WHEN THINGS GO WRONG – DISEASES AND DISORDERS OF THE HUMAN BRAIN

Edited by **Theo Mantamadiotis**

When Things Go Wrong – Diseases and Disorders of the Human Brain
Edited by Theo Mantamadiotis

Published by InTech
Janeza Trdine 9, 51000 Rijeka, Croatia

Publishing Process Manager Sandra Bakic

Technical Editor Teodora Smiljanic

Cover Designer InTech Design Team

Additional hard copies can be obtained from orders@intechopen.com

When Things Go Wrong – Diseases and Disorders of the Human Brain,
Edited by Theo Mantamadiotis
p. cm.
ISBN 978-953-51-0111-6

For EU product safety concerns:
IN TECH d.o.o., Prolaz Marije Krucifikse Kozulić 3, 51000 Rijeka, Croatia,
info@intechopen.com or visit our website at intechopen.com.

Meet the editor

Dr Theo Mantamadiotis, as a molecular cell biologist was able to combine his interests in the neurosciences and cancer biology. His specific research focuses on understanding molecular and cellular mechanisms regulating neural cell survival, growth and differentiation in the context of brain development and cancer. Much of Dr Mantamadiotis' work has focussed on a gene encoding the CREB transcription factor which has been implicated in a spectrum of neural functions ranging from learning and memory, drug addiction and neurodegeneration. In the words of Dr Mantamadiotis: "The most exciting part of the journey is just beginning, as our investigations on the role of CREB in brain have led us to investigate neural stem cells and their role in brain cancer".

Contents

Preface

From ancient times there was an appreciation, albeit controversial at the time, that mind and brain occupied the same space. Man's fascination with the human brain led to an evolution of thinking as to the origin of what we now call mental disorders. The Hippocratic Doctors of Ancient Greece left us with scripts describing the early philosophical-scientific thoughts of the nature of the brain and its ailments. In 'On the Sacred Disease' Hippocrates (400 B.C.) writes: "It ought to be generally known that the source of our pleasure, merriment, laughter and amusement, as of our grief, pain, anxiety and tears, is none other than the brain. It is specially the organ which enables us to think, see and hear, and to distinguish the ugly and the beautiful, the bad and the good, pleasant and unpleasant... It is the brain too which is the seat of madness and delirium, of the fears and frights which assail us" (Taken from Chadwick and Mann's translation, The medical works of Hippocrates, 1950, Oxford: Blackwells, pp. 179-189).

For all the advances in the medical and life sciences, the brain is perhaps still the least understood organ of the human body. This is in no way due to the lack of interest or research on the brain but rather due to the complexity of its structure at the macroscopic, microscopic and molecular levels. The brain provides clinicians and scientists a black box which is slowly being illuminated by the advances in understanding through advances in research. For patients suffering from disorders and diseases of the brain, the advances in brain research provide hope in the form of their own understanding of what is going wrong and in the form of advances in novel therapies which help alleviate debilitating symptoms.

In this book we have experts writing on various neuroscience topics ranging from mental illness, syndromes, compulsive disorders, brain cancer and advances in therapies and imaging techniques. Although diverse, the topics provide an overview of an array of diseases and their underlying causes, as well as advances in the treatment of these ailments. This book includes three chapters dedicated to neurodegenerative diseases, undoubtedly a group of diseases of huge socio-economic importance due to the number of people currently suffering from this type of disease but also the prediction of a huge increase in the number of people becoming afflicted.

The book also includes a chapter on the molecular and cellular aspects of brain cancer, a disease which is still amongst the least treatable of cancers.

Theo Mantamadiotis, PhD
Department of Pathology
The University of Melbourne,
Melbourne,
Australia

Laboratory of Physiology
Medical School
University of Patras
Rio-Patras, Greece

Part 1

Syndromes and Disorders

1

The Unique Properties of the Prefrontal Cortex and Mental Illness

Wen-Jun Gao*, Huai-Xing Wang, Melissa A. Snyder and Yan-Chun Li
Department of Neurobiology and Anatomy, Drexel University College of Medicine, Philadelphia, USA

1. Introduction

The prefrontal cortex (PFC) is part of the frontal lobes lying just behind the forehead and is one of the most important areas in the brain. This brain region is responsible for executive functions, which include mediating conflicting thoughts, making choices (between right and wrong or good and bad), predicting future events, and governing social and emotional control. All of the senses feed information to the PFC, which combines this information to form useful judgements. Further, it constantly contains active representation in working memory, as well as goals and contexts. The PFC is also the brain center most strongly implicated in conscience, human intelligence, and personality. Because of its critical role in executive functions, it is often referred to as the "CEO of the brain."

Unfortunately, the PFC is also one of the most susceptible regions to injury and environmental risk factors. As such, the PFC has been the focus of considerable scientific investigation, owing in part to the growing recognition that dysfunction of this region and related networks underlies many of the cognitive and behavioral disturbances associated with neuropsychiatric disorders such as schizophrenia, attention-deficit/hyperactivity disorder (ADHD), drug addiction, autism, and depression. Because all of these diseases are mental disorders related to psychiatric concerns, the prefrontal neuron has been called the "psychic cell" of the brain by the late neuroscientist Dr. Patricia Goldman-Rakic [1, 2]. She famously stated: "Santiago Ramón y Cajal might have envisioned, but likely could not have anticipated, the scientific advances that have allowed the functional validation of the existence of a "psychic cell" in the PFC and its extension to human cognition at the end of the 20th century [2]."

Scientific research on the PFC has been booming and great progress has been achieved since the late 1970s, especially after the "Decade of the Brain" began in 1990. As Dr. Goldman-Rakic stated: "This achievement rests not only on the shoulders of giants but on many small steps in the development of primate cognition, single and multiple unit recording in behaving monkeys, light and electron microscopic analysis of cortical circuitry no less than on the evolution of concepts about memory systems and parallel processing networks,

* Corresponding Author

among other advance." Indeed, compared to other neocortical regions, recent studies have reported that PFC has several distinct features that make this brain region special for its functions and associated diseases. First, the PFC is widely connected with many other brain regions, particularly those in the limbic system. A recent approach to PFC anatomy defines it on the basis of a combination of cortical types, topology and connectivity. Second, unlike primary sensory cortical regions, such as primary visual cortex (V1), primary auditory cortex (A1) and somatosensory cortex (S1), the PFC lacks direct sensory thalamocortical inputs. However, all of the salient sensory information is indirectly sent to the PFC through other associative cortical regions, such as the parietal cortex and temporal cortex. These characteristic connections make direct testing of PFC function in animals difficult and thus research is much delayed compared to other primary cortical areas. Third, the PFC is densely innervated by monoamine systems, especially the dopaminergic system. This can explain why many of the PFC functions are associated with the functions of dopamine system. Fourth, the PFC has special local circuitry designated for unique functions such as persistent activity for working memory. Fifth, because of these properties, the PFC is mainly associated with psychiatric disorders that are closely related to higher cognitive processes and emotions. The last and the most important is that the executive functions of the PFC develop to their full capabilities throughout the juvenile and adolescent period in humans. This higher brain region, unlike other primary cortical areas, exhibits delayed cortical development until young adulthood. During postnatal development, it gradually takes on its adult form as prefrontal neuron synapses are pruned to the adult level. Further, numerous data show that juvenile and adolescence are time periods of great vulnerability, with special sensitivity to environmental factors in humans, and eruption of neuropsychiatric disorders.

In this chapter, we will focus on the unique properties of PFC circuitry and development. Provide an overview of how during windows of vulnerability the maturation of this specific brain region and environmental factors initiate a series of events that render the PFC exceptionally susceptible to the development of neuropsychiatric disorders such as schizophrenia. Understanding the neurobiological basis is important in the development of more effective intervention strategies to treat or prevent these disorders.

2. The functions of the PFC are defined by its extensive connections with limbic system

The limbic system of the brain consists of many brain structures such as the hippocampal formation, amygdaloid complex, and nucleus accumbens. Limbic system structures are involved in emotions and motivations, particularly those related to survival such as fear, anger, pleasure, and sexual behavior. It is almost impossible to identify specific roles to definite structures, since psychological functions performed are not by single formations but by complexes of the interacting system. Overall, the limbic brain appears to be organized less in terms of precise physiological functions than in terms of elaboration and coordination of varied complexes of behavior [3, 4].

Recent findings in rodents and non-human primates suggest that divergent cognitive processes are carried out by anatomically distinct subregions of the PFC [5-7], although the extent to which these processes can be considered functionally homologous in different species remains controversial [8]. As part of the limbic system, the PFC is widely connected with many brain structures, particularly those in the Papez circuit. These wide connections make the PFC extremely responsive to stimulation such as emotion, stress, motivation, and learning and memory processes [6, 9-11].

2.1 PFC connections in the rat brain

The rat PFC is divided into the prelimbic, infralimbic, anterior cingulate, agranular insular cortices, and orbitofrontal areas [12-14]. Each of these subregions of the PFC appears to make individual contributions to emotional and motivational influences on behavior [15]. The PFC has complex functions such as working memory as well as attention, cognition, emotion and executive control [16]. The glutamatergic pyramidal neurons in the anterior cingulate cortex send descending projections to the nucleus accumbens core, the center for reward and emotional processing [13, 17, 18]. Additional descending projections from the PFC to nucleus accumbens, amygdala and other limbic brain regions appear to exert regulatory control over reward-seeking behavior. Therefore, the PFC is a key component of the limbic system with many inputs and outputs, and its heterogeneous cytoarchitectonic structure implies a complex functional organization.

The PFC can also be divided into dorsal and ventral divisions [14] and the attentional and emotional mechanisms appear to be segregated into dissociable prefrontal networks in the brain [16]. The reciprocal relationship between dorsal and ventral PFC may provide a neural substrate for cognitive - emotional interactions, and dysregulation in these systems is clearly related to various mental diseases [11]. It has been reported that the PFC is primarily connected with the mediodorsal thalamic nucleus with distinctions between the dorsal and ventral prefrontal cortices [14]. The dorsal PFC (prelimbic and anterior cingulate cortex) and ventral PFC (infralimbic area) appear to be differentiated with distinct afferent terminations. The dorsal PFC has connections with sensorimotor and association neocortex, while the ventral PFC shows strong connections with the amygdaloid complex and limbic association cortices. The ventral PFC projects heavily to the subcortical limbic structures, including the hypothalamic areas and septum, and of particular interest, the ventral PFC shows more powerful influences on brainstem monoaminergic cells than does the dorsal PFC.

2.2 Different structural features of the PFC in primate versus rodent

The PFC shows enormous variation across species in terms of cytoarchitectonics and connectivities, especially in the presence or absence of a granular zone and the existence of strong reciprocal connections from the mediodorsal nucleus of the thalamus [17, 19, 20]. One major problem about the PFC has been the long-standing debate over what constituents equivalent regions of the PFC between different species [8, 17, 19, 20]. In addition, unlike posterior and temporal regions of neocortex, the PFC receives highly organized indirect inputs from the basal ganglia via striatopallidal and striatonigral projections, and subsequently pallidothalamic and nigrothalamic neurons that project, in a parallel segregated manner, to different areas of the PFC in both rodents and primates [19, 21]. The PFC also receives extensive corticocortical inputs, for example, from parietal cortex and sensory cortical areas, as well as connections from subcortical structures such as the substantia nigra, ventral tegmental area, amygdala, lateral hypothalamus, and hippocampus [19].

The distinctive feature of primate PFC is the emergence of dysgranular and granular cortices, which are completely absent in the rodent. Some of the subregions in the primate PFC do not have a clear-cut homolog in rodents because the rat PFC is entirely agranular [4, 20, 22]. The primate PFC is often divided into different subregions, such as dorsolateral, ventrolateral, medial, and orbitofrontal. These subregions are extensively interconnected, with information to be shared within the PFC circuitries [23]. In addition, information from sensory cortices also converges to the PFC in multiple modalities [24]. Generally speaking, dorsolateral areas receive input from earlier sensory areas; whereas orbitofrontal areas receive inputs from advanced stages of sensory processing from every modality, including gustatory and olfactory [23, 25]. Thus, extrinsic and intrinsic connections make the PFC a site of multimodal convergence of information about the external environment. Furthermore, the PFC receives inputs that could inform it about internal mental states, such as motivation and emotion. As discussed above, orbital and medial PFC are closely connected with limbic structures such as the amygdala, hippocampus, and rhinal cortices [23], as well as the hypothalamus and other subcortical targets that are associated with autonomic responses [26]. Finally, outputs from the PFC, especially from the dorsolateral PFC, are directed to motor systems, and thus the PFC may form or control motor planning. Altogether, the PFC receives inputs that provide information about many external and internal variables, including those related to emotions and to cognitive functions, providing a potential anatomical substrate for the representation of mental states.

2.3 PFC-amygdala connection and interaction

The amygdala is a structurally and functionally heterogeneous group of nuclei lying in the anterior medial portion of the temporal lobe. The amygdala is most often discussed in the context of emotional processes; yet it is extensively interconnected with the PFC, especially with the orbitofrontal cortex and anterior cingulate cortex, as well as diffusely with other parts of the PFC [4, 27]. Sensory information enters the amygdala from visual, auditory, and somatosensory cortices, from the olfactory system, and from the perirhinal cortex and the parahippocampal gyrus [27]. Output from the amygdala is directed to a wide range of target structures, including the PFC, the striatum, sensory cortices, the hippocampus, the entorhinal cortex, and the basal forebrain, and to subcortical structures related to autonomic responses, hormonal responses, and startle [27]. Overall, the bidirectional communication between the amygdala and the PFC provides a potential basis for the integration of cognitive, emotional, and physiological processes into a unified representation of mental states [3, 15, 28].

3. Despite the widespread connections with the mediodorsal nucleus of the thalamus, the PFC lacks direct sensory thalamo-cortical connections

As discussed above, the PFC is mainly defined by projections from the mediodorsal nucleus of the thalamus [12, 14, 20]. Specifically, reciprocal and topographically organized connections between the medial PFC and various thalamic nuclei are well known [29-34]. A ventral to dorsal gradient in the PFC is corresponding to a medial to lateral gradient in the dorsal thalamus where the medial prefrontal cortex primarily projects to the midline, mediodorsal and intralaminar thalamus [3, 33, 34]. In general, the cortico-thalamic projections are largely reciprocated by thalamo-cortical fibers. The midline thalamic nuclei are largely involved in arousal and visceral functions while the intralaminar nuclei subserve orienting and attentional aspects of behavior [3, 14]. The limbic thalamus includes the anterior thalamus, which is part of the Papez circuit, and the mediodorsal thalamic nucleus. The mediodorsal nucleus is a major element within the thalamus of all mammals and undergoes a progressive expansion of cytoarchitectonic differentiation in higher animals, reaching its greatest development in human beings [35]. Importantly, this development parallels the development of the PFC. The mediodorsal thalamic nucleus projects to a large area of the frontal cortex in the rat, including the precentral area, anterior cingulate area, prelimbic area, orbital areas, and the insular areas [29, 36, 37].

Despite the widespread connections between the PFC and mediodorsal nucleus of the thalamus, unlike other sensory cortices, the PFC lacks direct afferent inputs from sensory thalamus. Therefore, research on the PFC is rather delayed compared to the studies on other cortical regions owing to the difficulty in making animal models or direct stimulation.

4. PFC receives rich monoaminergic, especially dopaminergic (DA), and cholinergic (ACh) innervations

Monoamines contribute to stable moods, and an excess or deficiency of monoamines cause several mood disorders. The PFC targets the main major forebrain cholinergic and monoaminergic systems, including noradrenaline (NA)-containing neurons in the pontine locus coeruleus, dopamine (DA)-containing neurons in the ventral tegmental area, serotonin (5-HT) neurons in the raphe nuclei and acetylcholine (ACh) neurons in the basal forebrain [5, 38, 39]. These systems act in turn to modulate cortical networks by influencing both excitatory and inhibitory synaptic transmissions as well as other cortical processing in the PFC [9, 38]. Neuromodulatory input to the PFC from these neuromodulatory systems could also convey information about internal state [40]. Further, the ascending monoaminergic (NA, DA and 5-HT) and ACh systems contribute to different aspects of performance on animal behaviors [40].

When considering the functions of the chemical modulatory inputs to the PFC, a general principle that has emerged in the past decade is the inverted U-shaped function, which links the efficiency of behavioral performance to the level of activity in the DA- and NE-ergic systems [40, 41]. The inverted-U dose response has been demonstrated with pharmacological agents in both animals [42-44] and humans [45]. A major advance in understanding the roles of the neuromodulatory systems is the in vivo measurement of ACh, DA, NA and 5-HT release in the PFC during behavioral tests [5, 46, 47]. This powerful approach directly links PFC functions with specific changes of individual neurotransmitter systems and their interactions in a behavioral task. It is possible that the neuromodulatory systems of the PFC are functionally specialized, and that each of them are engaged by different feedback circuits required for specific information processing. However, a better understanding of the role of each neuromodulator in different cognitive control processes is needed. It is also important to explore whether the regulatory signaling is distributed or localized within the different parts of the PFC neurons [48]. The PFC has a top-down regulatory control over the ascending modulatory systems of the brain, and that in turn, powerfully influences the neuromodulatory functions on the PFC [40, 41]. These projections widely innervate diverse forebrain regions, including the hippocampus, striatum, amygdala, and thalamus, as well as the entire neocortex. In turn, these neuromodulatory systems likely adjust signal-to-noise ratios in terminal domains to influence information processing and their conjoint activity, and consequently, to affect behaviors.

Among these ascending modulatory systems, the DA system is the most important one that plays a critical role in both normal cognitive process and neuropsychiatric pathologies associated with the PFC [49]. It has been known for several decades that the frontal lobe receives a major dopamine innervation. Furthermore, the PFC receives more DA innervations compared with other cortical regions. In contrast, all other ascending modulatory innervations are more evenly distributed among cortical regions. Researchers, however, have only recently been able to link dopamine afferents to specific cellular targets and neuronal circuits [49, 50]. Understanding the details of this linkage in prefrontal circuits may be important in resolving the various dilemmas concerning the mechanisms of dopamine action or cognitive processes, as well as the validity of the dopamine hypothesis of diseases like schizophrenia [51-54].

Accordingly, there have been considerable efforts by many groups to understand the cellular mechanisms of DA modulation in PFC neurons [49, 50, 55-60]. Although the results of these efforts sometimes lead to contradictions and controversies, these studies from both in vivo and in vitro experiments have provided some principal features and mechanisms of DA modulation in the PFC circuitry [49]. One principal feature of DA is that, as a neuromodulator, it is neither an excitatory nor an inhibitory neurotransmitter. It becomes apparent that DA's actions in PFC are regulatory and an optimal concentration of DA is required for normal operation of the PFC. Either too much or too little DA will result in serious mental problems that are associated with prefrontal cognitive functions. For example, hyperfunction of the dopaminergic system is believed to be related to several psychiatric disorders [50, 61]. Previous studies in both rats and primates indicate that excessive dopamine activity is detrimental to cognitive functions mediated by the PFC [62, 63]. DA's effects on the PFC depend on a variety of factors, especially activation of different dopamine receptors. There are at least five subtypes of dopamine receptors, D1, D2, D3, D4, and D5. The D1 and D5 receptors are members of the D1-like family of dopamine receptors, whereas the D2, D3 and D4 receptors are members of the D2-like family. The distinct inverted-U dose-response profiles of postsynaptic DA responses are contingent on the duration of DA receptor stimulation, the bidirectional effects following activation of D1 or D2 classes of receptors, the membrane potential state of the prefrontal neurons, and the history dependence of subsequent DA actions [49]. Based on these factors, a theory is proposed for DA's action in the PFC which suggests that DA acts to regulate the information held in working memory and then modulates the cognitive and executive performance of the PFC [49].

5. Unique PFC circuitry for persistent activity – The cellular basis/correlate for working memory

Working memory is the ability to hold an item of information transiently in mind in the service of comprehension, thinking, and planning [64-69]. It encompasses information retrieval, transient storage, and re-update/recycle processing. Thus working memory serves as a workspace for holding items of information in mind as they are recalled, manipulated, and/or associated to other ideas and incoming information. "Blackboard of the mind" has been a useful metaphor for the limited capacity and processing dynamics of the working memory mechanism [64, 69]. Information such as a rule or goal is held temporarily in working memory and used to guide behavior, attention or emotions, dependent on the PFC region(s) involved. In addition to the ability to transiently hold the information 'on-line" for working memory, the PFC is also able to represent information that is not currently in the environment through persistently activated recurrent networks of pyramidal neurons [70]. This process has been referred to as representational knowledge and is thought to be a fundamental component of abstract thought [69].

5.1 Persistent activity in primate studies

The circuitry underlying working memory or representational knowledge in the PFC has been most intensively studied in the past decades. In primates, visuospatial information is processed by the parietal association cortices, and fed forward to the dorsolateral PFC, where pyramidal cells excite each other to maintain information briefly in memory. A major advance in our understanding of PFC and working memory function came in the early 1970s. Electrophysiological studies revealed that neurons in the PFC become activated during the delay period of a delayed-response trial when a monkey recalled a visual stimulus that had been presented at the beginning of a trial [71, 72]. Patricia Goldman-Rakic and her colleagues [69] further discovered and elaborated the PFC microcircuitry subserving spatial working memory using anatomical tracing techniques and physiological recordings from monkeys performing an oculomotor spatial working memory task. They found that the dorsolateral PFC is key for spatial working memory, and many neurons in this region exhibit spatially tuned, persistent firing during the delay period in a spatial working memory task [73]. Goldman-Rakic posited that the delay-related firing arises from pyramidal cells with similar spatial characteristics exciting each other to maintain information in working memory. It quickly became evident that the persistent activity of these prefrontal neurons could be the cellular correlate of a mnemonic event for working memory.

5.2 Physiological and morphological properties of persistent activity

Then, what is the neural basis of persistent activity in the prefrontal neural circuitry? Are the prefrontal cortical circuitries specialized to generate persistent action potentials needed for working memory? What are the microcircuit properties that enable the PFC to subserve cognitive functions such as working memory and decision making in contrast to early sensory coding and processing in primary sensory areas? Although the mechanism remains elusive, a large body of evidence indicates that the PFC is both functionally and structurally specialized with unique properties differing from other cortical areas. It has been hypothesized that persistent activity is generated by sufficiently strong recurrent excitation among prefrontal neurons [69]. Specifically, prefrontal neurons that reside in layer II/III, contain extensive horizontal connections that are characteristic of recurrent connections [69]. Pyramidal cell networks interconnect on dendritic spines, exciting each other via postsynaptic N-Methyl-D-aspartate (NMDA) receptors. NMDA currents are particularly evident in the recurrent network of PFC circuitry [74], and seem to be necessary for delay-related firing in monkeys performing a working memory task [70].

In addition, neurons in the PFC circuitry exhibit distinct morphological properties. In an interesting study, the basal dendritic arbors of pyramidal cells in prefrontal areas of the macaque monkey were revealed by intracellular injection in fixed cortical slices and the spine density in the basal dendrites were quantified and compared with those of pyramidal cells in the occipital, parietal, and temporal lobes [75]. These analyses revealed that cells in the frontal lobe were significantly more spinous than those in the other lobes, having as many as 16 times more spines than cells in the primary visual area (V1), four times more those in area 7a, and 45% more than those in temporal cortex [75]. As each dendritic spine receives at least one excitatory input, the large number of spines reported in layer III pyramidal cells in the primate PFC suggests that they are capable of integrating a greater number of excitatory inputs than layer III pyramidal cells in the occipital, parietal, and temporal lobes. The ability to integrate a large number of excitatory inputs may be important for the sustained activity in the PFC and their role in memory and cognition [75-79]. In addition, Elston et al also presented evidence that the pyramidal cell phenotype varies markedly in the cortex of different anthropoid species. Regional and species differences in the size and number of bifurcations and spine density of the basal dendritic arbors cannot be explained by brain size. Instead, pyramidal cell morphology appears to accord with the specialized cortical function these cells perform. Cells in the PFC of humans are likely more branched and more spinous than those in the temporal and occipital lobes. Moreover, cells in the PFC of humans are more branched and more spinous than those in the PFC of macaque and marmoset monkeys. These results suggest that highly spinous and compartmentalized pyramidal cells (and the circuits they form) are required to perform complex cortical functions such as working memory and executive functions for comprehension, perception, and planning [77]. Because of the high density of dendritic spines in the PFC neurons [75, 76] and presumably more excitatory synapses in the recurrent circuitry in the PFC [80], the PFC is thought to be specialized to generate persistent action potentials (or persistent activity), the presumptive mechanism of working memory [81-87].

Furthermore, it has been appreciated that several types of interneurons reside in the PFC and interact with pyramidal cells. Using simultaneous recordings in monkeys, it has been revealed that the inhibitory interactions between neurons at different time points are relative to the cue presentation, delay interval and response period of a working memory task [88, 89]. These data indicate that pyramidal - interneuron interactions may be critical to the formation of memory fields in PFC [88]. The PFC network activity is 'tuned' by inhibitory GABAergic interneurons so that the contents of working memory are contained, specific and informative. For example, when pyramidal cells are active they excite GABAergic interneurons that suppress the firing of pyramidal cells in another microcircuit, and vice versa [88, 89]. These findings suggest an important role of inhibition in the PFC: controlling the timing of neuronal activities during cognitive operations and thereby shaping the temporal flow of information [90].

6. Delayed development or maturation of the PFC

6.1 Synaptogenesis, synaptic remodeling and maturation

Development is a complex process involving changes in white matter and the establishment of neuronal connections in the brain, both of which are influenced by genetic and environmental factors. Generally speaking, the development of the nervous system occurs through the interaction of several processes, some of which are completed before birth, while others continue into adulthood [91]. For example, proliferation and migration of cells mostly occurs during fetal development, although in postnatal development, the formation of neuronal circuits, along with neuronal death and the rapid formation and elimination of synapses, occurs in the cerebral cortex, including the PFC [92-95]. It is known that synaptic density in the brain increases with age, and it occurs as a result of trillions of neurological connections, commonly called "wiring." Neuronal firing creates a network that is permanently established with repetitive experiences. Connections no longer being used or relied upon are eliminated through a process called synaptic pruning. Although the development of neural connections in the brain is not fully understood, it is clear that the time courses of such neuronal and synaptic formation and elimination are considerably different across diverse cortical areas, with the PFC generally being one of the latest [96]. Therefore, the childhood development of the cerebral cortex may be characterized by neuronal death and the elimination of unused synapses during a defined time window such as adolescence. Synaptic density in the PFC reaches the net highest value at age 3.5 years, showing a level approximately 50% greater than that in adults but decreasing gradually through adolescence [96]. Developmental changes in cellular morphology have also been observed during early childhood, including expansion of the dendritic trees of the pyramidal neurons [97].

6.2 Delayed maturation of the PFC

PFC development in humans begins from the neural tube, which is an embryonic structure that eventually becomes the brain and spinal cord. PFC experiences one of the longest periods of development of any brain region, taking over two decades to reach full maturity in humans, i.e., PFC exhibits a significant delayed maturation compared to other brain regions [98-100]. As children explore their environments and begin to develop speech, motor skills, and a sense of themselves as separate human beings, the PFC undergoes rapid growth during infancy [101]. Several characteristic functions of the PFC, such as planning, reasoning, and language comprehension, change dramatically as a function of age throughout childhood and adolescence [102]. The processes involved in the development of these PFC functions have been debated for several decades at the level of both brain and behavior, and it has been established that changes in structural architecture and cognitive maturation occur concurrently throughout childhood development [103]. Complete frontal cortex development takes many years, and new functions are added well beyond the childhood years. Accumulating evidence suggests that early childhood appears to be comparably important for functional neural development of the PFC [104]. While the most dramatic structural changes in the healthy human brain are thought to occur in the perinatal period [96], there is a growing body of evidence suggesting that adolescence is also a period of substantial neurodevelopment [105]. Understanding the brain maturation over adolescence and early adulthood is particularly important, given that it is a peak period of neural reorganization that contributes to both normal variation and the onset of some major mental illnesses, such as schizophrenia [106, 107]. Despite support for pronounced changes in both the structure and function of the brain during adolescence, the relationship among these changes has not been fully examined.

6.3 Adolescence is a critical period for PFC maturation – Molecular and cellular alterations in the PFC circuitry

To encourage the establishment of new neuronal connections, the frontal lobe must be stimulated. While frontal cortex development is significantly influenced by genetics, environmental factors play a pivotal role. Children who are exposed to varied environments; encouraged to solve problems; challenged to reason; and engaged in different games, songs and memory tasks will benefit from these stimulations that facilitate the development of the PFC. Conversely, children with sensory processing disorders often struggle with the reasoning and decision making tasks controlled by the PFC, and damage to the PFC results in an inability to control impulses and learn from experiences with reward and punishment.

PFC development is thus characterized by maturational processes that span the period from early childhood through adolescence to adulthood [108, 109], but little is known whether and how developmental processes differ during these phases. In the past two decades, numerous studies have been focused on detail changes in the functional maturation of the PFC circuitry. For example, it is now clear that the underlying synaptic refinement process in the PFC is not completed until late adolescence and early adulthood [110, 111], which coincides with the period when symptoms of schizophrenia typically begin to emerge [112]. Indeed, our study indicated that the NMDA receptor subunit NR2B-to-NR2A shift does not occur during prefrontal development. The NMDA receptor-mediated currents in the recurrent synapses of the PFC exhibit a 2-fold longer decay time-constant and temporally summate a train of stimuli more effectively than those in the primary visual cortex [74]. Pharmacological experiments suggest a greater contribution by NR2B subunits at prefrontal synapses than in the visual cortex. Therefore, the biophysical properties of NMDA receptors in PFC may be critically important to the generation of slow reverberating dynamics required for cognitive computations [74]. However, the enriched NR2B subunit in the PFC appears to be a double-edged sword - important for normal working memory but easy to be targeted by detrimental stimulation. In addition, we also reported that parvalbumin-containing fast-spiking interneurons in the PFC undergo dramatic changes in glutamatergic receptors during the adolescent period, including both NMDA receptors and calcium-permeable AMPA receptors [113, 114]. Furthermore, Tseng and O'Donnell found significant changes in the susceptibility of interneurons to dopaminergic D2 receptor modulation during adolescence. Importantly, D2 agonists were effective only in adult but not in prepubertal animals [115]. Many other late occurring changes in GABAergic neurons, GABAergic neurotransmission and $GABA_A$ receptors have also been demonstrated [112, 116, 117]. Similarly, developmental trends have been reported for the dopaminergic [118] and glutamatergic systems [119] and for interactions of these neurotransmitters with GABAergic interneurons. It is possible that these prominent changes may make fast-spiking cells particularly sensitive and vulnerable to epigenetic or environmental stimulation, thus contributing to the onset of psychiatric disorders, including schizophrenia, bipolar disorder, and depression.

While these findings suggest important evidence on late-occurring anatomical and physiological modifications, the precise implications of these changes for coordinated network activity in the PFC are unknown. It is believed that these anatomical and physiological changes impact critically upon the functional properties of large-scale cortical networks [120, 121]. The alterations in GABAergic neurons during adolescence may be of particular relevance for synchronous oscillations because GABAergic interneurons and their interactions with excitatory neurotransmission have been shown to be critical for the generation of high-frequency oscillations [122-132]. Following early developmental periods, changes in the amplitude of neural oscillations and their synchronization continue until early adulthood, suggesting ongoing modifications in network properties. One of the most replicated findings is the alteration in resting-state oscillations. In the adult brain, resting-state activity is characterized by prominent alpha oscillations over occipital regions while low (delta, theta) and high (beta, gamma) frequencies are attenuated. During adolescence, there is a reduction in the amplitude of oscillations over a wide frequency range, particularly in the delta and theta band, while oscillations in the alpha and beta range become more prominent with age [133]. Interestingly, these changes occur more rapidly in posterior than in frontal regions and follow a linear trajectory until age 30 [133]. Alteration in the amplitude of oscillations is accompanied by modifications in the synchrony of resting-state oscillations. Thatcher et al investigated modifications in the coherence of beta oscillation in children and adolescents between 2 months and 16 years of age. During development, beta-band coherence increased over shorter distances while long-range coherence did not vary with age [134]. Uhlhaas et al further reported that until early adolescence, developmental improvements in cognitive performance were accompanied by increases in neural synchrony [121]. This developmental phase was followed by an unexpected decrease in neural synchrony that occurred during late adolescence and was associated with reduced performance. After this period of destabilization, a reorganization of synchronization patterns occurred with a pronounced increase in gamma-band power and in theta and beta phase synchrony. These findings provide evidence for the relationship between neural synchrony and late brain development that has important implications for the understanding of adolescence as a critical period of brain maturation [121].

7. Diseases associated with the development of PFC – Mental illness

7.1 What is a mental disorder?

Mental illness refers to a wide range of mental health disorders that affect people's mood, thinking and behavior. Examples of mental illness include schizophrenia, ADHD, depression, bipolar disorders, anxiety disorders, autism spectrum disorders, obsessive-compulsive disorder, eating disorders, and addictive behaviors. As repeatedly discussed above, the PFC plays a critical role in cognitive functions and cortical inhibition, especially for insight, judgment, the ability to inhibit inappropriate responses, and the ability to plan and organize for future events. Therefore, PFC dysfunction is greatly associated with disorders/deficits in cognitive and executive functions that are seen in most mental illnesses.

Many people have mental health concerns from time to time, but this only becomes a mental illness when clear signs and symptoms cause severe stress and affect people's ability to function properly. A mental illness can make people miserable and can cause problems in daily life, such as at work or in personal relationships. Signs and symptoms of mental illness vary, depending on the particular disorder. In most cases, mental illness symptoms can be managed with a combination of medications and counseling such as psychotherapy. Most major or serious mental illnesses tend to have symptoms that come and go, with periods in between when the person can lead a relatively normal life, i.e., episodic illness. The most common serious mental disorders are schizophrenia, bipolar disorder, and depression.

Although the exact cause of most mental illnesses is unknown, it is becoming clear that many of these conditions are caused by a combination of genetic, biological, psychological and environmental factors.

1. Genetics: Many mental illnesses have family histories, suggesting that the illnesses may be passed on from parents to children through specific genes. Many mental illnesses are linked to multiple problem genes that are still largely unknown. The disorder occurs from the interaction of these genes and other factors, such as psychological trauma and environmental stressors - which can influence or trigger the illness in a person who has inherited a susceptibility to the disease.
2. Biology: Mental illnesses have been linked to an abnormal balance of neurotransmitters, mis-wired neuronal connections in the network, and disrupted communications between neurons within the brain. When neuronal signals cannot be properly transmitted within the brain, particularly within the brain region such as PFC, signs and symptoms of a mental disorder will emerge.
3. Psychological trauma: Some mental illnesses may be triggered by psychological trauma suffered as a child, such as severe emotional, physical or sexual abuse, etc.
4. Environmental stressors or risk factors: Certain stressors or risk factors - such as a brain injury, dysfunctional family life, substance abuse, or a life threatening event - can trigger a disorder in a person who may be at risk for developing a mental illness.

7.2 Circuit basis for cognitive dysfunction in mental illness

The cognitive operations of the PFC are especially vulnerable to physiological, genetic and environmental factors. They can be altered by changes in arousal state such as fatigue or stress [135] and are profoundly impaired in most mental illnesses [40, 136-139]. However, it is unknown how these functions are affected. There are many questions that need to be answered. Specifically, for example, what are the specific genes that are involved in a mental disorder such as schizophrenia or depression? There are some high risk genes identified for an individual disease. However, it is unclear how these identified genes interact to other factors and how these susceptible genes are triggered by aforementioned psychological trauma or environmental risk factors, and consequently result in a domino effect in the brain. A large body of evidence indicates that the onset of a mental disorder is triggered by a risk factor but the pathological process of a mental illness is complex and unclear. Apparently, many mental illnesses are associated with impaired brain development, especially broken PFC circuitry.

As discussed above, PFC cognitive functions rely on networks of interconnected pyramidal cells [1, 2, 69], as well as GABAergic interneurons [112, 116, 140]. Recent studies reveals that neuronal connections in the PFC network are influenced by powerful molecular events that determine whether a network is connected or disconnected at a given moment, thus determining the strength of cognitive abilities [70]. These mechanisms provide great flexibility, but also confer vulnerabilities and limit mental capacity. A remarkable number of genetic and/or environmental insults to these molecular signaling cascades are associated with cognitive disorders such as schizophrenia [77, 138, 139, 141-144], ADHD [145, 146], depression [100, 101, 147-149], and autism spectrum disorder [150-155]. These insults can dysregulate network connections in the PFC and weaken its capabilities in cognitive control. It is evident that many genetic and environmental insults would have an impact on signaling molecules within PFC networks [70] and its highly linked limbic systems.

Alterations in PFC circuitry are therefore associated with a variety of cognitive disorders, ranging from mild PFC impairment (e.g. anxiety disorder, depression, normal aging) to severe deficits (e.g., schizophrenia, bipolar disorder, Alzheimer's disease).

The question is that what causes a circuit disorder? Mental disorders such as schizophrenia and mood and anxiety disorders are mostly diseases of early life; their onset tends to occur during adolescence or early adulthood, when the brain is still developing. Because of page limits and the complex etiology and pathological process in different mental disorders, it is not possible for us to describe all aforementioned mental illnesses in detail in this chapter. So next we use schizophrenia as an example to illustrate the role of PFC in this devastating disorder.

7.3 Disrupted development of PFC circuitry in schizophrenia

Schizophrenia is a disorder of cognitive neurodevelopment with characteristic abnormalities in working memory attributed, at least in part, to alterations in the circuitry of the PFC. Schizophrenia is associated with altered PFC circuits, arising from both developmental insults in utero, and continuing in the mature brain, for example with impaired neural circuitry and synaptic connectivity in late adolescence and adulthood. Various environmental exposures from conception through adolescence increase risk for the illness, possibly by altering the developmental trajectories of prefrontal cortical circuits.

Several lines of evidence support the notion that a substantial reorganization of cortical connections takes place during adolescence in humans. A review of neurobiological abnormalities in schizophrenia indicates that the neurobiological parameters that undergo peripubertal regressive changes may be abnormal in this disorder. An excessive pruning of the prefrontal corticocortical, and corticosubcortical synapses, perhaps involving the excitatory glutamatergic inputs to pyramidal neurons, may underlie schizophrenia [99, 106]. Several developmental trajectories, which are related to early brain insults as well as genetic factors affecting postnatal neurodevelopment, could lead to the illness. These models would have heuristic value and may be consistent with several known facts of the schizophrenic illness, such as its onset in adolescence. For example, a person with schizophrenia usually experiences a psychotic break in early adulthood, which is a time when the number of cortical synapses is being pruned. The disorder might result from the excessive loss of synapses in a critical cortical pathway when the normal process overshoots.

Although psychosis always emerges in late adolescence or early adulthood, we still do not understand all of the changes in normal or abnormal development prior to and during this period. It is particularly unclear what factors alter the excitatory-inhibitory synaptic balance in the juvenile and what changes induce the onset of cognitive dysfunction. Current studies suggest that problems related to schizophrenia are evident much earlier. The emerging picture from genetic and epigenetic studies indicates that early brain development is affected. Many of the structural variants associated with schizophrenia implicate that neurodevelopmental genes or epigenetic factors are involved with neuronal development [156-159]. A remarkable number of genetic insults in schizophrenia involve proteins found at prefrontal synapses. There are well-established genetic changes associated with NMDA receptor signaling [160-162], DA [51, 163-165], GABA [112, 116, 140, 166], and α7 nicotinic receptors [167-170]. More recently, a number of high-risk genes are found to be associated with schizophrenia [171]. Four out of the top 10 risk gene variants most strongly associated with schizophrenia are directly involved in DA-ergic systems, including the catechol-o-methyltransferase gene (COMT) [142, 172-177], neuregulin 1 (NRG1) [178, 179], disrupted in schizophrenia 1 protein (DISC1) [157, 180], and dystrobrevin-binding protein 1 (dysbindin) [181-184]. Many of these gene variants are involved in brain development, such as reelin, or influence more ubiquitous brain transmitters such as glutamate or GABA [171, 184-189]. These postnatal developmental trajectories of neural circuits in the PFC identify the sensitive adolescent period for vulnerability to schizophrenia [112].

Furthermore, recent data from developmental cognitive neuroscience highlight the profound changes in the organization and function of PFC networks during the transition from adolescence to adulthood. While previous studies have focused on the development of neuronal components in gray matter, as well as axonal fibers and myelination in white matter [190], recent evidence suggests that brain maturation during adolescence extends to fundamental changes in the properties of cortical circuits that in turn promote the precise temporal coding of neural activity. Specifically, schizophrenia is associated with impaired neuronal synchronized activity that occurred during PFC maturation, suggesting an important role of adolescent brain development for the understanding, treatment, and prevention of the disorder [120].

These findings, although intriguing, are limited in that they do not reveal the changes before psychosis. At present, the diagnosis of schizophrenia is based primarily on the symptoms and signs of psychosis. Recently, it has been proposed that schizophrenia may progress through four stages: from risk to prodrome to psychosis and to chronic disability [191]. Obviously, the key to prevent or forestall the disorder is to detect early stages of risk and prodrome. Therefore identification of novel biomarkers, new cognitive tools, as well as subtle clinical features is urgently needed for early diagnosis and treatment [191, 192]. Animal studies, particularly developmental models, will certainly help to reveal the neurodevelopmental trajectory of schizophrenia, yield disease mechanisms, and eventually offer opportunities for the development of new treatments. As Thomas Insel pointed out in a recent review of schizophrenia [191]: "This 'rethinking' of schizophrenia as a neurodevelopmental disorder, which is profoundly different from the way we have seen this illness for the past century, yields new hope for prevention and cure over the next two decades."

8. Summary

The cognitive and executive functions of the prefrontal cortex (PFC) develop to their full capabilities throughout the juvenile and adolescent period in humans. The PFC is critical for cognitive functions and cortical inhibition, especially for insight, judgment, the ability to inhibit inappropriate responses, and the ability to plan and organize for the future. This higher brain region, unlike other primary cortical areas, exhibits unique connectivity and delayed cortical maturation. During postnatal development, it gradually takes on its adult form as prefrontal neuron synapses are pruned and neuronal connections are reformatted to adult level. Further, numerous data show that juvenile and adolescence are time periods of great vulnerability, with special sensitivity to risk environmental factors, and eruption of neuropsychiatric disorders. We have provided an overview of the unique properties and connectivity of the PFC circuitry and alterations during the juvenile and adolescent development under both normal and abnormal conditions. Understanding the neurobiological basis is important in the development of more effective intervention strategies to treat or prevent mental disorders such as schizophrenia.

9. Acknowledgement

This study was supported by grant R01MH232395 to W.-J Gao from the National Institutes of Health, USA.

10. Conflict of interest

The authors claim no financial conflicts of interest.

11. References

[1] Goldman-Rakic, P.S., *The "psychic" neuron of the cerebral cortex.* Ann N Y Acad Sci, 1999. 868: p. 13-26.

[2] Goldman-Rakic, P.S., *The "psychic cell" of Ramon y Cajal.* Prog Brain Res, 2002. 136: p. 427-34.

[3] Ray, R.D. and D.H. Zald, *Anatomical insights into the interaction of emotion and cognition in the prefrontal cortex.* Neurosci Biobehav Rev, 2011.

[4] Salzman, C.D. and S. Fusi, *Emotion, cognition, and mental state representation in amygdala and prefrontal cortex.* Annu Rev Neurosci, 2010. 33: p. 173-202.

[5] Dalley, J.W., R.N. Cardinal, and T.W. Robbins, *Prefrontal executive and cognitive functions in rodents: neural and neurochemical substrates.* Neurosci Biobehav Rev, 2004. 28(7): p. 771-84.

[6] Wilson, C.R.E., et al., *Functional localization within the prefrontal cortex: missing the forest for the trees?* Trends in Neurosciences, 2010. 33(12): p. 533-540.

[7] Kesner, R.P., *Subregional analysis of mnemonic functions of the prefrontal cortex in the rat.* Psychobiology, 2000. 28(2): p. 219-228.

[8] Brown, V.J. and E.M. Bowman, *Rodent models of prefrontal cortical function.* Trends Neurosci, 2002. 25(7): p. 340-3.

[9] Arnsten, A.F., *Catecholamine regulation of the prefrontal cortex.* J Psychopharmacol, 1997. 11(2): p. 151-62.

[10] Aultman, J.M. and B. Moghaddam, *Distinct contributions of glutamate and dopamine receptors to temporal aspects of rodent working memory using a clinically relevant task.* Psychopharmacology (Berl), 2001. 153(3): p. 353-64.

[11] Benes, F.M., *Amygdalocortical circuitry in schizophrenia: from circuits to molecules.* Neuropsychopharmacology, 2010. 35(1): p. 239-257.

[12] Morgane, P.J., J.R. Galler, and D.J. Mokler, *A review of systems and networks of the limbic forebrain/limbic midbrain.* Progress in Neurobiology, 2005. 75(2): p. 143-160.

[13] Cardinal, R.N., et al., *Emotion and motivation: the role of the amygdala, ventral striatum, and prefrontal cortex.* Neurosci Biobehav Rev, 2002. 26(3): p. 321-52.

[14] Heidbreder, C.A. and H.J. Groenewegen, *The medial prefrontal cortex in the rat: evidence for a dorso-ventral distinction based upon functional and anatomical characteristics.* Neurosci Biobehav Rev, 2003. 27(6): p. 555-79.

[15] LeDoux, J.E., *Emotion Circuits in the Brain.* Annual Review of Neuroscience, 2000. 23(1): p. 155-184.

[16] Goldman-Rakic, P.S., *The physiological approach: functional architecture of working memory and disordered cognition in schizophrenia.* Biol Psychiatry, 1999. 46(5): p. 650-61.

[17] Ongur, D. and J.L. Price, *The organization of networks within the orbital and medial prefrontal cortex of rats, monkeys and humans.* Cereb Cortex, 2000. 10(3): p. 206-19.
[18] Ongur, D., A.T. Ferry, and J.L. Price, *Architectonic subdivision of the human orbital and medial prefrontal cortex.* J. Comp. Neurol., 2003. 460: p. 425-449.
[19] Groenewegen, H.J., C.I. Wright, and H.B. Uylings, *The anatomical relationships of the prefrontal cortex with limbic structures and the basal ganglia.* J Psychopharmacol, 1997. 11(2): p. 99-106.
[20] Preuss, T.M., *Do rats have prefrontal cortex? The Rose-Woolsey-Akert program reconsidered.* Journal of Cognitive Neuroscience, 1995. 7(1): p. 1-24.
[21] Haber, S.N., et al., *The orbital and medial prefrontal circuit through the primate basal ganglia.* J Neurosci, 1995. 15(7 Pt 1): p. 4851-67.
[22] Price, J.L., *Definition of the orbital cortex in relation to specific connections with limbic and visceral structures and other cortical regions.* Ann N Y Acad Sci, 2007. 1121: p. 54-71.
[23] Carmichael, S.T. and J.L. Price, *Connectional networks within the orbital and medial prefrontal cortex of macaque monkeys.* J. Comp. Neurol., 1996. 371: p. 179-207.
[24] Romanski, L.M. and P.S. Goldman-Rakic, *An auditory domain in primate prefrontal cortex.* Nat Neurosci, 2002. 5(1): p. 15-6.
[25] Romanski, L.M., J.F. Bates, and P.S. Goldman-Rakic, *Auditory belt and parabelt projections to the prefrontal cortex in the rhesus monkey.* J Comp Neurol, 1999. 403(2): p. 141-57.
[26] Ongur, D., X. An, and J.L. Price, *Prefrontal cortical projections to the hypothalamus in macaque monkeys.* J Comp Neurol, 1998. 401(4): p. 480-505.
[27] Stefanacci, L. and D.G. Amaral, *Some observations on cortical inputs to the macaque monkey amygdala: an anterograde tracing study.* J Comp Neurol, 2002. 451(4): p. 301-23.
[28] Maeng, L.Y., J. Waddell, and T.J. Shors, *The prefrontal cortex communicates with the amygdala to impair learning after acute stress in females but not in males.* J. Neurosci., 2010. 30(48): p. 16188-16196.
[29] Krettek, J.E. and J.L. Price, *The cortical projections of the mediodorsal nucleus and adjacent thalamic nuclei in the rat.* J Comp Neurol, 1977. 171(2): p. 157-91.
[30] Ferron, A., et al., *Inhibitory influence of the mesocortical dopaminergic system on spontaneous activity or excitatory response induced from the thalamic mediodorsal nucleus in the rat medial prefrontal cortex.* Brain Res, 1984. 302(2): p. 257-65.
[31] Vertes, R.P., *Analysis of projections from the medial prefrontal cortex to the thalamus in the rat, with emphasis on nucleus reuniens.* J Comp Neurol, 2002. 442(2): p. 163-87.
[32] Vertes, R.P., *Differential projections of the infralimbic and prelimbic cortex in the rat.* Synapse, 2004. 51(1): p. 32-58.
[33] Ray, J.P. and J.L. Price, *The organization of projections from the mediodorsal nucleus of the thalamus to orbital and medial prefrontal cortex in macaque monkeys.* J Comp Neurol, 1993. 337(1): p. 1-31.
[34] Ray, J.P. and J.L. Price, *The organization of the thalamocortical connections of the mediodorsal thalamic nucleus in the rat, related to the ventral forebrain-prefrontal cortex topography.* J Comp Neurol, 1992. 323(2): p. 167-97.
[35] van Eden, C.G., A. Rinkens, and H.B. Uylings, *Retrograde degeneration of thalamic neurons in the mediodorsal nucleus after neonatal and adult aspiration lesions of the medial prefrontal cortex in the rat. Implications for mechanisms of functional recovery.* Eur J Neurosci, 1998. 10(5): p. 1581-9.

[36] Rotaru, D.C., G. Barrionuevo, and S.R. Sesack, *Mediodorsal thalamic afferents to layer III of the rat prefrontal cortex: Synaptic relationships to subclasses of interneurons.* J. Comp. Neurol., 2005. 490(3): p. 220-238.

[37] Negyessy, L., J. Hamori, and M. Bentivoglio, *Contralateral cortical projection to the mediodorsal thalamic nucleus: origin and synaptic organization in the rat.* Neuroscience, 1998. 84(3): p. 741-53.

[38] Robbins, T.W., *Chemical neuromodulation of frontal-executive functions in humans and other animals.* Exp Brain Res, 2000. 133(1): p. 130-8.

[39] Bjorklund, A. and S.B. Dunnett, *Dopamine neuron systems in the brain: an update.* Trends in Neurosciences, 2007. 30(5): p. 194-202.

[40] Robbins, T.W. and A.F. Arnsten, *The neuropsychopharmacology of fronto-executive function: monoaminergic modulation.* Annu Rev Neurosci, 2009. 32: p. 267-87.

[41] Robbins, T.W., *Chemistry of the mind: Neurochemical modulation of prefrontal cortical function.* The Journal of Comparative Neurology, 2005. 493(1): p. 140-146.

[42] Granon, S., et al., *Enhanced and impaired attentional performance after infusion of D1 dopaminergic receptor agents into rat prefrontal cortex.* J. Neurosci., 2000. 20(3): p. 1208-1215.

[43] Zahrt, J., et al., *Supranormal stimulation of D1 dopamine receptors in the rodent prefrontal cortex impairs spatial working memory performance.* J Neurosci, 1997. 17(21): p. 8528-35.

[44] Vijayraghavan, S., et al., *Inverted-U dopamine D1 receptor actions on prefrontal neurons engaged in working memory.* Nat Neurosci, 2007. 10(3): p. 376-384.

[45] Gibbs, S.E. and M. D'Esposito, *A functional magnetic resonance imaging study of the effects of pergolide, a dopamine receptor agonist, on component processes of working memory.* Neuroscience, 2006. 139(1): p. 359-71.

[46] Dalley, J.W., et al., *Distinct changes in cortical acetylcholine and noradrenaline efflux during contingent and noncontingent performance of a visual attentional task.* J Neurosci, 2001. 21(13): p. 4908-14.

[47] Dalley, J.W., et al., *Specific abnormalities in serotonin release in the prefrontal cortex of isolation-reared rats measured during behavioural performance of a task assessing visuospatial attention and impulsivity.* Psychopharmacology (Berl), 2002. 164(3): p. 329-40.

[48] Arnsten, A.F. and B.M. Li, *Neurobiology of executive functions: catecholamine influences on prefrontal cortical functions.* Biol Psychiatry, 2005. 57(11): p. 1377-84.

[49] Seamans, J.K. and C.R. Yang, *The principal features and mechanisms of dopamine modulation in the prefrontal cortex.* Prog Neurobiol, 2004. 74(1): p. 1-58.

[50] Li, Y.C. and W.J. Gao, *GSK-3beta activity and hyperdopamine-dependent behaviors.* Neurosci Biobehav Rev, 2011. 35: p. 645-654.

[51] Howes, O.D. and S. Kapur, *The dopamine hypothesis of schizophrenia: version III--the final common pathway.* Schizophr Bull, 2009. 35(3): p. 549-62.

[52] O'Donnell, P., *Adolescent maturation of cortical dopamine.* Neurotox Res, 2010. 18(3-4): p. 306-12.

[53] Goto, Y., S. Otani, and A.A. Grace, *The Yin and Yang of dopamine release: a new perspective.* Neuropharmacology, 2007. 53(5): p. 583-587.

[54] Li, Y.-C., et al., *D2 receptor overexpression in the striatum leads to a deficit in inhibitory transmission and dopamine sensitivity in mouse prefrontal cortex.* Proceedings of the National Academy of Sciences, 2011. 108(29): p. 12107-12112.
[55] Gao, W.J. and P.S. Goldman-Rakic, *Selective modulation of excitatory and inhibitory microcircuits by dopamine.* Proc Natl Acad Sci U S A, 2003. 100(5): p. 2836-41.
[56] Gao, W.J., L.S. Krimer, and P.S. Goldman-Rakic, *Presynaptic regulation of recurrent excitation by D1 receptors in prefrontal circuits.* Proc Natl Acad Sci U S A, 2001. 98(1): p. 295-300.
[57] Gao, W.J., Y. Wang, and P.S. Goldman-Rakic, *Dopamine modulation of perisomatic and peridendritic inhibition in prefrontal cortex.* J Neurosci, 2003. 23(5): p. 1622-30.
[58] Li, Y.C., et al., *Dopamine D1 receptor-mediated enhancement of NMDA receptor trafficking requires rapid PKC-dependent synaptic insertion in the prefrontal neurons.* J Neurochem, 2010. 114: p. 62-73.
[59] Hu, J.-L., et al., *Dopamine D1 receptor-mediated NMDA receptor insertion depends on Fyn but not Src kinase pathway in prefrontal cortical neurons.* Molecular Brain 2010. 3(20): p. 1-14.
[60] Li, Y.C., et al., *Activation of glycogen synthase kinase-3 beta is required for hyperdopamine and D2 receptor-mediated inhibition of synaptic NMDA receptor function in the rat prefrontal cortex.* J Neurosci, 2009. 29(49): p. 15551-63.
[61] Gainetdinov, R.R. and M.G. Caron, *Monoamine transporters: from genes to behavior.* Annual Rev Pharmacol Toxicol, 2003. 43(1): p. 261-284.
[62] Murphy, B.L., et al., *Increased dopamine turnover in the prefrontal cortex impairs spatial working memory performance in rats and monkeys.* Proc Natl Acad Sci U S A, 1996. 93(3): p. 1325-9.
[63] Murphy, B.L., et al., *Dopamine and spatial working memory in rats and monkeys: pharmacological reversal of stress-induced impairment.* J Neurosci, 1996. 16(23): p. 7768-75.
[64] Goldman-Rakic, P.S., *Regional and cellular fractionation of working memory.* Proc Natl Acad Sci U S A, 1996. 93(24): p. 13473-80.
[65] Baddeley, A., *Working memory.* Science, 1992. 255(5044): p. 556-9.
[66] Baddeley, A., *Recent developments in working memory.* Curr Opin Neurobiol, 1998. 8(2): p. 234-8.
[67] Baddeley, A., *Working memory.* Curr Biol, 2010. 20(4): p. R136-40.
[68] Baddeley, A., *Working Memory: Theories, Models, and Controversies.* Annual Review of Psychology, 2010.
[69] Goldman-Rakic, P.S., *Cellular basis of working memory.* Neuron, 1995. 14(3): p. 477-85.
[70] Arnsten, A.F., et al., *Dynamic Network Connectivity: A new form of neuroplasticity.* Trends Cogn Sci, 2010. 14(8): p. 365-75.
[71] Fuster, J.M. and G.E. Alexander, *Neuron activity related to short-term memory.* Science, 1971. 173(997): p. 652-4.
[72] Kubota, K. and H. Niki, *Prefrontal cortical unit activity and delayed alternation performance in monkeys.* J Neurophysiol, 1971. 34(3): p. 337-47.
[73] Funahashi, S., C.J. Bruce, and P.S. Goldman-Rakic, *Mnemonic coding of visual space in the monkey's dorsolateral prefrontal cortex.* J Neurophysiol, 1989. 61(2): p. 331-49.

[74] Wang, H.X., et al., *A specialized NMDA receptor function in layer 5 recurrent microcircuitry of the adult rat prefrontal cortex.* Proc. Nat. Acad. Sci. U. S. A. , 2008. 105(43): p. 16791-16796.

[75] Elston, G.N., *Pyramidal cells of the frontal lobe: all the more spinous to think with.* J Neurosci, 2000. 20(18): p. RC95.

[76] Elston, G.N., *Cortex, cognition and the cell: new insights into the pyramidal neuron and prefrontal function.* Cereb Cortex, 2003. 13(11): p. 1124-38.

[77] Elston, G.N., R. Benavides-Piccione, and J. DeFelipe, *The pyramidal cell in cognition: a comparative study in human and monkey.* J Neurosci, 2001. 21(17): p. RC163.

[78] Elston, G.N., R. Benavides-Piccione, and J. Defelipe, *A study of pyramidal cell structure in the cingulate cortex of the macaque monkey with comparative notes on inferotemporal and primary visual cortex.* Cereb Cortex, 2005. 15(1): p. 64-73.

[79] Elston, G.N. and J. DeFelipe, *Spine distribution in cortical pyramidal cells: a common organizational principle across species.* Prog Brain Res, 2002. 136: p. 109-33.

[80] Wang, Y., et al., *Heterogeneity in the pyramidal network of the medial prefrontal cortex.* Nat Neurosci, 2006. 9: p. 534-542.

[81] Barak, O. and M. Tsodyks, *Persistent activity in neural networks with dynamic synapses.* PLoS Computational Biology, 2007. 3(2): p. e35.

[82] Brunel, N., *Persistent activity and the single-cell frequency-current curve in a cortical network model.* Network, 2000. 11(4): p. 261-80.

[83] Compte, A., et al., *Synaptic mechanisms and network dynamics underlying spatial working memory in a cortical network model.* Cereb Cortex, 2000. 10(9): p. 910-23.

[84] Curtis, C.E. and M. D'Esposito, *Persistent activity in the prefrontal cortex during working memory.* Trends in Cognitive Sciences, 2003. 7(9): p. 415-423.

[85] Durstewitz, D., J.K. Seamans, and T.J. Sejnowski, *Neurocomputational models of working memory.* Nat Neurosci, 2000. 3 Suppl: p. 1184-91.

[86] McCormick, D.A., et al., *Persistent cortical activity: mechanisms of generation and effects on neuronal excitability.* Cereb Cortex, 2003. 13(11): p. 1219-31.

[87] Wang, X.-J., *Synaptic basis of cortical persistent activity: the importance of NMDA receptors to working memory.* J Neurosci, 1999. 19(21): p. 9587-603.

[88] Constantinidis, C., G.V. Williams, and P.S. Goldman-Rakic, *A role for inhibition in shaping the temporal flow of information in prefrontal cortex.* Nat Neurosci, 2002. 5(2): p. 175-80.

[89] Constantinidis, C. and P.S. Goldman-Rakic, *Correlated discharges among putative pyramidal neurons and interneurons in the primate prefrontal cortex.* J Neurophysiol, 2002. 88(6): p. 3487-97.

[90] Wang, X.J., et al., *Division of labor among distinct subtypes of inhibitory neurons in a cortical microcircuit of working memory.* Proc Natl Acad Sci U S A, 2004. 101(5): p. 1368-73.

[91] Rakic, P., *Evolution of the neocortex: a perspective from developmental biology.* Nat Rev Neurosci, 2009. 10(10): p. 724-735.

[92] Goldman-Rakic, P.S., *Development of cortical circuitry and cognitive function.* Child Dev, 1987. 58(3): p. 601-22.

[93] Lenroot, R.K. and J.N. Giedd, *Brain development in children and adolescents: insights from anatomical magnetic resonance imaging.* Neurosci Biobehav Rev, 2006. 30(6): p. 718-29.

[94] Kuboshima-Amemori, S. and T. Sawaguchi, *Plasticity of the primate prefrontal cortex.* Neuroscientist, 2007. 13(3): p. 229-40.

[95] Rakic, P., et al., *Concurrent overproduction of synapses in diverse regions of the primate cerebral cortex.* Science, 1986. 232(4747): p. 232-5.

[96] Huttenlocher, P.R. and A.S. Dabholkar, *Regional differences in synaptogenesis in human cerebral cortex.* J Comp Neurol, 1997. 387(2): p. 167-78.

[97] Mrzljak, L., et al., *Neuronal development in human prefrontal cortex in prenatal and postnatal stages.* Prog Brain Res, 1990. 85: p. 185-222.

[98] Tsujimoto, S., *The prefrontal cortex: functional neural development during early childhood.* Neuroscientist, 2008. 14(4): p. 345-58.

[99] Gonzalez-Burgos, G., et al., *Functional maturation of excitatory synapses in layer 3 pyramidal neurons during postnatal development of the primate prefrontal cortex.* Cereb Cortex, 2008. 18: p. 626-637.

[100] Davey, C.G., M. Yucel, and N.B. Allen, *The emergence of depression in adolescence: development of the prefrontal cortex and the representation of reward.* Neurosci Biobehav Rev, 2008. 32(1): p. 1-19.

[101] Andersen, S.L. and M.H. Teicher, *Stress, sensitive periods and maturational events in adolescent depression.* Trends Neurosci, 2008. 31(4): p. 183-91.

[102] Davidson, M.C., et al., *Development of cognitive control and executive functions from 4 to 13 years: evidence from manipulations of memory, inhibition, and task switching.* Neuropsychologia, 2006. 44(11): p. 2037-78.

[103] Casey, B.J., J.N. Giedd, and K.M. Thomas, *Structural and functional brain development and its relation to cognitive development.* Biological Psychology, 2000. 54(1-3): p. 241-57.

[104] Diamond, A. and P.S. Goldman-Rakic, *Comparison of human infants and rhesus monkeys on Piaget's AB task: evidence for dependence on dorsolateral prefrontal cortex.* Exp Brain Res, 1989. 74(1): p. 24-40.

[105] Sisk, C.L. and D.L. Foster, *The neural basis of puberty and adolescence.* Nat Neurosci, 2004. 7(10): p. 1040-7.

[106] Keshavan, M.S., S. Anderson, and J.W. Pettegrew, *Is schizophrenia due to excessive synaptic pruning in the prefrontal cortex? The Feinberg hypothesis revisited.* J Psychiatr Res, 1994. 28(3): p. 239-65.

[107] Adriani, W. and G. Laviola, *Windows of vulnerability to psychopathology and therapeutic strategy in the adolescent rodent model.* Behav Pharmacol, 2004. 15(5-6): p. 341-52.

[108] Lewis, D.A., *Development of the prefrontal cortex during adolescence: insights into vulnerable neural circuits in schizophrenia.* Neuropsychopharmacol, 1997. 16(6): p. 385-98.

[109] Arnsten, A.F. and R.M. Shansky, *Adolescence: vulnerable period for stress-induced prefrontal cortical function?* Ann N Y Acad Sci, 2004. 1021: p. 143-7.

[110] Bourgeois, J.P., P.S. Goldman-Rakic, and P. Rakic, *Synaptogenesis in the prefrontal cortex of rhesus monkeys.* Cereb Cortex, 1994. 4(1): p. 78-96.

[111] Woo, T.U., et al., *Peripubertal refinement of the intrinsic and associational circuitry in monkey prefrontal cortex.* Neuroscience, 1997. 80(4): p. 1149-58.

[112] Hoftman, G.D. and D.A. Lewis, *Postnatal developmental trajectories of neural circuits in the primate prefrontal cortex: identifying sensitive periods for vulnerability to schizophrenia.* Schizophr Bull, 2011. 37(3): p. 493-503.

[113] Wang, H.X. and W.J. Gao, *Development of calcium-permeable AMPA receptors and their correlation with NMDA receptors in fast-spiking interneurons of rat prefrontal cortex.* J Physiol, 2010. 588: p. 2823-2838.

[114] Wang, H.X. and W.J. Gao, *Cell type-specific development of NMDA receptors in the interneurons of rat prefrontal cortex.* Neuropsychopharmacol, 2009. 34(8): p. 2028-40.

[115] Tseng, K.Y. and P. O'Donnell, *Dopamine modulation of prefrontal cortical interneurons changes during adolescence.* Cereb Cortex, 2007. 17(5): p. 1235-1240.

[116] Lewis, D.A., et al., *Postnatal development of prefrontal inhibitory circuits and the pathophysiology of cognitive dysfunction in schizophrenia.* Ann N Y Acad Sci, 2004. 1021: p. 64-76.

[117] Vincent, S.L., L. Pabreza, and F.M. Benes, *Postnatal maturation of GABA-immunoreactive neurons of rat medial prefrontal cortex.* J Comp Neurol, 1995. 355(1): p. 81-92.

[118] O'Donnell, P., *Adolescent onset of cortical disinhibition in schizophrenia: insights from animal models.* Schizophr Bull, 2011. 37(3): p. 484-92.

[119] Anderson, S.A., et al., *Synchronous development of pyramidal neuron dendritic spines and parvalbumin-immunoreactive chandelier neuron axon terminals in layer III of monkey prefrontal cortex.* Neuroscience, 1995. 67(1): p. 7-22.

[120] Uhlhaas, P.J., *The adolescent brain: implications for the understanding, pathophysiology, and treatment of schizophrenia.* Schizophr Bull, 2011. 37(3): p. 480-3.

[121] Uhlhaas, P.J., et al., *The development of neural synchrony reflects late maturation and restructuring of functional networks in humans.* Proceedings of the National Academy of Sciences, 2009. 106(24): p. 9866-9871.

[122] Sohal, V.S., et al., *Parvalbumin neurons and gamma rhythms enhance cortical circuit performance.* Nature, 2009. 459(7247): p. 698-702.

[123] Cobb, S.R., et al., *Synchronization of neuronal activity in hippocampus by individual GABAergic interneurons.* Nature, 1995. 378(6552): p. 75-8.

[124] Whittington, M.A., R.D. Traub, and J.G. Jefferys, *Synchronized oscillations in interneuron networks driven by metabotropic glutamate receptor activation.* Nature, 1995. 373(6515): p. 612-5.

[125] Wang, X.J. and G. Buzsaki, *Gamma oscillation by synaptic inhibition in a hippocampal interneuronal network model.* J Neurosci, 1996. 16(20): p. 6402-13.

[126] Traub, R.D., J.G. Jefferys, and M.A. Whittington, *Simulation of gamma rhythms in networks of interneurons and pyramidal cells.* J Comput Neurosci, 1997. 4(2): p. 141-50.

[127] Csicsvari, J., et al., *Fast network oscillations in the hippocampal CA1 region of the behaving rat.* J Neurosci, 1999. 19(16): p. RC20.

[128] Csicsvari, J., et al., *Oscillatory coupling of hippocampal pyramidal cells and interneurons in the behaving Rat.* J Neurosci, 1999. 19(1): p. 274-87.

[129] Bartos, M., et al., *Fast synaptic inhibition promotes synchronized gamma oscillations in hippocampal interneuron networks.* Proc Natl Acad Sci U S A, 2002. 99(20): p. 13222-7.

[130] Vida, I., M. Bartos, and P. Jonas, *Shunting inhibition improves robustness of gamma oscillations in hippocampal interneuron networks by homogenizing firing rates.* Neuron, 2006. 49(1): p. 107-117.

[131] Fuchs, E.C., et al., *Recruitment of Parvalbumin-Positive Interneurons Determines Hippocampal Function and Associated Behavior.* Neuron, 2007. 53(4): p. 591-604.

[132] Woo, T.U., K. Spencer, and R.W. McCarley, *Gamma oscillation deficits and the onset and early progression of schizophrenia.* Harvard Review of Psychiatry, 2010. 18(3): p. 173-89.

[133] Whitford, T.J., et al., *Brain maturation in adolescence: concurrent changes in neuroanatomy and neurophysiology.* Human Brain Mapping, 2007. 28(3): p. 228-37.

[134] Thatcher, R.W., D.M. North, and C.J. Biver, *Development of cortical connections as measured by EEG coherence and phase delays.* Human Brain Mapping, 2008. 29(12): p. 1400-15.
[135] Arnsten, A.F.T., *Stress signalling pathways that impair prefrontal cortex structure and function.* Nat Rev Neurosci, 2009. 10(6): p. 410-422.
[136] Weinberger, D.R., K.F. Berman, and R.F. Zec, *Physiologic dysfunction of dorsolateral prefrontal cortex in schizophrenia. I. Regional cerebral blood flow evidence.* Arch. Gen. Psychiatry, 1986. 43: p. 114-124.
[137] Fumagalli, F., et al., *Stress during development: Impact on neuroplasticity and relevance to psychopathology.* Progress in Neurobiology, 2007. 81(4): p. 197-217.
[138] Roberts, R.C., *Schizophrenia in Translation: Disrupted in Schizophrenia (DISC1): Integrating Clinical and Basic Findings.* Schizophrenia Bulletin, 2007. 33(1): p. 11-15.
[139] Hains, A.B. and A.F. Arnsten, *Molecular mechanisms of stress-induced prefrontal cortical impairment: implications for mental illness.* Learn Mem, 2008. 15(8): p. 551-64.
[140] Benes, F.M. and S. Berretta, *GABAergic interneurons: implications for understanding schizophrenia and bipolar disorder.* Neuropsychopharmacol, 2001. 25(1): p. 1-27.
[141] Weinberger, D.R., *Schizophrenia and the frontal lobe.* Trends Neurosci, 1988. 11(8): p. 367-70.
[142] Weinberger, D.R., et al., *Prefrontal neurons and the genetics of schizophrenia.* Biol Psychiatry, 2001. 50(11): p. 825-44.
[143] Harrison, P.J., *Schizophrenia: a disorder of neurodevelopment?* Curr Opin Neurobiol, 1997. 7(2): p. 285-9.
[144] Lewis, D.A. and G. Gonzalez-Burgos, *Neuroplasticity of neocortical circuits in schizophrenia.* Neuropsychopharmacol, 2008. 33(1): p. 141-65.
[145] Levy, F. and M. Farrow, *Working memory in ADHD: prefrontal/parietal connections.* Curr Drug Targets, 2001. 2(4): p. 347-52.
[146] Arnsten, A.F., *Catecholamine influences on dorsolateral prefrontal cortical networks.* Biol Psychiatry, 2011. 69(12): p. e89-99.
[147] Myers-Schulz, B. and M. Koenigs, *Functional anatomy of ventromedial prefrontal cortex: implications for mood and anxiety disorders.* Mol Psychiatry, 2011.
[148] Kolb, B., S. Pellis, and T.E. Robinson, *Plasticity and functions of the orbital frontal cortex.* Brain Cogn, 2004. 55(1): p. 104-15.
[149] Lyons, D.M., *Stress, depression, and inherited variation in primate hippocampal and prefrontal brain development.* Psychopharmacol Bull, 2002. 36(1): p. 27-43.
[150] Mundy, P., M. Gwaltney, and H. Henderson, *Self-referenced processing, neurodevelopment and joint attention in autism.* Autism, 2010. 14(5): p. 408-29.
[151] Shalom, D.B., *The medial prefrontal cortex and integration in autism.* Neuroscientist, 2009. 15(6): p. 589-98.
[152] Hill, E.L., *Executive dysfunction in autism.* Trends Cogn Sci, 2004. 8(1): p. 26-32.
[153] Sabbagh, M.A., *Understanding orbitofrontal contributions to theory-of-mind reasoning: implications for autism.* Brain Cogn, 2004. 55(1): p. 209-19.
[154] Bachevalier, J. and K.A. Loveland, *The orbitofrontal-amygdala circuit and self-regulation of social-emotional behavior in autism.* Neurosci Biobehav Rev, 2006. 30(1): p. 97-117.
[155] Zikopoulos, B. and H. Barbas, *Changes in prefrontal axons may disrupt the network in autism.* J Neurosci, 2010. 30(44): p. 14595-609.

[156] Walsh, T., et al., *Rare structural variants disrupt multiple genes in neurodevelopmental pathways in schizophrenia.* Science, 2008. 320(5875): p. 539-43.

[157] Niwa, M., et al., *Knockdown of DISC1 by in utero gene transfer disturbs postnatal dopaminergic maturation in the frontal cortex and leads to adult behavioral deficits.* Neuron, 2010. 65(4): p. 480-9.

[158] Costa, E., et al., *Epigenetic targets in GABAergic neurons to treat schizophrenia.* Adv Pharmacol, 2006. 54: p. 95-117.

[159] Crow, T.J., *How and why genetic linkage has not solved the problem of psychosis: review and hypothesis.* Am J Psychiatry, 2007. 164(1): p. 13-21.

[160] Farber, N.B., J.W. Newcomer, and J.W. Olney, *The glutamate synapse in neuropsychiatric disorders. Focus on schizophrenia and Alzheimer's disease.* Prog Brain Res, 1998. 116: p. 421-37.

[161] Coyle, J.T., G. Tsai, and D. Goff, *Converging evidence of NMDA receptor hypofunction in the pathophysiology of schizophrenia.* Ann N Y Acad Sci, 2003. 1003: p. 318-27.

[162] Marek, G.J., et al., *Glutamatergic (N-methyl-D-aspartate receptor) hypofrontality in schizophrenia: too little juice or a miswired brain?* Mol Pharmacol, 2010. 77(3): p. 317-26.

[163] Seeman, P., *All roads to schizophrenia lead to dopamine supersensitivity and elevated dopamine D2 receptors.* CNS Neurosci Ther, 2011. 17(2): p. 118-132.

[164] Remington, G., O. Agid, and G. Foussias, *Schizophrenia as a disorder of too little dopamine: implications for symptoms and treatment.* Expert Rev Neurother, 2011. 11(4): p. 589-607.

[165] Artigas, F., *The prefrontal cortex: a target for antipsychotic drugs.* Acta Psychiatr Scand, 2010. 121(1): p. 11-21.

[166] Gonzalez-Burgos, G., T. Hashimoto, and D.A. Lewis, *Alterations of cortical GABA neurons and network oscillations in schizophrenia.* Curr Psychiatry Rep, 2010. 12(4): p. 335-44.

[167] Martin, L.F. and R. Freedman, *Schizophrenia and the alpha7 nicotinic acetylcholine receptor.* Int Rev Neurobiol, 2007. 78: p. 225-46.

[168] Leonard, S., et al., *Nicotinic receptor function in schizophrenia.* Schizophr Bull, 1996. 22(3): p. 431-45.

[169] Lyon, E.R., *A review of the effects of nicotine on schizophrenia and antipsychotic medications.* Psychiatr Serv, 1999. 50(10): p. 1346-50.

[170] Mansvelder, H.D., et al., *Nicotinic modulation of neuronal networks: from receptors to cognition.* Psychopharmacology (Berl), 2005: p. 1-14.

[171] Allen, N.C., et al., *Systematic meta-analyses and field synopsis of genetic association studies in schizophrenia: the SzGene database.* Nat Genet, 2008. 40(7): p. 827-34.

[172] Tan, H.Y., J.H. Callicott, and D.R. Weinberger, *Prefrontal cognitive systems in schizophrenia: towards human genetic brain mechanisms.* Cogn Neuropsychiatry, 2009. 14(4-5): p. 277-98.

[173] Tunbridge, E.M., P.J. Harrison, and D.R. Weinberger, *Catechol-o-Methyltransferase, cognition, and psychosis: Val158Met and beyond.* Biological Psychiatry, 2006. 60(2): p. 141-151.

[174] Savitz, J., M. Solms, and R. Ramesar, *The molecular genetics of cognition: dopamine, COMT and BDNF.* Genes Brain Behav, 2006. 5(4): p. 311-28.

[175] Harrison, P.J. and D.R. Weinberger, *Schizophrenia genes, gene expression, and neuropathology: on the matter of their convergence.* Mol Psychiatry, 2005. 10(1): p. 40-68.
[176] Cannon, T.D., *The inheritance of intermediate phenotypes for schizophrenia.* Curr Opin Psychiatry, 2005. 18(2): p. 135-40.
[177] Bilder, R.M., et al., *The catechol-O-methyltransferase polymorphism: relations to the tonic - phasic dopamine hypothesis and neuropsychiatric phenotypes.* Neuropsychopharmacology, 2004. 29(11): p. 1943-61.
[178] Kato, T., et al., *Transient exposure of neonatal mice to neuregulin-1 results in hyperdopaminergic states in adulthood: implication in neurodevelopmental hypothesis for schizophrenia.* Mol Psychiatry, 2011. 16(3): p. 307-320.
[179] Roy, K., et al., *Loss of erbB signaling in oligodendrocytes alters myelin and dopaminergic function, a potential mechanism for neuropsychiatric disorders.* PNAS, 2007. 104(19): p. 8131-8136.
[180] Lipina, T.V., et al., *Enhanced dopamine function in DISC1-L100P mutant mice: implications for schizophrenia.* Genes Brain Behav, 2010. 9(7): p. 777-89.
[181] Ji, Y., et al., *Role of dysbindin in dopamine receptor trafficking and cortical GABA function.* Proceedings of the National Academy of Sciences, 2009. 106(46): p. 19593-19598.
[182] Papaleo, F. and D.R. Weinberger, *Dysbindin and Schizophrenia: it's dopamine and glutamate all over again.* Biol Psychiatry, 2011. 69(1): p. 2-4.
[183] Iizuka, Y., et al., *Evidence that the BLOC-1 protein dysbindin modulates dopamine D2 receptor internalization and signaling but not D1 internalization.* J. Neurosci., 2007. 27(45): p. 12390-12395.
[184] Papaleo, F., B.K. Lipska, and D.R. Weinberger, *Mouse models of genetic effects on cognition: Relevance to schizophrenia.* Neuropharmacology, 2011.
[185] Shi, J., E.S. Gershon, and C. Liu, *Genetic associations with schizophrenia: meta-analyses of 12 candidate genes.* Schizophr Res, 2008. 104(1-3): p. 96-107.
[186] Hahn, C.G., et al., *Altered neuregulin 1-erbB4 signaling contributes to NMDA receptor hypofunction in schizophrenia.* Nat Med, 2006. 12(7): p. 824-8.
[187] Guidotti, A., et al., *Epigenetic GABAergic targets in schizophrenia and bipolar disorder.* Neuropharmacology, 2011. 60(7-8): p. 1007-1016.
[188] Kundakovic, M., et al., *The reelin and GAD67 promoters are activated by epigenetic drugs that facilitate the disruption of local repressor complexes.* Mol Pharmacol, 2009. 75(2): p. 342-54.
[189] Guidotti, A., et al., *Characterization of the action of antipsychotic subtypes on valproate-induced chromatin remodeling.* Trends Pharmacol Sci, 2009. 30(2): p. 55-60.
[190] Woo and Crowell, *Targeting synapses and myelin in the prevention of schizophrenia.* Schizophrenia Research, 2005. 73(2-3): p. 193-207.
[191] Insel, T.R., *Rethinking schizophrenia.* Nature, 2010. 468(7321): p. 187-193.
[192] Lieberman, J.A., L.F. Jarskog, and D. Malaspina, *Preventing clinical deterioration in the course of schizophrenia: the potential for neuroprotection.* J Clin Psychiatry, 2006. 67(6): p. 983-90.

2

Neurocognitive Aspects of Tourette Syndrome and Related Disorders

Marc E. Lavoie and Kieron P. O'Connor
Cognitive and Social Psychophysiology Laboratory,
FRSQ Research Team on Obsessive-Compulsive Spectrum,
Fernand-Seguin Research Center of the Louis-H Lafontaine Hospital,
Department of Psychiatry,
University of Montreal, Québec,
Canada

1. Introduction

1.1 The challenge of characterizing Gilles de la Tourette Syndrome

One of the top priorities, for current research in Gilles de la Tourette Syndrome (GTS), is to disentangle the intricate interactions between regions of the frontal cortex and the basal ganglia. This approach will reveal how these interactions act in concert to regulate motor, emotional, and cognitive action plans (Keen-Kim & Freimer, 2006; Leckman, 2002; State, 2011). Another key issue is the understanding of these brain mechanisms with GTS in the presence of obsessive-compulsive disorders (OCD) (Gaze, Kepley, & Walkup, 2006). The heuristic value of our proposed approach resides in the fact that cognitive and cerebral functions are two salient features easily quantified with non invasive protocols. As proposed by Swain et al., (Swain, Scahill, Lombroso, King, & Leckman, 2007) *"a determined effort to explore the electrophysiology of this disorder using EEG/MEG recordings is our next best step"*. We will first review the current state of the literature regarding specific cerebral structures underlying GTS symptoms. Secondly, we will expose a strategy to integrate brain imaging, electrophysiology and neuropsychology in the exploration of the GTS brain in action. Third, we will investigate clinical and phenomenological aspects of comorbidity in GTS patients. We will thus, expose a functional method based on multimodal assessments to characterize the relationship between tic expression, brain activity and different levels of cognitive processing such as motor activation, memory and emotions.

1.2 Definition

In 1885, Dr. Georges Gilles de la Tourette described nine patients with motor and vocal tics, some of which had echo phenomena (a tendency to repeat things said to them) and coprolalia (utterances of obscene phrases) (Gilles de la Tourette, 1885). This syndrome is currently classified in the DSM-IV-TR (APA, 2000) with disorders first diagnosed in infancy, childhood or adolescence. The essential features are the presence of *simple* or *complex* multiple motor tics and one or more vocal tics. *Simple tics* are defined as repetitive non-voluntary contractions of functionally related groups of skeletal muscles in one or more parts of the body including blinking, cheek twitches and head or knee jerks among others (Leckman et al., 1997; Shapiro & Shapiro, 1986). *Complex tics* may take the form of self-inflicted repetitive actions such as nail biting, hair pulling, head slapping, teeth grinding or tense-release hand gripping cycles. Tics appear many times a day with onset longer than a year and prior to 18 years old.

1.3 Genetics in GTS

Since the first systematic report of tics in the 19th century by Itard (Itard, 1825) and later by Gilles de la Tourette (Gilles de la Tourette, 1885), generational transmission of the disease was suspected. More than one century later, genetic factors in GTS remain hypothetical. A large twin study showed concordance rates that are three to four times higher for monozygotic than to dizygotic twins (Price, Leckman, Pauls, Cohen, & Kidd, 1986). Studies investigating affected families with GTS suggests that the trait is inherited in an autosomal dominant pattern with variable expression (Eapen, Pauls, & Robertson, 1993; Alsobrook & Pauls, 1997). Analysis of vertical transmission patterns in families has revealed that OCD and GTS may share some underlying genetic vulnerabilities (Pauls, 1992). The pattern of comorbidity and other evidence indicates that GTS genes may be responsible for a spectrum of disorders, including OCD and Attention Deficit Hyperactivity Disorder (ADHD) even if OCD and ADHD can equally exist with their own etiologies. The inherited trait may not cause any disorder or may manifest as GTS, chronic multiple tic disorder, ADHD and/or OCD (Keen-Kim & Freimer, 2006). In a comprehensive review, Pauls (2003), underlined that genetic factors play an important role in the manifestation of GTS and that several genes are important with some possibly having major effect; and several regions of the genome have been identified as potential locations of these susceptibility genes.

More specifically, sequencing of SLIT and TRK like family member 1 (SLITRK1), revealed a single base deletion as well as two independent occurrences of a mutation called the var321 (Abelson et al., 2005), likely associated with GTS. SLITRK1 expression was confirmed in cortical striatal circuits, which is consistent with regions implicated in GTS pathology (Stillman et al., 2009). An animal model of SLITRK1 deficiency shows altered noradrenergic function phenotype related to alpha-agonists, which are used in the treatment of Tourette syndrome (Katayama et al., 2010). However, the SLITRK1 gene expression in GTS remain under question since other research was not able to replicate these results in human (Scharf et al., 2008). Other candidate genes have been tested with mixed or equivocal results such as genes related to dopamine and serotonin transporters, glycine receptor, 5q33-q35 neuroreceptors, adrenergic receptors, methyl-CpG binding protein 2, and human leukocyte antigen (Keen-Kim & Freimer, 2006; Pauls, 2003).

In brief, GTS is a genetically complex disorder that probably arises with multiple genes interacting with environmental components. Recent development could certainly show promises for success in finding the responsible genes and sequence variants, resulting in better targeted treatments.

1.4 Epidemiology and prevalence

Depending on the sample characteristics, between 0.15% and 1.1% of all children have GTS and boys outnumber girls by at least 4:1 (Kadesjo & Gillberg, 2000), with the most severe period of tic severity occurring at 10 years old (Leckman et al., 1998), followed by a decrease until the adult age with approximately 40% eventually becoming symptom-free (Burd et al., 2001). Although whether tics disappear or adapt in adults remains controversial (Pappert, Goetz, Louis, Blasucci, & Leurgans, 2003). Tics are also sensitive to a number of exacerbating factors including everyday psychosocial stress, anxiety, emotional excitement, and fatigue (Findley et al., 2003). Once considered very rare, the incidence of GTS in adults is about 0.1-1% (Leckman et al., 1998). The lifetime prevalence of GTS in adults is not known, but estimates vary between 5% and 10% of the population. In a recent study, O'Connor (2005) found a self-report life-time prevalence rate of 8%. Other recent estimates have placed the prevalence of GTS at 1% and chronic tic disorders at 10% of the population (Robertson, 2003; Robertson & Stern, 2000).

1.5 Secondary distress caused by tics

Tics are rarely life-threatening except in cases where they may provoke auto-mutilation. Psychosocial distress however can be considerable and can involve secondary phobias, depressions, social anxieties and worries over self-image, and relationship problems. In our estimation of the interference of tic and habit disorders in daily activities, we found problems ranging from unemployment, marital conflict, interpersonal difficulties, employer relations, travel restrictions, problems attending social or public functions, performance worries (e.g. about driving, speaking, teaching, dancing, sport) all of which were perceived (by the affected person) to be a result of the tic habit (O'Connor, 2005; O'Connor et al., 2001). People with tics often experience low self-esteem and are (or become) hyperattentive to the judgment of others with consequent low self-satisfaction (Thibert, Day, & Sandor, 1995).

1.6 Comorbidity and associated disorders

The presence of tic symptoms alone is often the exception rather than the rule (Scahill, Sukhodolsky, Williams, & Leckman, 2005) and the expression of tics is a constituent part of a larger mosaic of collateral symptoms. Comorbidity is defined as an additional coexisting diagnosable problem distinct from the principal complaint. So, in addition to this clinical picture defined herein, GTS often appears in association with other psychopathologies, typically referred to as the "GTS+" group (Robertson, 2003). Freeman *et al.*, (2000) established that anger control problems, sleep difficulties, coprolalia, and self-injurious behavior, reached high levels in GTS individuals with comorbidity. Large epidemiological studies also showed that the most frequent comorbidity in GTS is ADHD in children and OCD in adults, affecting each about 50% of GTS patients (Alsobrook & Pauls, 2002; Freeman, 2007; Freeman et al., 2000). Studies are frequently compromised because of not factoring out comorbidity. There are however challenges in detecting and diagnosing comorbidities in GTS. For instance, early research (e.g., (Shapiro & Shapiro, 1992) argued that high rates of comorbidity of GTS with OCD result from mistaking impulsion for compulsion particularly in the case of complex tics, and this may explain the wide range in prevalence estimation. Another difficulty is that the multiple forms of tics (phonic, motor, sensory, cognitive, simple, complex) can be mistaken for symptoms of other disorders.

1.7 Externalizing and aggressive behavior in GTS

The challenge of characterizing GTS *per se* is often confounded by externalizing symptoms that superimpose on tics and there is a clear consensus on the importance of considering these symptoms. Stephens and Sandor (1999) found that conduct disorder was significantly higher in the GTS+ comorbidity group than in the GTS-only or control groups, with more problems reported in older children. These findings provide evidence that aggressive behavior observed in children with GTS may be associated with comorbidity, independently of tic severity or age. Consistently, Carter (2000) demonstrated that children with GTS+comorbidity showed more behavior problems and poorer social adaptation than children with GTS only or unaffected controls. Children with GTS only were not significantly different from controls on most measures of externalizing behaviors and social adaptation, but did exhibit more internalizing symptoms. Moreover, tic symptom severity was not associated with social, behavioral, or emotional functioning among children with GTS, even after stratifying by medication status. These findings suggest that much of the social and behavioral dysfunction in children with GTS could be ADHD or OCD-specific and children with GTS alone may have a very different social-emotional profile than those with GTS+comorbidity. The impact of such comorbidity is especially evident in children where the co-occurrence of GTS with OCD, particularly in the presence of ADHD, increase the likelihood of explosive behavior (Budman, Bruun, Park, & Olson, 1998; Budman, Rockmore, Stokes, & Sossin, 2003). In explosive behavior, the child, for no apparent reason and for a brief period, flies into a state of seemingly uncontrollable, sometimes aggressive, anger, only to resume a normal demeanor a few minutes later. But rage and explosive behavior may be an emotional tic, similar in form and onset to motor tics and hence form part of GTS (Budman, Bruun, Park, Lesser, & Olson, 2000). The difficulty for diagnosis is that non-tic features of GTS may nonetheless be characteristic of GTS rather than other problems. For example, motor restlessness which is a symptom of sensorimotor activation accompanying GTS is also a symptom of ADHD. Clearly, an important step in clarifying diagnosis and the role of comorbidity is to develop a coherent account of the various manifestations of GTS, and in particular the precise form and function of tics.

1.8 The consequence of collateral symptoms in GTS

The impact of comorbidity in GTS touches on clinical manifestations and management. Cases with comorbidity are likely to show more severe symptoms, show poorer prognosis, and are more likely to be treatment-resistant (Leclerc, Forget, & O'Connor, 2008). Children with GTS and comorbidity, in particular OCD and ADHD, show more behavioral problems and poorer psychosocial adaptation whether at school or in other domains. There is also the question of what problem to treat first, and whether treating one problem impacts on the treatment of other problems. For example, treatment strategies for treating hyperactivity involve medication, which can (at least temporarily) exacerbate tics. In addition, particularly in children, it is frequently comorbid behavioral problems (e.g., explosive outbursts) which are most disruptive for the family. The presence of other problems in GTS also adds to feelings of stress, inability to cope and low self-esteem. A further consequence is that existence of at least one comorbidity increase the probability of further comorbidity, such as OCD and hyperactivity, that substantially increases the risk of concurrent explosive outbursts (Budman et al., 2003).

The clinician is frequently confronted with the issue of which problem to treat first. Usually there are multiple comorbidities and their assessment are often unreliable, in part because it is unclear which comorbidities are distinct from GTS or part of the same problem. For example, all the comorbidities in GTS have distinct tic-like features. Explosive outbursts may be viewed as emotional tics and what appear as OCD-like behaviors in GTS may in fact be complex tics, and hyperactivity may be a by-product of the heightened sensorimotor activation often found in GTS (O'Connor, 2002).

Tics are usually preceded by "premonitory urges," described by patients as growing tension of the ticcing muscle or as increased anxiety, which is temporarily relieved after performance of the tic (Leckman, Bloch, Scahill, & King, 2006). These manifestations are very similar to OCD, in which subjects feel increased anxiety and discomfort until certain compulsions are performed (King & Scahill, 2001). More precisely, the manifestation of OCD symptoms is characterized by recurrent intrusive thoughts (*e.g.* obsession) accompanied by repetitive, seemingly purposeful behaviors (*e.g.* compulsion), sufficiently severe to interfere with daily functioning. OCD appears in In half of GTS (Apter et al., 1992) in comparison with 3-4% in the non-GTS adult population (Karno, Golding, Sorenson, & Burnam, 1988; Zohar et al., 1992). Three main questions arise from these findings. How to discriminate OCD characteristics from typical symptoms of GTS? How to objectively characterize expression of motor tics in GTS and GTS+OCD? And finally, how to characterize these comorbid groups with neurocognitive measures. This will constitute one of the primary focuses of the current chapter.

2. Neurobiological basis of Tourette

2.1 Can we identify specific cerebral structures underlying GTS symptoms?

Studies using magnetic resonance imaging (MRI), have identified minor reduction in the putamen and the caudate nuclei when confounding variables such as sex, age, OCD, attention-deficit hyperactivity disorders (ADHD) and streptococcal infection were taken into account (Peterson et al., 2000). Other MRI and positron emission tomography (PET) studies consistently reported volumetric and metabolic reductions in lentiform (Braun et al., 1995; Eidelberg et al., 1997) and caudate nuclei (Bloch, Leckman, Zhu, & Peterson, 2005; Hyde et al., 1995; Stoetter et al., 1992). The basal ganglia are not the sole cerebral structures involved in the pathogenesis of GTS. An extensive investigation (Peterson et al., 2007) comparing a large sample of GTS and controls aged between 6-63 years old, showed increased volumes of the head and medial surface of the hippocampus and the dorsal and ventral surfaces of the amygdala. Volumes of these subregions declined with age in the GTS group but not in controls, so the sub-regions were larger in GTS children, but significantly smaller in GTS adults than in the control group. In children and adults, volumes in these subregions correlated inversely with the severity of tic, suggesting that enlargement of these structures have a neuro-modulatory effect on tics. In addition to these networks, motor and sensorimotor cortices have showed metabolic increases associated with heightened activation in premotor cortex and supplementary motor area (SMA) with PET imaging (Braun et al., 1993; Eidelberg et al., 1997; Stoetter et al., 1992). Cortical thinning in sensorimotor areas was also correlated with tic severity and was most prominent in ventral portions of the homunculi that control the facial, orolingual and laryngeal muscles commonly involved in tic expressions (Sowell et al., 2008). In a recent review of neuroimaging studies, Sheppard et al., (1999) underlined that GTS patients may develop clinical levels of OCD and/or ADHD since all three disorders involve neuropathology of the Basal-Ganglia Thalamo Cortical (BGTC) pathways. For instance, GTS patients may have a dysfunction in sensorimotor and limbic BGTC circuits; OCD in the prefrontal and limbic BGTC pathways; and ADHD in the sensorimotor, orbitofrontal, and limbic BGTC circuits.

In summary, the most recent volumetric observations in structural brain imaging suggest that complex networks related to sensorimotor functions are involved in GTS rather than a defined region of interest. The next important question is to address the functional problem of how these altered cerebral networks affect cognitive processing in GTS.

2.2 Neuropsychology of GTS

A comprehensive understanding of this syndrome requires a multidimensional approach, ranging from clinical psychology and psychiatry to neurology and cognitive neuroscience. For instance, several studies have uncovered cognitive specificities in GTS such as deficit in learning for mathematics and written language (Brookshire, Butler, Ewing-Cobbs, & Fletcher, 1994; Como, 2001), verbal fluency (Bornstein, 1991b; Brookshire et al., 1994) and nonverbal memory (Harris et al., 1995; Lavoie, Thibault, Stip, & O'Connor, 2007; Schuerholz, Baumgardner, Singer, Reiss, & Denckla, 1996). Other investigations proposed that GTS children achieved normal performances on tasks evaluating abstract concepts (Bornstein, 1990; Bornstein & Baker, 1991; Braun et al., 1993; Harris et al., 1995; Schuerholz et al., 1996; Yeates, 1994), planning and response inhibition (Ozonoff & Jensen, 1999) as well as verbal fluency (Braun et al., 1993; Mahone, Koth, Cutting, Singer, & Denckla, 2001), whilst, on the other hand, others proposed several types of executive function impairments (Baron-Cohen, Cross, Crowson, & Robertson, 1994; Bornstein, King, & Carroll, 1983; Brookshire et al., 1994; Schuerholz et al., 1996; Sutherland, Kolb, Schoel, Whishaw, & Davies, 1982). Additional investigations have reported abnormalities with motor skills tasks like the Purdue and Groove Pegboard (PGP) in children (Bornstein, 1990; Hagin, Beecher, Pagano, & Kreeger, 1982), pre-adolescents (Bornstein, 1991a) and adults (O'Connor, Lavoie, Stip, Borgeat, & Laverdure, 2008). Perhaps the most interesting observation is the finding that poorer performances on the PGP, during childhood, predicted worse adulthood tic severity and psychosocial functioning (Bloch, Sukhodolsky, Leckman, & Schultz, 2006).

2.3 Integration of neuropsychology and functional imaging

Individuals with GTS do not necessarily have a characteristic neuropsychological profile which distinguishes them clearly from other psychiatric groups. The large array of behavioral problems in GTS touches various cognitive functions and the apparent lack of consistency in the neuropsychological results could be due to methodological problems considering that, in some cases; studies did not include a control group and often included small samples. Another possible confounding element could be related to the lack of sensitivity of the neuropsychological tests to tap subtle abnormalities often present in these groups. One solution is to adapt neuropsychological tasks to functional magnetic resonance imaging (fMRI) in order to record live brain activity during tic generation or cognitive and motor processing. In an elegant study using fMRI with GTS adults, Peterson et al. (1998) compared brain activity during blocks of time, during which tics were voluntarily suppressed or not suppressed. During tic suppression, prefrontal cortical, thalamic and basal ganglia areas were activated and less activation corresponded with higher tic severity which was consistent with volumetric studies.

In addition, there is often a problem in planning and execution of motor action in GTS. One of the first fMRI study investigating motor functions in GTS showed heightened activation in premotor cortex and SMA during a finger tapping task (Biswal et al., 1998). This, however, could depend on a non selective overactivity of the motor system or on a problem in modulating effort. To address that question, Serrien et al., (Serrien et al., 2002) showed that the SMA of the GTS patients have small or greatly reduced activation when executing a manipulative task as compared with a baseline condition. Nonetheless, cortical areas involved in movement preparation were continuously activated. It was hypothesized that the constant activation of SMA may explain the involuntary urges to move, preventing an accurate planning of voluntary behavior. These first fMRI results suggest that the problem may not be unidirectional with over- or under- activation of motor-related brain networks, but can also relate to a problem of modulation of effortfull and goal directed behavior. This also suggests a deficit not only in motor response inhibition but also in cognitive control. Recent brain imaging findings seems to point towards greater activation of bilateral frontostriatal regions in GTS, which accompanied poorer performance on the Stroop, a well known task of cognitive inhibition. This finding implied that greater activation of the frontostriatal system helps to maintain task performance in individuals with GTS (Marsh, Zhu, Wang, Skudlarski, & Peterson, 2007). Another study (Baym, Corbett, Wright, & Bunge, 2008) confirmed that GTS children exhibit increased activation in the direct pathway through the basal ganglia, as well as increased activation in the prefrontal cortex and the subthalamic nucleus during an inhibition control task. In that study, higher tic severity was associated with enhanced activation of dopaminergic nuclei, cortical, striatal and thalamic regions (*i.e.* direct pathway) and with greater engagement of the subthalamic nucleus area, suggestive of a compensatory mechanism.

In summary, findings from both neuropsychology and neuroimaging suggest the presence of a dysfunction in a cortico-striatal-thalamo-cortical (CSTC) circuit loop. More precisely, recent findings pinpoint a chronic overactivation in cerebral regions associated with motor processing. Finally it seems that a problem of cognitive inhibition is present, which is likely to interfere with accurate planning and execution of voluntary movements. The next challenge is to seek integration of these functional neuroimaging results with real-time information processing in GTS.

2.4 Cognitive electrophysiology and experimental neuropsychology in GTS

As demonstrated in the previous section, despite recent developments in the understanding of GTS, most hypotheses consider the behavioral, cognitive and neurobiological levels independently, whereas an integrative model of GTS, that combines all levels of functioning, would address the relationship between these levels. Tools for such multi-level research would require sensitivity to high-speed cognitive processing, which changes in a matter of milliseconds in synchrony with a specific time-lock event. One solution is the Event-Related Potentials (ERP) which are cortical electrical deflections derived from the time-locked averaged EEG signal and labeled by their polarity and temporal ranges in milliseconds (i.e.

P300) (Sarason, Johnson, & Siegel, 1978). An initial study, using an auditory oddball[1] paradigm, with GTS patients, has shown an abnormal N100/P200 complex, while finding an intact P300 (Van de Wetering, Martens, Fortgens, Slaets, & van Woerkom, 1985) so, suggesting a deficit in attention and vigilance, but with intact memory updating processes. Other studies found larger N100 amplitude to both target and non-target stimuli, proposing that GTS patients allocate more attention than controls in processing both relevant and non relevant stimuli (van Woerkom, Roos, & van Dijk, 1994). However, recent findings with an auditory-visual oddball (Johannes, Wieringa, Nager et al., 2001), found a reduced amplitude of the P300 indicating an increased interference of visual task demands with auditory target perception, which suggested a deficit in cognitive control in GTS patients.

Despite these interesting results, it has been unclear whether this particular problem is associated with a cognitive control deficit and/or with a core motor deficit interfering with cognitive control. An alternative hypothesis is that these results are not only the reflection of a deficit *per se*, but represent instead a mechanism that acts to overcome a motor inhibition problem. For instance, the readiness potential (RP) activation was consistently larger over frontal and smaller over central areas in the GTS group (see Rothenberger et al., 1982; 1986) supporting a possible frontal compensation hypothesis. However, the extent to which the motor preparation is linked with actual cerebral activity has not been systematically analyzed. In one of our earlier studies, we showed that patients with chronic tic disorder failed to show any relationship between reaction times and cortical activation (*i.e.* RP) during a fore period reaction time task (O'Connor, Lavoie, Robert, Stip, & Borgeat, 2005). This finding supports the possibility that people with tic disorders may not be able to modulate cortical activation optimally when planning and executing motor responses.

The caveat with the RP, nonetheless, resides in its high variability, probably reflecting overlapping of non motor as well as motor activity. Also, its early onset may implicate general anticipatory processes rather than the specific cortical preparation preceding movement (Trevena & Miller, 2002). To circumvent this problem, the Lateralized Readiness Potential (LRP), which has its generators in the primary motor cortex (Requin & Riehle, 1995), the SMA (Rektor, 2002) and the basal ganglia (Rektor et al., 2003) represents an excellent candidate measure of motor processing, that could be affected in GTS. Specifically, the LRP has been shown to be a marker of selective motor activation, representing the differential engagement of the left and right motor cortices in the preparation and initiation of motor responses (Coles, 1989; Kutas & Donchin, 1980). The LRP could be analyzed time-locked to the stimulus or to the response, reflecting two levels of processing (at premotor or at the motor level). Using LRPs, the team of Johannes et coll. (Johannes, Wieringa et al., 2001b) failed to show any response-specific difference to GTS patients. In this paradigm, however, stimulus-locked LRPs were pooled across conditions and the peak amplitude was analyzed as a non-specific measure of motor processing, which may have reduced its sensitivity to detect any subtle motor processing differences. To resolve this limitation, we investigated LRPs in GTS adults across diverse conditions of stimulus-response interference (Thibault, O'Connor, Stip, & Lavoie, 2008). GTS groups showed faster response times and

---

[1] During the oddball task, a train of rare stimuli is presented among frequent ones. The task is to identify rare-targets among frequents. This normally triggers the P300 component, which shows larger amplitude to the rare than to the frequent stimuli.

earlier LRP onset to the incompatible condition, which was correlated with tic severity. These findings support the hypothesis of faster motor program retrieval, congruent with the hypothesis of a neuro-modulatory mechanism. This allow a compensation mechanism to achieve normal or above normal motor performance (Biswal et al., 1998; Eidelberg et al., 1997). Interestingly, these results are consistent with observations that, for instance, activities that require focused attention and fine motor dexterity, such as playing a musical instrument are frequently associated with the momentary disappearance of tics (Swain et al., 2007).

In sum, previous ERP studies showed, first, that people with GTS may not be able to modulate cortical activation optimally, when planning and executing motor responses, and secondly, they need to compensate to achieve normal of better performances. However, some results are contradictory and could be related to the presence of other symptomatic elements or to an erroneous diagnosis. To understand the specificity of other findings, we propose to propose to take into account more thoroughly the presence of other conditions often associated with problems of inhibition.

2.5 The puzzling problem of inhibition in GTS and OCD

Even if earlier findings are consistent with an inhibitory dysfunction hypothesis in GTS, there are a lot of inconsistencies in the literature and many studies find no evidence of such deficit in children (Channon, Pratt, & Robertson, 2003; Ozonoff & Jensen, 1999) and adults (Channon, Flynn, & Robertson, 1992; Channon et al., 2003; Ray Li, Hsu, Wang, & Ko, 2006). What could be the reason of these inconsistencies? To address that point, Ozonoff and collaborators (Ozonoff, Strayer, McMahon, & Filloux, 1998) suggested that inhibitory deficits could be largely caused by the presence of comorbid disorders that often arise in GTS. Indeed, the authors found no performance difference between relatively pure GTS (without comorbidity) and control children, in a negative priming task. In fact, only the GTS+ADHD and/or the OCD showed signs of an inhibitory dysfunction compared to controls and GTS without comorbidity. Again, the comorbidity factor appears very important in altering the neurocognitive profile of GTS. In general, frequent comorbidities between GTS and OCD, along with behavioral similarities between them, leads several researchers to propose that they might share common neurophysiological bases (Pauls, 1992; Pauls, Alsobrook, Goodman, Rasmussen, & Leckman, 1995; Pauls, Towbin, Leckman, Zahner, & Cohen, 1986; Sheppard et al., 1999).

2.6 Inhibitory function and attention in dissociating GTS and OCD

However, there are several points that discriminate GTS and OCD. Brain imaging investigations suggest that both GTS and OCD could be initially provoked by a default in inhibitory functions, caused by a metabolic reduction in basal ganglia structures projecting to either the prefrontal and primary motor cortices in GTS, or the orbitofrontal cortex and the anterior cingulate cortex in OCD (Menzies et al., 2008; Mink, 2001; Saxena, Brody, Schwartz, & Baxter, 1998; Sheppard et al., 1999). The prefrontal cortex plays an important role in the ability to orchestrate thought and action in accordance with internal goals and the means to achieve them (Miller & Cohen, 2001), while the primary motor cortex is responsible for simple static or repetitive movements as well as complex preprogrammed or spontaneous purposeful movements (Lassen & Ingvar, 1990). The orbitofrontal cortex appears to be fundamentally critical for outcome-guided behavior and also for facilitating changes in behavior in the face of unexpected outcomes (*e.g.* habit reversal) (Murray, O'Doherty, & Schoenbaum, 2007). In OCD, alteration of this circuit could be responsible for functional deficit in procedural memory as assessed by the pursuit rotor test (Roth, Baribeau, Milovan, O'Connor, & Todorov, 2004), while another study found that these functions were well preserved in GTS (Marsh, Alexander, Packard, Zhu, & Peterson, 2005). Common problems associated with both OCD as well as GTS may stem from their difficulty to inhibit interference from non-relevant cues. For instance, a semantic inhibition task revealed that GTS and OCD groups were consistently disadvantaged in the more demanding inhibition conditions compared to matched controls (Rankins, Bradshaw, & Georgiou-Karistianis, 2006). This difficulty to inhibit interference could also rely on a problem of overfocused attention particularly salient in OCD (Savage et al., 1994). This hypothesis was confirmed in ERP research, where attention-related components peaked at a faster latency in OCD (Towey et al., 1990; Towey et al., 1993; Towey et al., 1994) than in a control group, which was not found with GTS (Johannes, Wieringa, Nager et al., 2001; Johannes et al., 2002; van Woerkom, Fortgens, Rompel-Martens, & Van de Wetering, 1988).

2.7 Inhibitory and sensorimotor integration specificity in GTS

Even if these findings underline differences and similarities in GTS and OCD, only few ERP investigations have compared pure GTS, comorbid GTS+OCD and pure OCD in the same experiment. One of our recent study (Thibault et al., 2008) focusing on comorbidity in GTS, showed a normal P200, whilst the P300 amplitude was clearly affected by the occurrence of clinical symptoms. The OCD and the GTS+OCD group showed reduced rare-target P300 amplitude, mainly in the right anterior region, but otherwise did not differ significantly from each other. The target P300 amplitude was also negatively correlated with OCD, which confirmed numerous findings reported in OCD (Beech, Ciesielski, & Gordon, 1983; Malloy, Rasmussen, Braden, & Haier, 1989; Miyata et al., 1998; Morault, Guillem, Bourgeois, & Paty, 1998; Morault, Bourgeois, Laville, Bensch, & Paty, 1997; Oades, Dittmann-Balcar, Schepker, Eggers, & Zerbin, 1996; Sanz, Molina, Martin-Loeches, Calcedo, & Rubia, 2001; Thibault et al., 2008; Towey et al., 1994). Conversely, participants suffering from GTS showed larger target P300 amplitude, positively correlated with tic frequency. These results suggest that OCD and GTS symptoms have opposing influences on the P300 amplitude during a non-motor oddball task. During a motor inhibition task, however, the profile was different. Inhibitory mechanisms were investigated in a go-nogo task to assess whether sensorimotor integration processes are similar in GTS and OCD (Johannes, Wieringa et al., 2001a). Results showed that the 'no-go' were associated with a frontal shift of the so-called NGA[2] in the GTS, but not in the OCD group. With a comparable STOP-task, we also found results similar to Johannes et al., (Johannes, Wieringa et al., 2001a) where the NGA related to the stop/inhibition was larger over frontal areas in the GTS group even in the absence of OCD comorbidity (Thibault et al., 2009). This finding led to the hypothesis that an overactivated frontal inhibitory function is specific to GTS patients.

[2] The No-Go anteriorization (NGA) is a frontally distributed ERP more prominent in response to response inhibition at approximately 400 ms post-stimulus. It represents a subtraction (*e.g.* voltage subtraction between go and no-go ERPs).

2.8 GTS children growing up: A model of inhibition and developmental neuroplasticity

GTS is characterized by its fluctuating nature over time, and its developmental trajectory needs to be considered. Through longitudinal studies, certain hypotheses have underlined cerebral anomalies associated with symptoms persistency in adulthood. Peterson and collaborators (Peterson et al.2001) proposed that because it is present in every age group, the hypometabolism of the caudate nucleus could constitute a feature of GTS. Moreover, the volume decrease of the putamen, the internal globus pallidus and prefrontal areas, as well as the increase of volume of premotor areas, are uniquely present among adults, which suggests that they are associated with specific pathological mechanisms contributing to the maintenance or inhibition of symptoms among sub-groups of adult with significant symptoms of GTS persisting during adulthood. Among these individuals, there seems to be a failure of cerebral plasticity mechanisms that allows compensating the presence of tics by an overactivation of a motor inhibition process. Unlike adults, children with GTS have a larger orbitofrontal volume (Peterson 2001; Peterson et al.2001; Spessot, Plessen, & Peterson, 2004), which would constitute an adaptive plasticity in response to the expression of tics which, in turn, would help to inhibit them more easily. With the maturation of the prefrontal cortex during adolescence, this mechanism could gain strength and explain the symptom decrease during adolescence and early adulthood. Among adults with persisting symptoms, this prefrontal compensation could not occur. The decrease in volume of the putamen and globus pallidus, and thus the increase in volume of the premotor area, would only be secondary to this compensation.

These neurodevelopmental observations are compatible with current cognitive-behavioral models (O'Connor, 2002; O'Connor, 2005; O'Connor et al., 2009). If the evolution and fluctuation of symptoms is related to a form of cerebral plasticity, then we propose that cognitive-behavioral treatment (CBT) will, in turn, improve symptoms as well as favoring neurophysiological changes corresponding to a normalization of cerebral function, a phenomenon which has recently been observed by our team (Branet, Hosatte-Ducassy, O'Connor, & Lavoie, 2010; Lavoie, Imbriglio, Stip & O'Connor, 2011; O'Connor 2005; O'Connor et al. 2001; O'Connor et al. 2008).

3. Treatment approaches with Gilles de la Tourette Syndrome

3.1 Pharmacological treatments

Pharmacological treatments remain the intervention of choice to help people with GTS. Various treatments have been proposed to help patients, but the majority of prescription drugs as much among adults as among children with GTS, show a variable response, even sometimes on the same individual. From the beginning, let us mention that no drug can lead to the complete remission of this syndrome and the dosage is usually graduated according to the presence of the dominant tic or behavioral symptoms. Because of the dominant hypothesis of tics as a problem of the motor CSTC circuit and the dopaminergic system, dopamine antagonist neuroleptics are routinely the main treatment. Therefore, many researchers have observed that pharmacological agents that trigger an increase (agonist) in dopaminergic functions will exacerbate tics (Golden, 1974; Price, Leckman, Pauls, Cohen, & Kidd, 1986; Riddle, Hardin, Towbin, Leckman, & Cohen, 1987), whereas those that bring a decrease (antagonist) of the dopaminergic action tend to reduce the tic frequency (Shapiro et al., 1989; Lombroso et al., 1995).

Haloperidol (neuroleptic) and clonidine (antihypertensive) are currently the favored medication for the management of tics (Bruun & Budman, 1996; Dion, Annable, Sandor, & Chouinard, 2002; Gilbert et al., 2004; Scahill, Leckman, Schultz, Katsovich, & Peterson, 2003). Among children and teenagers, controlled trials have shown that the frequency of tics decreases by 50% after the use of haloperidol or pimozide (Sallee, Nesbitt, Jackson, Sine, & Sethuraman, 1997). However, typical antipsychotics like Haldol may cause extrapyramidal signs, characterized by involuntary movements, impatience and a need to constantly move and significant trembling among other symptoms. Atypical drug therapy or drug combinations are reserved for more complex cases as well as in the presence of associated disorders. However, side effects also occur in approximately 80% of individuals, and only 20-30% of patients afflicted with GTS continue pharmacological treatment for an extended period (Peterson, Campise, & Azrin, 1994). The effectiveness of risperidone (atypical neuroleptic) has progressively been proven to reduce tics, despite the possibility of significant long term side effects, such as an increased risk of hyperglycemia and diabetes (see review of Lavenstein, 2003). Other pharmacological agents (antidepressants or other neuroleptics) can provide positive results in reducing tics, but these results are often inconsistent and generally come from unique cases, non randomized trials (Pringsheim & Marras, 2009).

In addition, the consumption of psychostimulants (e.g. methylphenidate) was not recommended given the increase in tics in children with concomitant ADHD (Bremness & Sverd, 1979; Golden, 1974; Golden, 1977). However, the majority of recent studies showed that psychostimulants decrease ADHD symptoms without involving much of an increase in the long-term tics (for a review see Erenberg, 2006). Furthermore, other studies have shown that the tic increase due to psychostimulants, is no longer visible after approximately 18 weeks of treatment, so challenging the restriction on the use of psychostimulants among children with GTS and ADHD (Debes, Hjalgrim, & Skov, 2009). However, it is the caregiver's responsibility to inform the family of the possible secondary effects of psychostimulants.

3.2 Cognitive-behavioral treatment

Alternative treatments have shown some success with tic management, including hypnotism, relaxation, muscle feedback, awareness training, negative reinforcement, response prevention and massed practice (Bergin et al., 1998; Azrin et Peterson, 1988; 1990). Therapeutic interventions target not only tic symptoms, but also coping strategies that can modify the unique impact that GTS symptoms may have on an individual's well being (Petersen and Cohen, 1998). The most compelling treatment medium for managing the tics themselves seems to be behavioral treatment, in particular 'habit reversal' (HR) (Azrin and Peterson, 1988). This package involves multiple stages, including relaxation, awareness, contingency training and positive reinforcement of not ticcing and the crucial element of practice of a competitive antagonistic response. This latter technique involves tensing the muscle antithetical and incompatible with the tic-implicated muscle. Awareness training and competing response training seem the most crucial elements of the program (Miltenberger et al., 1988), which can be applied to both tics and habit disorders. Three developmentally normal adolescents with chronic hair pulling were treated with a simplified HR procedure and resulted in an immediate reduction to near-zero levels of hair pulling, with one to three booster sessions required to maintain these levels (Rapp et al., 1998). Azrin & Peterson (1988) report an improvement of between 64-100% in several studies using this method in populations with both simple tics and/or GTS. Peterson & Azrin, (1992) compared the efficacy of awareness, relaxation, and HR in six participants using a within participants design. HR produced the largest overall reduction in tics (55%) and led to the largest reduction in total tics (95%) for any individual, but there was no significant difference between treatments. In an initial wait-list controlled treatment trial, a cognitive-behavioral package based on HR showed significant post treatment clinical improvement for 52% of the adult patients (O'Connor et al. 2001).

However, these results were collected during experiments with small numbers of participants from various populations affected with chronic tics, GTS or habit disorders. Recently this type of behavioral therapy was evaluated in a multi-site randomized controlled trial which followed 126 children between 9 to 17 years-old afflicted with GTS or chronic tic (Piacentini et al., 2010). In this study, all children were randomly assigned to 8 sessions of behavioral therapy during 10 weeks or to equivalent support and education therapy sessions. The sessions of behavioral treatment helped to significantly decrease the tic symptoms in comparison with the support therapy (in 53% vs. 19% of cases respectively) with, in addition, the effects lasting 6 months in 87% of cases.

3.3 Multilevel treatment of GTS: Integrating cognitive, behavioral and neurophysiological findings

Over the last 10 years, our group has conducted a number of studies exploring the cognitive behavioral and psychophysiological manifestations of motor activation in GTS/Chronic Tic, with the aim of linking the multi-level processes evoking tic onset with behavioral management procedures (Lavoie et al., 2008; Leclerc et al., 2008; O'Connor, 2005; O'Connor, 2005; O'Connor et al., 2009; O'Connor et al., 2005; O'Connor et al., 2008; Thibault et al., 2009). As part of the research program, we developed a style of planning questionnaire (STOP) which measures style of planning in everyday life. The STOP has now been validated and has good reliability and discriminates between tic disorder and controls, (O'Connor, 2005). Its three main factors are: overactivity, overpreparation and overrigidity in planning action. The results suggest that all GTS show elevated scores on the first two factors. In addition, the overactivity subscale correlates highly with the Tourette symptom global subscale of motor restlessness.

These experimental and clinical findings have led to elaboration of a cognitive behavioral/psychophysiological model of treatment (O'Connor, 2005) which proposes: 1) an over-active style of planning that prevents optimal preparation for action; 2) this style leads to problems regulating arousal/inhibition processes particularly under circumstances where regulation is open-looped, controlled, and has unpredictable parameters; 3) such high levels of motor activation create tension and frustration and are likely to evoke ticcing; 4) hence a CBT package which addresses the cognitive psychophysiological sources of motor activation who will reduce background tension and prevent tic onset. Whereas traditional HR targets solely the tic implicated muscle in a competing response, an important additional component in our CBT program is modification of excessive overall motor activation, by targeting cognitive and behavioural/physiological sources creating tension. An initial study using this CBT program demonstrated its efficacy on 47 chronic tic and 43 habit disorder (other manual impulse disorder, e.g., hair pulling, nail biting, teeth grinding) receiving a 4-month treatment program. Thirty-eight (22 chronic tic TD, 16 habit disorders) were placed on a wait-list control group, which subsequently received treatment. The treatment approach combined awareness training, relaxation (including modification of a tension-producing overactive style of action), and habit-reversal training, with more general cognitive restructuring of anticipations linked to ticcing. Sixty-five percent of completers reported between 75 and 100% control over the tic. At 2-year follow-up, 52% rated 75-100% control. There were also significant changes post-treatment in measures of self-esteem, anxiety, depression and style of planning action (O'Connor et al., 2001). The majority of participants in this study were diagnosed with light to moderate symptoms.

3.4 Cognitive-behavioral treatment and his impact on brain plasticity

A strong relationship has also been found between symptom reduction following a CBT and brain glucose metabolism in patients with OCD. Using PET imaging, Baxter et al., (Baxter, 1990) found a decrease in the glucose metabolic rate in the right head of the caudate nucleus when OCD was treated successfully with fluoxetine or CBT. A further investigation (Brody et al., 1998) suggested that subjects with differing patterns of metabolism preferentially respond to CBT versus medication. Left orbital-frontal cortex metabolism alone was selected as predicting treatment response in the CBT treated group.

Our team also found interesting impact of CBT on motor dexterity (O'Connor et al., 2008) as well as comparable effect of CBT on those receiving or not medication (O'Connor et al., 2009). One recent research also showed not only behavioral, but also electro-cortical effects post CBT. Thus, before treatment, GTS patients showed reduced electrophysiological response in comparison with the control group during a motor inhibition task. Following CBT administration, this response was normalized concomitantly with decrease of tics frequency (Lavoie et al., 2011). Despite the innovation and evolutionary character of this model, more studies are nonetheless necessary in order to validate the foundation and the efficiency of this intervention program to better assist clinicians in an innovative way.

In sum, CBT and pharmacotherapy focusing on motor regulation can lead to significant clinical improvement in GTS. Brain imaging results after CBT and/or pharmacotherapy in patients with OCD also suggest strong relationships between altered brain activity and symptoms reduction.

4. Conclusion

GTS is a complex neuropsychiatric disorder that affects more people than previously thought. In the last decade, past research has made progress in the treatment of this syndrome, but many questions remain open. Why many patients failed to respond to current treatment? Why are they often misdiagnosed? Are the symptoms really disappearing in adults? These questions can only be approached with a multidisciplinary team combining neurologist, psychologist and neuroscientist from different background. So, a unidisciplinary approach disallows integrating the cognitive, structural and the functional levels of cerebral functioning. Structured interviews are valuable to follow up on clinical states, but they only yield superficial or indirect information on brain functioning. A coherent model of GTS from a single approach is unlikely, since this pathology is multifaceted. A cognitive-behavioral approach links impairments with the clinical expression of the illness that will impact on therapeutic strategies. However, it provides little information about the cerebral roots of the disease. Neuropsychology allows valid inferences about discrete anomalies, but inferences are mainly based on our knowledge of focal lesions, not on functional disorders. Brain imaging is appropriate for identifying localized metabolic abnormalities. However, it is limited by its low temporal resolution that does not take account of the real-time dynamics of the neurocognitive mechanisms involved in the cascade of information processing (Logothetis, 2008). ERPs provide clues to the cerebral activation underlying cognitive processes. But the activity recorded over the scalp might also reflect deeper subcortical activity, which can be only extrapolated or modeled through the analysis of multiple generating sources at best. Moreover, the scalp distribution has often been neglected in clinical studies, so losing both spatial and temporal resolution. However, the ERP approach might still be insufficient because the main limitation will always reside in its low spatial resolution even with a larger electrode array. An alternative will be the use of fMRI cluster techniques to seed dipoles into the EEG head model. Another important point is to anchor both measures to behavioral and neurocognitive expressions of GTS and OCD. As a result, other associated symptoms are often underestimated in populations of GTS, leading to incorrect diagnostic or treatment. To address that issue, we propose in depth neuropsychological evaluation as well as brain activity recordings in order to characterize a particular profile pertaining to GTS and/or OCD groups. The potential benefit of the current approach will be to extract a complete profile allowing prediction of symptom development or treatment success.

From a clinical perspective, effective and individualized therapeutic action should not only include the modification of motor symptoms and inhibition, but should also include cognitive strategies to deal with tics. It is necessary to broaden our conception of GTS in order to see it not only as a neurological, but also as a psychobiological syndrome, because a multifactorial treatment induces a maximal effect on many levels and helps to decrease and to better manage the frequency and intensity of the symptoms. This approach needs to combine nonetheless both cognitive and behavioral perspective, while taking into account physiological aspects that can also exacerbate the behavioral reaction.

In conclusion, two considerations seem fundamental for the development of specialized interventions for GTS in the near future. First, integrating psychophysiological technology as an instrument of treatment: these new possibilities can support cognitive and behavioral management through learning self-controlled strategies. Second, the dissemination of study results on alternate interventions or other front lines must be done. Finally, treatments for GTS symptoms, empirically acknowledged to be effective, should be known by the public and be more accessible.

5. Acknowledgments

This work was supported in part by a Canadian Institutes of Health Research (CIHR) operating grant (MOP57936), a Fonds pour la Recherche en Santé du Québec (FRSQ), team research grant *(Subvention à la recherche en santé mentale -FRSQ # 20573).*

6. References

Abelson, J. F., Kwan, K. Y., O'Roak, B. J., Baek, D. Y., Stillman, A. A., Morgan, T. M., et al. (2005). Sequence variants in SLITRK1 are associated with Tourette's syndrome. *Science, 310*(5746), 317-320.

Alsobrook, J. P., 2nd, & Pauls, D. L. (2002). A factor analysis of tic symptoms in Gilles de la Tourette's syndrome. *Am J Psychiatry, 159*(2), 291-296.

APA. (2000). *Diagnostic and Statistical Manual of Mental Disorders (4th edition-Text revision ed.)*. Washington DC: American Psychiatric Association.

Apter, A., Pauls, D. L., Bleich, A., Zohar, A. H., Kron, S., Ratzoni, G., et al. (1992). A population-based epidemiological study of Tourette syndrome among adolescents in Israel. *Adv Neurol, 58*, 61-65.

Baron-Cohen, S., Cross, P., Crowson, M., & Robertson, M. (1994). Can children with Gilles de la Tourette syndrome edit their intentions? *Psychol.Med., 24*(1), 29-40.

Baxter, L. R. (1990). Brain imaging as a tool in establishing a theory of brain pathology in obsessive compulsive disorder. *J Clin Psychiatry, 51 Suppl*, 22-25; discussion 26.

Baym, C. L., Corbett, B. A., Wright, S. B., & Bunge, S. A. (2008). Neural correlates of tic severity and cognitive control in children with Tourette syndrome. *Brain, 131*(Pt 1), 165-179.

Beech, H. R., Ciesielski, K. T., & Gordon, P. K. (1983). Further observations of evoked potentials in obsessional patients. *Br J Psychiatry, 142*, 605-609.

Biswal, B., Ulmer, J. L., Krippendorf, R. L., Harsch, H. H., Daniels, D. L., Hyde, J. S., et al. (1998). Abnormal cerebral activation associated with a motor task in Tourette syndrome. *American Journal of Neuroradiology, 19*(8), 1509-1512.

Bloch, M. H., Leckman, J. F., Zhu, H., & Peterson, B. S. (2005). Caudate volumes in childhood predict symptom severity in adults with Tourette syndrome. *Neurology, 65*(8), 1253-1258.

Bloch, M. H., Sukhodolsky, D. G., Leckman, J. F., & Schultz, R. T. (2006). Fine-motor skill deficits in childhood predict adulthood tic severity and global psychosocial functioning in Tourette's syndrome. *J Child Psychol Psychiatry, 47*(6), 551-559.

Bornstein, R. A. (1990). Neuropsychological performance in children with Tourette's syndrome. *Psychiatry Research, 33*(1), 73-81.

Bornstein, R. A. (1991a). Neuropsychological correlates of obsessive characteristics in Tourette syndrome. *J.Neuropsychiatry Clin.Neurosci., 3*(2), 157-162.

Bornstein, R. A. (1991b). Neuropsychological performance in adults with Tourette's syndrome. *Psychiatry Res, 37*(3), 229-236.

Bornstein, R. A., & Baker, G. B. (1991). Neuropsychological performance and urinary phenylethylamine in Tourette's syndrome. *J Neuropsychiatry Clin Neurosci, 3*(4), 417-421.

Bornstein, R. A., King, G., & Carroll, A. (1983). Neuropsychological abnormalities in Gilles de la Tourette's syndrome. *J.Nerv.Ment.Dis., 171*(8), 497-502.

Branet, I., Hosatte-Ducassy, C., O'Connor, K. P., & Lavoie, M. E. (2010). Motor processing and brain activity are related to cognitive-behavioral improvement in chronic tic and habit disorders. *International Journal of Psychophysiology., 77*(3).

Braun, A. R., Randolph, C., Stoetter, B., Mohr, E., Cox, C., Vladar, K., et al. (1995). The functional neuroanatomy of Tourette's syndrome: an FDG-PET Study. II: Relationships between regional cerebral metabolism and associated behavioral and cognitive features of the illness. *Neuropsychopharmacology, 13*(2), 151-168.

Braun, A. R., Stoetter, B., Randolph, C., Hsiao, J. K., Vladar, K., Gernert, J., et al. (1993). The functional neuroanatomy of Tourette's syndrome: an FDG-PET study. I. Regional changes in cerebral glucose metabolism differentiating patients and controls. *Neuropsychopharmacology, 9*(4), 277-291.

Bremness, A. B., & Sverd, J. (1979). Methylphenidate-induced Tourette syndrome: case report. *American Journal of Psychiatry, 136*, 1334-1335.

Brody, A. L., Saxena, S., Schwartz, J. M., Stoessel, P. W., Maidment, K., Phelps, M. E., et al. (1998). FDG-PET predictors of response to behavioral therapy and pharmacotherapy in obsessive compulsive disorder. *Psychiatry Res, 84*(1), 1-6.

Brookshire, B. L., Butler, I. J., Ewing-Cobbs, L., & Fletcher, J. M. (1994). Neuropsychological characteristics of children with Tourette syndrome : Evidence for a nonverbal learning disability?, *Journal of Clinical and Experimental Neuropsychology* (Vol. 16, pp. 289-302).

Bruun, R. D., & Budman, C. L. (1996). Risperidone as a treatment for Tourette's syndrome. *Journal of clinical psychiatry., 57*(1), 29-31.

Budman, C. L., Bruun, R. D., Park, K. S., Lesser, M., & Olson, M. (2000). Explosive outbursts in children with Tourette's disorder. *J Am Acad Child Adolesc Psychiatry, 39*(10), 1270-1276.

Budman, C. L., Bruun, R. D., Park, K. S., & Olson, M. E. (1998). Rage attacks in children and adolescents with Tourette's disorder: a pilot study. *J Clin Psychiatry, 59*(11), 576-580.

Budman, C. L., Rockmore, L., Stokes, J., & Sossin, M. (2003). Clinical phenomenology of episodic rage in children with Tourette syndrome. *J Psychosom Res, 55*(1), 59-65.

Burd, L., Kerbeshian, P. J., Barth, A., Klug, M. G., Avery, P. K., & Benz, B. (2001). Long-term follow-up of an epidemiologically defined cohort of patients with Tourette syndrome. *J Child Neurol, 16*(6), 431-437.

Carter, A. S., O'Donnell, D. A., Schultz, R. T., Scahill, L., Leckman, J. F., & Pauls, D. L. (2000). Social and emotional adjustment in children affected with Gilles de la Tourette's syndrome: associations with ADHD and family functioning. Attention Deficit Hyperactivity Disorder. *J Child Psychol Psychiatry, 41*(2), 215-223.

Channon, S., Flynn, D., & Robertson, M. M. (1992). Attentional deficits in Gilles de la Tourette syndrome, *Neuropsychiatry, Neuropsychology and Behavioral Neurology* (Vol. 5, pp. 170-177).

Channon, S., Pratt, P., & Robertson, M. M. (2003). Executive function, memory, and learning in Tourette's syndrome. *Neuropsychology, 17*(2), 247-254.

Coles, M. G. (1989). Modern mind-brain reading: psychophysiology, physiology, and cognition. *Psychophysiology, 26*(3), 251-269.

Como, P. G. (2001). Neuropsychological function in Tourette syndrome (pp. 103-111).

Debes, N., Hjalgrim, H., & Skov, L. (2009). The Presence of Attention-Deficit Hyperactivity Disorder (ADHD) and Obsessive-Compulsive Disorder Worsen Psychosocial and Educational Problems in Tourette Syndrome. *J Child Neurol.*

Dion, Y., Annable, L., Sandor, P., & Chouinard, G. (2002). Risperidone in the treatment of tourette syndrome: a double-blind, placebo-controlled trial. *J Clin Psychopharmacol, 22*(1), 31-39.

Eidelberg, D., Moeller, J. R., Antonini, A., Kazumata, K., Dhawan, V., Budman, C., et al. (1997). The metabolic anatomy of Tourette's syndrome. *Neurology, 48*(4), 927-934.

Erenberg, G. (2006). The Relationship Between Tourette Syndrome, Attention Deficit Hyperactivity Disorder, and Stimulant Medication: A Critical Review. *Seminars in Pediatric Neurology 12*, 217-221.

Findley, D. B., Leckman, J. F., Katsovich, L., Lin, H., Zhang, H., Grantz, H., et al. (2003). Development of the Yale Children's Global Stress Index (YCGSI) and its application in children and adolescents ith Tourette's syndrome and obsessive-compulsive disorder. *J Am Acad Child Adolesc Psychiatry, 42*(4), 450-457.

Freeman, R. D. (2007). Tic disorders and ADHD: answers from a world-wide clinical dataset on Tourette syndrome. *Eur Child Adolesc Psychiatry, 16 Suppl 1*, 15-23.

Freeman, R. D., Fast, D. K., Burd, L., Kerbeshian, J., Robertson, M. M., & Sandor, P. (2000). An international perspective on Tourette syndrome: selected findings from 3,500 individuals in 22 countries. *Dev Med Child Neurol, 42*(7), 436-447.

Gaze, C., Kepley, H. O., & Walkup, J. T. (2006). Co-occurring psychiatric disorders in children and adolescents with Tourette syndrome. *J Child Neurol, 21*(8), 657-664.

Gilbert, D. L., Bansal, A. S., Sethuraman, G., Sallee, F. R., Zhang, J., Lipps, T., et al. (2004). Association of cortical disinhibition with tic, ADHD, and OCD severity in Tourette syndrome. *Movement disorders 19*(4), 416-425.

Gilles de la Tourette, G. (1885). Étude sur une affection nerveuse caractérisée par de l'incoordination motrice accompagnée d'écholalie et de coprolalie. *Archives de Neurologie, 9*, 19-42; 158-200.

Golden, G. S. (1974). Gilles de la Tourette's syndrome following methylphenidate administration. *Developmental Medicine and Child Neurology 16:*, 76-78.

Golden, G. S. (1977). The effect of central nervous system stimulants on Tourette syndrome. *Ann.Neurol., 2*(1), 69-70.

Hagin, R. A., Beecher, R., Pagano, G., & Kreeger, H. (1982). Effects of Tourette syndrome on learning. *Advances in neurology, 35*, 323-328.

Harris, E. L., Schuerholz, L. J., Singer, H. S., Reader, M. J., Brown, J. E., Cox, C., et al. (1995). Executive function in children with Tourette syndrome and/or attention deficit hyperactivity disorder. *J.Int.Neuropsychol.Soc., 1*(6), 511-516.

Hyde, T. M., Stacey, M. E., Coppola, R., Handel, S. F., Rickler, K. C., & Weinberger, D. R. (1995). Cerebral morphometric abnormalities in Tourette's syndrome: a quantitative MRI study of monozygotic twins. *Neurology, 45*(6), 1176-1182.

Itard, J. (1825). Mémoire sur quelques fonctions involontaires des appareils de la locomotion, de la préhension et de la voix. *Archives Générales de Médecine, 8*, 385-407.

Johannes, S., Wieringa, B. M., Mantey, M., Nager, W., Rada, D., Muller-Vahl, K. R., et al. (2001a). Altered inhibition of motor responses in Tourette syndrome and Obsessive-Compulsive disorder. *Acta Neurologica Scandinavica, 104*, 36-43.

Johannes, S., Wieringa, B. M., Mantey, M., Nager, W., Rada, D., Muller-Vahl, K. R., et al. (2001b). Altered inhibition of motor responses in Tourette Syndrome and Obsessive-Compulsive Disorder. *Acta Neurol Scand, 104*(1), 36-43.

Johannes, S., Wieringa, B. M., Nager, W., Muller-Vahl, K. R., Dengler, R., & Munte, T. F. (2001). Electrophysiological measures and dual-task performance in Tourette syndrome indicate deficient divided attention mechanisms. *European journal of neurology 8*(3), 253-260.

Johannes, S., Wieringa, B. M., Nager, W., Muller-Vahl, K. R., Dengler, R., & Munte, T. F. (2002). Excessive action monitoring in Tourette syndrome. *J.Neurol., 249*(8), 961-966.

Kadesjo, B., & Gillberg, C. (2000). Tourette's disorder: epidemiology and comorbidity in primary school children. *J Am Acad Child Adolesc Psychiatry, 39*(5), 548-555.

Karno, M., Golding, J. M., Sorenson, S. B., & Burnam, M. A. (1988). The epidemiology of obsessive-compulsive disorder in five US communities. *Arch Gen Psychiatry, 45*(12), 1094-1099.

Katayama, K., Yamada, K., Ornthanalai, V. G., Inoue, T., Ota, M., Murphy, N. P., et al. (2010). Slitrk1-deficient mice display elevated anxiety-like behavior and noradrenergic abnormalities. *Mol Psychiatry, 15*(2), 177-184.

Keen-Kim, D., & Freimer, N. B. (2006). Genetics and epidemiology of Tourette syndrome. *J Child Neurol, 21*(8), 665-671.

King, R. A., & Scahill, L. (2001). Emotional and behavioral difficulties associated with Tourette syndrome. *Adv Neurol, 85*, 79-88.

Kutas, M., & Donchin, E. (1980). Preparation to respond as manifested by movement-related brain potentials. *Brain Research, 202*(1), 95-115.

Lassen, N. A., & Ingvar, D. H. (1990). Brain regions involved in voluntary movements as revealed by radioisotopic mapping of CBF or CMR-glucose changes. *Rev Neurol (Paris), 146*(10), 620-625.

Lavenstein, B. L. (2003). Treatment approaches for children with Tourette's syndrome. *Curr Neurol Neurosci Rep, 3*(2), 143-148.

Lavoie, M. E., Imbriglio, T. V., Baltazar, L., Thibault, G., Stip, E., & O'Connor, K. P. (2008). Evaluation and treatment of Tourette syndrome: An integrated behavioral and neurocognitive approach focusing on motor processing. *Journal of the International Neuropsychological Society, 14*(Suppl. 1), 273.

Lavoie, M. E., Imbriglio, T. V., Stip, E., & O'Connor, K. P. (2011). Neurocognitive changes following cognitive-behavioral treatment in the Tourette syndrome and chronic tic disorder. *International Journal of Cognitive Psychotherapy, 4*(2), 34-50.

Lavoie, M. E., Thibault, G., Stip, E., & O'Connor, K. P. (2007). Memory and executive functions in adults with Gilles de la Tourette syndrome and chronic tic disorder. *Cognit Neuropsychiatry, 12*(2), 165-181.

Leckman, J. F. (2002). Tourette's syndrome. *Lancet, 360*(9345), 1577-1586.

Leckman, J. F., Bloch, M. H., Scahill, L., & King, R. A. (2006). Tourette syndrome: the self under siege. *J Child Neurol, 21*(8), 642-649.

Leckman, J. F., Peterson, B. S., Anderson, G. M., Arnsten, A. F., Pauls, D. L., & Cohen, D. J. (1997). Pathogenesis of Tourette's syndrome. *Journal of child psychology and psychiatry, and allied disciplines, 38*(1), 119-142.

Leckman, J. F., Zhang, H., Vitale, A., Lahnin, F., Lynch, K., Bondi, C., et al. (1998). Course of tic severity in Tourette syndrome: the first two decades. *Pediatrics, 102*(1 Pt 1), 14-19.

Leclerc, J., Forget, J., & O'Connor, K. P. (2008). *Quand le corps fait à sa tête - Le syndrome de Gilles de la Tourette.* Montréal: Multimondes.

Logothetis, N. K. (2008). What we can do and what we cannot do with fMRI. *Nature, 453*(7197), 869-878.

Mahone, E. M., Koth, C. W., Cutting, L., Singer, H. S., & Denckla, M. B. (2001). Executive function in fluency and recall measures among children with Tourette syndrome or ADHD. *J.Int.Neuropsychol.Soc., 7*(1), 102-111.

Malloy, P., Rasmussen, S., Braden, W., & Haier, R. J. (1989). Topographic evoked potential mapping in obsessive-compulsive disorder: evidence of frontal lobe dysfunction. *Psychiatry Res, 28*(1), 63-71.

Marsh, R., Alexander, G. M., Packard, M. G., Zhu, H., & Peterson, B. S. (2005). Perceptual-motor skill learning in Gilles de la Tourette syndrome. Evidence for multiple procedural learning and memory systems. *Neuropsychologia, 43*(10), 1456-1465.

Marsh, R., Zhu, H., Wang, Z., Skudlarski, P., & Peterson, B. S. (2007). A developmental fMRI study of self-regulatory control in Tourette's syndrome. *Am J Psychiatry, 164*(6), 955-966.

Menzies, L., Chamberlain, S. R., Laird, A. R., Thelen, S. M., Sahakian, B. J., & Bullmore, E. T. (2008). Integrating evidence from neuroimaging and neuropsychological studies of obsessive-compulsive disorder: the orbitofronto-striatal model revisited. *Neurosci Biobehav Rev, 32*(3), 525-549.

Miller, E. K., & Cohen, J. D. (2001). An integrative theory of prefrontal cortex function. *Annu Rev Neurosci, 24*, 167-202.

Mink, J. W. (2001). Basal ganglia dysfunction in Tourette's syndrome: a new hypothesis. *Pediatr Neurol, 25*(3), 190-198.

Miyata, A., Matsunaga, H., Kiriike, N., Iwasaki, Y., Takei, Y., & Yamagami, S. (1998). Event-related potentials in patients with obsessive-compulsive disorder. *Psychiatry Clin Neurosci, 52*(5), 513-518.

Morault, P., Guillem, F., Bourgeois, M., & Paty, J. (1998). Improvement predictors in obsessive-compulsive disorder. An event-related potential study. *Psychiatry Res, 81*(1), 87-96.

Morault, P. M., Bourgeois, M., Laville, J., Bensch, C., & Paty, J. (1997). Psychophysiological and clinical value of event-related potentials in obsessive-compulsive disorder. *Biol Psychiatry, 42*(1), 46-56.

Murray, E. A., O'Doherty, J. P., & Schoenbaum, G. (2007). What we know and do not know about the functions of the orbitofrontal cortex after 20 years of cross-species studies. *J Neurosci, 27*(31), 8166-8169.

O'Connor, K. P. (2002). A cognitive-behavioral/psychophysiological model of tic disorders. *Behaviour Research and Therapy, 40*, 1113-1142.

O'Connor, K. P. (2005). *Cognitive-behavioral management of tic disorders.* . New York: John Wiley.

O'Connor, K. P. (2005). Testing the cognitive-psychophysiological model: validation of a style of planning action (STOP) as a discriminator between tic disorder, obsessive-compulsive disorder and generalized anxiety. In *Cognitive-Behavioral Management of tic disorders* (pp. 65-73.). Chichester: John-Wiley and Sons.

O'Connor, K. P., Brault, M., Robillard, S., Loiselle, J., Borgeat, F., & Stip, E. (2001). Evaluation of a cognitive-behavioural program for the management of chronic tic and habit disorders. *Behav.Res.Ther., 39*(6), 667-681.

O'Connor, K. P., Laverdure, A., Taillon, A., Stip, E., Borgeat, F., & Lavoie, M. (2009). Cognitive behavioral management of Tourette's syndrome and chronic tic disorder in medicated and unmedicated samples. *Behav Res Ther, 47*(12), 1090-1095.

O'Connor, K. P., Lavoie, M. E., Robert, M., Stip, E., & Borgeat, F. (2005). Brain-behavior relations during motor processing in chronic tic and habit disorder. *Cognitive and behavioral neurology 18*(2), 79-88.

O'Connor, K. P., Lavoie, M. E., Stip, E., Borgeat, F., & Laverdure, A. (2008). Cognitive-behaviour therapy and skilled motor performance in adults with chronic tic disorder. *Neuropsychol Rehabil, 18*(1), 45-64.

Oades, R. D., Dittmann-Balcar, A., Schepker, R., Eggers, C., & Zerbin, D. (1996). Auditory event-related potentials (ERPs) and mismatch negativity (MMN) in healthy children and those with attention-deficit or tourette/tic symptoms. *Biological Psychology, 43*(2), 163-185.

Ozonoff, S., & Jensen, J. (1999). Brief report: specific executive function profiles in three neurodevelopmental disorders. *J.Autism Dev.Disord., 29*(2), 171-177.

Ozonoff, S., Strayer, D. L., McMahon, W. M., & Filloux, F. (1998). Inhibitory deficits in Tourette syndrome: a function of comorbidity and symptom severity. *J.Child Psychol.Psychiatry, 39*(8), 1109-1118.

Pappert, E. J., Goetz, C. G., Louis, E. D., Blasucci, L., & Leurgans, S. (2003). Objective assessments of longitudinal outcome in Gilles de la Tourette's syndrome. *Neurology, 61*(7), 936-940.

Pauls, D. L. (1992). The genetics of obsessive compulsive disorder and Gilles de la Tourette's syndrome. *Psychiatr Clin North Am, 15*(4), 759-766.

Pauls, D. L. (2003). An update on the genetics of Gilles de la Tourette syndrome. *J Psychosom Res, 55*(1), 7-12.

Pauls, D. L., Alsobrook, J. P., 2nd, Goodman, W., Rasmussen, S., & Leckman, J. F. (1995). A family study of obsessive-compulsive disorder. *Am J Psychiatry, 152*(1), 76-84.

Pauls, D. L., Towbin, K. E., Leckman, J. F., Zahner, G. E., & Cohen, D. J. (1986). Gilles de la Tourette's syndrome and obsessive-compulsive disorder. Evidence supporting a genetic relationship. *Arch Gen Psychiatry, 43*(12), 1180-1182.

Peterson, B., Riddle, M. A., Cohen, D. J., Katz, L. D., Smith, J. C., Hardin, M. T., et al. (1993). Reduced basal ganglia volumes in Tourette's syndrome using three-dimensional reconstruction techniques from magnetic resonance images. *Neurology, 43*(5), 941-949.

Peterson, B. S. (2001). Neuroimaging studies of Tourette syndrome: a decade of progress. *Adv.Neurol., 85*, 179-196.

Peterson, B. S., Choi, H. A., Hao, X., Amat, J. A., Zhu, H., Whiteman, R., et al. (2007). Morphologic features of the amygdala and hippocampus in children and adults with Tourette syndrome. *Arch Gen Psychiatry, 64*(11), 1281-1291.

Peterson, B. S., Leckman, J. F., Tucker, D., Scahill, L., Staib, L., Zhang, H., et al. (2000). Preliminary findings of antistreptococcal antibody titers and basal ganglia volumes in tic, obsessive-compulsive, and attention deficit/hyperactivity disorders. *Arch Gen Psychiatry, 57*(4), 364-372.

Peterson, B. S., Skudlarski, P., Anderson, A. W., Zhang, H., Gatenby, J. C., Lacadie, C. M., et al. (1998). A functional magnetic resonance imaging study of tic suppression in Tourette syndrome. *Arch.Gen.Psychiatry, 55*(4), 326-333.

Peterson, B. S., Staib, L., Scahill, L., Zhang, H., Anderson, C., Leckman, J. F., et al. (2001). Regional brain and ventricular volumes in Tourette syndrome. *Archives of General Psychiatry, 58*(5), 427-440.

Piacentini, J., Woods, D. W., Scahill, L., Wilhelm, S., Peterson, A. L., Chang, S., et al. (2010). Behavior therapy for children with Tourette disorder: a randomized controlled trial. *Jama, 303*(19), 1929-1937.

Price, R. A., Leckman, J. F., Pauls, D. L., Cohen, D. J., & Kidd, K. K. (1986). Gilles de la Tourette's syndrome: tics and central nervous system stimulants in twins and nontwins. *Neurology, 36*(2), 232-237.

Pringsheim, T., & Marras, C. (2009). Pimozide for tics in Tourette's syndrome. *Cochrane Database Syst Rev*(2), CD006996.

Rankins, D., Bradshaw, J. L., & Georgiou-Karistianis, N. (2006). The semantic Simon effect in Tourette's syndrome and obsessive-compulsive disorder. *Brain Cogn, 61*(3), 225-234.

Ray Li, H.-L., Hsu, Y.-P., Wang, H.-S., & Ko, N.-C. (2006). Motor Response Inhibition in Children With Tourette's Disorder. *The Journal of Neuropsychiatry and Clinical Neurosciences, 18*(3), 417.

Rektor, I. (2002). Scalp-recorded Bereitschaftspotential is the result of the activity of cortical and subcortical generators--a hypothesis. *Clinical Neurophysiology, 113*(12), 1998-2005.

Rektor, I., Kaiiovsky, P., Bares, M., Brazdil, M., Streitova, H., Klajblova, H., et al. (2003). A SEEG study of ERP in motor and premotor cortices and in the basal ganglia. *Clinical Neurophysiology, 114*(3), 463-471.

Requin, J., & Riehle, A. (1995). Neural correlates of partial transmission of sensorimotor information in the cerebral cortex. *Acta Psychologica, 90*(1-3), 81-95.

Robertson, M. M. (2003). Diagnosing Tourette syndrome: is it a common disorder? *J Psychosom Res, 55*(1), 3-6.

Robertson, M. M., & Stern, J. S. (2000). Gilles de la Tourette syndrome: symptomatic treatment based on evidence. *Eur Child Adolesc Psychiatry, 9 Suppl 1*, I60-75.

Roth, R. M., Baribeau, J., Milovan, D., O'Connor, K., & Todorov, C. (2004). Procedural and declarative memory in obsessive-compulsive disorder. *J Int Neuropsychol Soc, 10*(5), 647-654.

Rothenberger, A. K., S. (1982). Bereitschaftpotential in children with multiple tics and Gilles de la Tourette syndrome. In A. Rothenberger. (Ed.), *Event-Related Potentials in children* (pp. 257-270). Amsterdam: Elsevier Biomedical Press.

Rothenberger, A. K., S., Schenk, G.K., Zerbin, D. Voss, M. (1986). Movement-Related potentials in children with hypermotoric behaviour. In R. Z. W.C. McCallum., F.Denoth. (Ed.), *Cerebral Psychophysiology: studies in event-related potentials* (Vol. EEG suppl 38, pp. 496-499.). Amsterdam: Elsevier Science publishers, B.V. (Biomedical division).

Sanz, M., Molina, V., Martin-Loeches, M., Calcedo, A., & Rubia, F. J. (2001). Auditory P300 event related potential and serotonin reuptake inhibitor treatment in obsessive-compulsive disorder patients. *Psychiatry Res, 101*(1), 75-81.

Sarason, I. G., Johnson, J. H., & Siegel, J. M. (1978). Assessing the impact of life changes: development of the Life Experiences Survey. *J Consult Clin Psychol, 46*(5), 932-946.

Savage, C. R., Weilburg, J. B., Duffy, F. H., Baer, L., Shera, D. M., & Jenike, M. A. (1994). Low-level sensory processing in obsessive-compulsive disorder: an evoked potential study. *Biol Psychiatry, 35*(4), 247-252.

Saxena, S., Brody, A. L., Schwartz, J. M., & Baxter, L. R. (1998). Neuroimaging and frontal-subcortical circuitry in obsessive-compulsive disorder. *Br J Psychiatry Suppl*(35), 26-37.

Scahill, L., Leckman, J. F., Schultz, R. T., Katsovich, L., & Peterson, B. S. (2003). A placebo-controlled trial of risperidone in Tourette syndrome. *Neurology, 60*(7), 1130-1135.

Scahill, L., Sukhodolsky, D. G., Williams, S. K., & Leckman, J. F. (2005). Public health significance of tic disorders in children and adolescents. *Adv Neurol, 96*, 240-248.

Scharf, J. M., Moorjani, P., Fagerness, J., Platko, J. V., Illmann, C., Galloway, B., et al. (2008). Lack of association between SLITRK1var321 and Tourette syndrome in a large family-based sample. *Neurology, 70*(16 Pt 2), 1495-1496.

Schuerholz, L. J., Baumgardner, T. L., Singer, H. S., Reiss, A. L., & Denckla, M. B. (1996). Neuropsychological status of children with Tourette's syndrome with and without attention deficit hyperactivity disorder. *Neurology, 46*(4), 958-965.

Serrien, D. J., Nirkko, A. C., Loher, T. J., Lovblad, K. O., Burgunder, J. M., & Wiesendanger, M. (2002). Movement control of manipulative tasks in patients with Gilles de la Tourette syndrome. *Brain, 125*(Pt 2), 290-300.

Shapiro, A. K., & Shapiro, E. (1992). Evaluation of the reported association of obsessive-compulsive symptoms or disorder with Tourette's disorder. *Compr Psychiatry, 33*(3), 152-165.

Shapiro, E., & Shapiro, A. K. (1986). Semiology, nosology and criteria for tic disorders. *Revue neurologique, 142*(11), 824-832.

Sheppard, D. M., Bradshaw, J. L., Purcell, R., & Pantelis, C. (1999). Tourette's and comorbid syndromes: obsessive compulsive and attention deficit hyperactivity disorder. A common etiology? *Clin Psychol Rev, 19*(5), 531-552.

Sowell, E. R., Kan, E., Yoshii, J., Thompson, P. M., Bansal, R., Xu, D., et al. (2008). Thinning of sensorimotor cortices in children with Tourette syndrome. *Nat Neurosci, 11*(6), 637-639.

Spessot, A. L., Plessen, K. J., & Peterson, B. S. (2004). Neuroimaging of developmental psychopathologies: the importance of self-regulatory and neuroplastic processes in adolescence. *Ann N Y Acad Sci, 1021*, 86-104.

State, M. W. The genetics of Tourette disorder. *Curr Opin Genet Dev, 21*(3), 302-309.

Stephens, R. J., & Sandor, P. (1999). Aggressive behaviour in children with Tourette syndrome and comorbid attention-deficit hyperactivity disorder and obsessive-compulsive disorder. *Can J Psychiatry, 44*(10), 1036-1042.

Stillman, A. A., Krsnik, Z., Sun, J., Rasin, M. R., State, M. W., Sestan, N., et al. (2009). Developmentally regulated and evolutionarily conserved expression of SLITRK1 in brain circuits implicated in Tourette syndrome. *J Comp Neurol, 513*(1), 21-37.

Stoetter, B., Braun, A. R., Randolph, C., Gernert, J., Carson, R. E., Herscovitch, P., et al. (1992). Functional neuroanatomy of Tourette syndrome. Limbic-motor interactions studied with FDG PET. *Advances in neurology, 58*, 213-226.

Sutherland, R. J., Kolb, B., Schoel, W. M., Whishaw, I. Q., & Davies, D. (1982). Neuropsychological assessment of children and adults with Tourette syndrome: a comparison with learning disabilities and schizophrenia. *Adv.Neurol., 35*, 311-322.

Swain, J. E., Scahill, L., Lombroso, P. J., King, R. A., & Leckman, J. F. (2007). Tourette syndrome and tic disorders: a decade of progress. *J Am Acad Child Adolesc Psychiatry, 46*(8), 947-968.

Thibault, G., Felezeu, M., O'Connor, K. P., Todorov, C., Stip, E., & Lavoie, M. E. (2008). Influence of comorbid obsessive-compulsive symptoms on brain event-related potentials in Gilles de la Tourette syndrome. *Prog Neuropsychopharmacol Biol Psychiatry, 32*(3), 803-815.

Thibault, G., O'Connor, K. P., Stip, E., & Lavoie, M. E. (2009). Electrophysiological manifestations of stimulus evaluation, response inhibition and motor processing in Tourette syndrome patients. *Psychiatry Res, 167*(3), 202-220.

Thibault, G., O'Connor, K. P., Stip, E., & Lavoie, M. E. (2008). Electrophysiological manifestations of stimulus evaluation, response inhibition and motor processing in Tourette syndrome patients. Psychiatry Research. *Psychiatry Research, doi:10.1016/j.psychres.*, 2008.2003.2021.

Thibert, A. L., Day, H. I., & Sandor, P. (1995). Self-concept and self-consciousness in adults with Tourette syndrome. *Can J Psychiatry, 40*(1), 35-39.

Towey, J., Bruder, G., Hollander, E., Friedman, D., Erhan, H., Liebowitz, M., et al. (1990). Endogenous event-related potentials in obsessive-compulsive disorder. *Biol Psychiatry, 28*(2), 92-98.

Towey, J., Bruder, G., Tenke, C., Leite, P., DeCaria, C., Friedman, D., et al. (1993). Event-related potential and clinical correlates of neurodysfunction in obsessive-compulsive disorder. *Psychiatry Res, 49*(2), 167-181.

Towey, J. P., Tenke, C. E., Bruder, G. E., Leite, P., Friedman, D., Liebowitz, M., et al. (1994). Brain event-related potential correlates of overfocused attention in obsessive-compulsive disorder. *Psychophysiology, 31*(6), 535-543.

Trevena, J. A., & Miller, J. (2002). Cortical movement preparation before and after a conscious decision to move. *Conscious Cogn, 11*(2), 162-190; discussion 314-125.

Van de Wetering, B. J., Martens, C. M., Fortgens, C., Slaets, J. P., & van Woerkom, T. C. (1985). Late components of the auditory evoked potentials in Gilles de la Tourette syndrome. *Clinical neurology and neurosurgery, 87*(3), 181-186.

van Woerkom, T. C., Fortgens, C., Rompel-Martens, C. M., & Van de Wetering, B. J. (1988). Auditory event-related potentials in adult patients with Gilles de la Tourette's syndrome in the oddball paradigm. *Electroencephalography and Clinical Neurophysiology, 71*(6), 443-449.

van Woerkom, T. C., Roos, R. A., & van Dijk, J. G. (1994). Altered attentional processing of background stimuli in Gilles de la Tourette syndrome: a study in auditory event-related potentials evoked in an oddball paradigm. *Acta Neurol.Scand., 90*(2), 116-123.

Yeates, K. O., Bornstein, R. A. (1994). Attention deficit disorder and neuropsychological functioning in children with Tourette's syndrome. *Neuropsychology, 8*, 65-74.

Zohar, A. H., Ratzoni, G., Pauls, D. L., Apter, A., Bleich, A., Kron, S., et al. (1992). An epidemiological study of obsessive-compulsive disorder and related disorders in Israeli adolescents. *J Am Acad Child Adolesc Psychiatry, 31*(6), 1057-1061.

3

How Much Serotonin in the CNS is Too Much?

Rui Tao and Zhiyuan Ma
Charles E. Schmidt College of Medicine,
Florida Atlantic University, Boca Raton, Florida
USA

1. Introduction

Serotonin syndrome is a neurological disorder primarily associated with inappropriate uses of serotonin (5HT)-promoting drugs such as serotonin reuptake inhibitors (SRIs), monoamine oxidase inhibitors (MAOIs) or 3,4-Methylenedioxymethamphetamine (MDMA; Ecstasy) (Boyer *et al.*, 2005; Karunatilake *et al.*, 2006; Parrott, 2002; Paruchuri *et al.*, 2006). The primary cause of the syndrome is due to a global increase in brain 5HT that can potentially activate all 14 subtypes of 5HT receptors (5HTRs). However, the significance of a given subtype contributing to the syndrome is not always the same, depending on the amount of 5HT evoked by drugs, physical condition of individual patients and even surrounding environment of drug administration. As a result, signs and symptoms of the syndrome vary widely (Mills, 1995) and thus it has been a clinical challenge to get an accurate diagnosis. Regardless, it has been recognized since the early 1990's that the symptoms of the syndrome can be generally classified into 3 categories: mental state changes (mood swings), neuromuscular hyperactivity and autonomic dysfunction (Sternbach, 1991). Among these, neuromuscular and autonomic symptoms can be replicated in laboratory animals with MAOIs combined with 5HT precursors (Shioda *et al.*, 2004) or by MDMA at a high dose (Baumann *et al.*, 2008a; Spanos *et al.*, 1989). Thus, despite many difficulties in clinical research, there are some great advances in understanding serotonin syndrome thanks primarily to preclinical investigation in animals.

Increasingly, the term "serotonin toxicity" has been used interchangeably with "serotonin syndrome", particularly in MDMA-related research. This is mainly because MDMA at high doses could cause a reduction in 5HT content in the brain and possibly axonal degeneration (Bhide *et al.*, 2009; Malberg *et al.*, 1998). In fact, there is no evidence indicating that clinically relevant doses of MDMA could produce such effects or neural death although symptoms of the serotonin syndrome occur [details reviewed by (Baumann *et al.*, 2007)]. For these reasons, we avoided using the term "toxicity" in this study unless more relevant information such as levels of cell death is available in clinically relevant literature.

To address the neurological mechanisms underlying the cause of the serotonin syndrome, we will review recent research findings on how brain 5HT is dynamically altered following administration of 5HT-promoting drugs by which the syndrome would potentially be evoked. We will focus on preclinical data since most experimental studies have been carried out in rodents (rats and mice). If available, human data will also be included for comparison. Additionally, the role of 5HT receptors in the syndrome will be discussed.

2. Dynamic changes in 5HT concentration

Despite the fact that brain 5HT can be found in both the extracellular space and intracellular compartments, extracellular 5HT ($5HT_{ext}$) is the one involved in neurotransmission in the brain. Therefore, first of all, we will review literature on $5HT_{ext}$ concentration which, in normal physiological condition, likely reflects the functional activity of serotonergic neurons. While an exhaustive literature review is not possible here, representative findings will be mentioned. For comparison, data related to the level of intracellular 5HT (non-functional component) will also be included. Pharmacologically, 5HT-promoting drugs could elevate $5HT_{ext}$ to a particular level resulting in improvement of behavioral response and enhancement of mood without causing mental impairment. Thus in the second part we will review literature concerning the $5HT_{ext}$ level at which behavioral incompetence can be improved in response to 5HT-promoting drugs. Thirdly, we will seek evidence of the upper limit level (threshold) above which $5HT_{ext}$ is too high in the brain and potentially causes serotonin syndrome.

2.1 Normal extracellular concentration

As a neurotransmitter, 5-hydroxytryptamine (5HT; serotonin) is synthesized mainly at serotonergic axon terminals through decarboxylation of 5-hydroxytryptophan (5HTP). Newly synthesized 5HT molecules have three possible fates as follows:

1. uptaken into vesicles by the vesicular monoamine transporter-2 (VMAT2);
2. oxidized by mitochondrial monoamine oxidase-A (MAO_A) followed by aldehyde dehydrogenase into 5-hydroxyindoleacetic acid (5HIAA);
3. spontaneously released into the extracellular space through reverse transporters (Gobbi *et al.*, 1993).

However, newly synthesized 5HT molecules are not randomly or equally directed into these three pathways. Their direction depends on physical states and drug action properties in medication. Normally, almost all newly synthesized 5HT molecules are uptaken into vesicles against concentration gradients. Although little is known about mammalian vesicular 5HT concentrations, it is suggested that there may be 270 mM (or 1.6×10^{23} molecules) in a 5HT vesicle of leech Retzius neurons (Bruns *et al.*, 1995; Bruns *et al.*, 2000). Generally in the brain, over 99% of 5HT molecules are stored in the synaptic vesicles. The amount of intracellular 5HT can be estimated using homogenized brain tissue. However, it should be kept in mind that the measures likely mirror intensity of serotonergic innervations (Schaefer *et al.*, 2008), but not functional activity of neurotransmission. Nevertheless, the range of 5HT content in the homogenized brain is 1 to 2 nmol/g (Baumann *et al.*, 2008a; Bhide *et al.*, 2009; Grahame-Smith *et al.*, 1974; Malberg *et al.*, 1998). Moreover, raphe nuclei have relatively higher contents than other regions (Adell *et al.*, 1991b). A similar range is also found in the mouse brain (Kim *et al.*, 2005; Pothakos *et al.*, 2010).

Several studies have been carried out to measure 5HT content in the post-mortem human brain (Parsons *et al.*, 1992; Seidl *et al.*, 1999). In general, 5HT is in the range of 100-400 pmol/g in homogenized tissues. Since 5HT is rapidly oxidized in the post-mortem tissues, it is likely that the level in the human brain is underestimated.

$5HT_{ext}$ is considered to be critical in maintaining mood and other affective functioning, although normally its quantity is estimated to be less than 0.1% of total 5HT molecules in the brain. At such an exceptionally small quantity, it is a challenge to determine its level in the human brain. Indeed, a PubMed search indicates that there are no relevant data available in clinical literature. Despite this, highly sensitive approaches have been developed during the last two decades for laboratory animals. For instance, Adell et al. measured $5HT_{ext}$ in the cerebrospinal fluid using conventional microdialysis in the frontal cortex, striatum, hypothalamus, hippocampus, inferior colliculus and raphe nuclei, demonstrating that the $5HT_{ext}$ concentration is in the range of 0.5-2 nM (Adell *et al.*, 1991b). Thus, it appears that $5HT_{ext}$ in the rat cerebrospinal fluid is at a low nanomolar level. The same conclusion has been obtained by other studies with rats and mice using zero-net-flux microdialysis (Calcagno *et al.*, 2007; Gardier *et al.*, 2003; Mathews *et al.*, 2004; Tao *et al.*, 2000).

The amount and distribution of 5HT into the extracellular space are strongly implicated in many mental health problems. Physiologically, the concentration of $5HT_{ext}$ is constantly kept at a state of equilibrium, involving balanced regulation between spontaneous release, $5HT_{1A}R$ feedback inhibition and reuptake mechanisms as demonstrated in *in vitro* and *in vivo* studies (Becquet *et al.*, 1990; Blier, 2001; Sharp *et al.*, 2007; Wolf *et al.*, 1986). On the other hand, the intracellular concentration in vesicles of serotonergic terminals is at a high milimolar (mM) level, implying that intracellular 5HT can rapidly elevate $5HT_{ext}$ to an extraordinary level in response to 5HT-promoting drugs. The response scale can be widely variable, ranging from a several-fold to a hundred-fold increase (Rutter *et al.*, 1995; Shioda *et al.*, 2004; Zhang *et al.*, 2009). Neuropharmacological investigations into the level of increased $5HT_{ext}$ in response to 5HT-promoting drugs will be highlighted in the next section.

2.2 Therapeutic elevation of $5HT_{ext}$ by 5HT-promoting drugs

Many psychoactive drugs used for patients are able to elevate $5HT_{ext}$ in the brain. To better elucidate the neuropharmacological range of extracellular concentrations, we focus on three categories of 5HT-promoting drugs: serotonin reuptake inhibitors (SRIs), monoamine oxidase inhibitors (MAOIs) and 5-hydroxyl-L-tryptophan (5HTP). SRIs, which elevate $5HT_{ext}$ by blocking 5HT from reentering synapses, have been widely prescribed for decades in the treatment for depression, anxiety and posttraumatic stress disorders. MAOIs are one of the oldest classes of antidepressants, functioning by increasing extravesicular accumulation and spontaneous release. 5HTP is the immediate precursor compound for 5HT synthesis, which is considered to be an important supplemental ingredient in promoting mood (Parker *et al.*, 2011). Altogether, $5HT_{ext}$ elevation is a part of critical mechanisms in the course of medical treatment for some mental diseases although other mechanisms based on 5HT receptors and relevant intracellular signaling pathways are also involved (Sharp *et al.*, 2011). Up to date, laboratory methods for determining 5HT release in patients' brain are not available and thus how much elevation of $5HT_{ext}$ is sufficient to improve mental health in humans is unknown. The relevant knowledge on the neurochemical effects of these drugs is almost exclusively obtained by animal studies.

There are several members in the SRI family, mainly (but not exhaustively) comprising selective serotonin reuptake inhibitors (SSRIs), serotonin-norepinephrine reuptake inhibitors (SNRIs) and tricyclic antidepressants (TCAs). For the last three decades a variety of approaches and treatment protocols have been examined on animals to investigate how these drugs act against 5HT reuptake into serotonergic neurons in the brain. It appears that acute systemic injection could cause only a regionally selective increase in $5HT_{ext}$ (Beyer *et al.*, 2008; David *et al.*, 2003; Rutter *et al.*, 1993). At clinically relevant doses, the maximum elevation evoked by SRIs is relatively low, at less than a 2-fold increase over baseline. This is because a global increase in $5HT_{ext}$ by systemic injection would activate autoreceptors, namely $5HT_{1A}$Rs in the raphe that inhibit discharge-dependent release of 5HT at axon terminals, limiting further increase in $5HT_{ext}$ (Hervas *et al.*, 2000; Rutter *et al.*, 1995). Related to this, $5HT_{1A}$R activation is associated with an acute anxiogenic response to SRIs, examined with laboratory animals (Birkett *et al.*, 2011; Greenwood *et al.*, 2008). Indeed, SRIs are known to have an anxiogenic effect in human patients, particularly during the early period of medical treatments (Bigos *et al.*, 2008; Browning *et al.*, 2007).

Functional activity of $5HT_{1A}$Rs can be to some extent desensitized after long-term use of SRIs (Blier, 2001; El Mansari *et al.*, 2005). As a result, $5HT_{ext}$ elevation is slightly augmented, most likely to around 2-3 fold (Dawson *et al.*, 2000) but never more than 5-fold (Popa *et al.*, 2010). It has been hypothesized that the desensitized $5HT_{1A}$Rs together with increased $5HT_{ext}$, but neither alone, are two required elements for antidepressant treatments (Sharp *et al.*, 2011). If the hypothesis were correct, it suggests that a less than 5-fold increase in $5HT_{ext}$ in the cerebrospinal fluid is sufficient for therapeutic purposes.

In addition to $5HT_{1A}$Rs, other 5HTR subtypes could also affect the profile of SRIs, thereby affecting $5HT_{ext}$ elevation (Cremers *et al.*, 2007; Jongsma *et al.*, 2005). This suggests that in the brain there exist several other feedback loops that collectively influence the effect of systemic administration of SRIs. In other words, the response to SRIs would be stronger after eliminating the feedback inhibition. Consistently, the $5HT_{ext}$ response is much higher when the drug administration is locally applied in the brain (Adell *et al.*, 1991a; Hervas *et al.*, 2000; Tao *et al.*, 2000). The maximum elevation of $5HT_{ext}$ by local application of SRIs can be 5-fold but less than 10-fold compared to the baseline. Although $5HT_{ext}$ at this level appears to be higher than normal, there is no evidence in the literature suggesting that a 5 to 9-fold increase is associated with mental impairment such as serotonin syndrome.

It should be kept in mind that 5HT reuptake is one of the important pathways for 5HT metabolism. Drugs that have an effect on the reuptake would likely pose the risk of interrupting the integrity of 5HT metabolism, particularly after long-term use (Moret *et al.*, 1992; Stenfors *et al.*, 2001). It has been shown that chronic treatment with SRIs could reduce the amount of intracellular 5HT in some brain regions (Bianchi *et al.*, 2002).Whether the reduced 5HT content would ultimately affect the SRI-evoked increase in $5HT_{ext}$, or whether such chronic effects on intracellular 5HT are associated with the development of treatment resistance in patients is not clear.

MAOIs are a family of drugs that inhibit the activity of the MAO-A isozyme located at the outer membrane of mitochondria in the synapses of serotonergic axon terminals, causing 5HT accumulation in the cytoplasmic compartment (Evrard *et al.*, 2002; Ferrer *et al.*, 1994). Since the capacity of the cytoplasm to retain 5HT molecules is very limited while the synaptic vesicles are full, the only place for the extravesicular 5HT is to "spillover" into the extracellular space. The mechanisms for 5HT spill are not fully elucidated, most likely through the transmitter carriers that reversely transport extravesicular 5HT into the extracellular space (Gobbi *et al.*, 1993; Silva *et al.*, 2008). Thus, administration of MAOIs would cause an increase in $5HT_{ext}$ proportional to that of intracellular content (Ferrer *et al.*, 1994). Normally, a maximum increase following systemic injection is about 1 to 2-fold above baseline (Ferrer *et al.*, 1994; Rollema *et al.*, 2011). The effect of MAOIs is also determined by dietary ingredients, particularly of those containing the 5HT precursor such as tryptophan and 5HTP. It deserves separate mention that one of the major side effects in the clinical use of the old generation of MAOIs is hypertensive crisis, known as "cheese reaction" due to ingestion with tyramine-rich diets. The good news is that the new generation of MAOIs has fewer side effects relevant to diet [reviewed by (Wimbiscus *et al.*, 2010)]. Clinically, while SRIs are widely used as a first-line drug to treat depression, MAOIs are often recommended for treatment-resistant depression. Similar to SRIs, the MAOI-evoked increase in $5HT_{ext}$ is also regulated by feedback mechanisms involving $5HT_{1A}Rs$ (Lanteri *et al.*, 2009). Hence, like SRIs, MAOIs could produce relatively higher increases after the feedback inhibition mechanism is eliminated or desensitized, showing a maximum elevation of 5-10 fold above baseline (Tao *et al.*, 1994).

5HTP is recommended as a supplemental treatment for depression and menopausal hot flush (Curcio *et al.*, 2005; Shaw *et al.*, 2002). In humans, the typical 5HTP dosage is 100-300 mg/day or 1-3 mg/kg [reviewed by (Turner *et al.*, 2006)]. In animals, 5HTP alone has no measurable effect on $5HT_{ext}$ except at a dose of 40 mg/kg and higher (Gartside *et al.*, 1992; Perry *et al.*, 1993). This implies that newly synthesized 5HT molecules can be rapidly metabolized before being accumulated in the extravesicular compartment for spontaneous release. Specifically, $5HT_{ext}$ was increased by 2-fold in response to 75 mg/kg 5HTP (Nakatani *et al.*, 2008) and by 4-fold in response to 100 mg/kg 5HTP (Gartside *et al.*, 1992). In summary, it appears that 5HTP alone at the dose typically used in human patients has little contribution to the level of $5HT_{ext}$.

2.3 Dangerous elevations of $5HT_{ext}$: How much $5HT_{ext}$ in the CNS is too much?

There is no doubt that $5HT_{ext}$ is tightly regulated as a part of a homeostasis, most likely remaining at a low nanomolar concentration for maintaining normal brain function. Hypothetically, $5HT_{ext}$ in patients with major depression is too low to stimulate 5HTRs for an affective process (such as mood). 5HT-promoting drugs could lift mood by correcting $5HT_{ext}$ level. The amount of 5HT elevation is still at a small scale when drugs are used alone (monopharmacy).

However, $5HT_{ext}$ could be overcorrected, particularly with medication switching and polypharmacy. Despite decades of clinical research, no laboratory test is available to monitor brain $5HT_{ext}$ during a drug regimen in humans. It has been suggested that the change in brain $5HT_{ext}$ is in parallel with changes in peripheral 5HT or 5HIAA measured in blood and urine (Alvarez *et al.*, 1999; Bianchi *et al.*, 2002; Celada *et al.*, 1993). Metabolically, peripheral 5HT is however independent of 5HT in the CNS; how changes in peripheral 5HT in response to drug treatment can infer changes in the brain is not yet established.

Although drug treatments would improve the mental well-being scale in terms of psychological evaluations, patients are at risk of having "excessive" $5HT_{ext}$ in the brain. Thus, it is important to determine a safe range of $5HT_{ext}$ concentrations for improvement in mood without causing excessive $5HT_{ext}$-induced side effects. It has been suggested by *in vivo* microdialysis studies that 5HT-promoting drugs usually produce a nanomolar (nM) level of

$5HT_{ext}$ in the cerebrospinal fluid. However, *in vitro* studies demonstrate that a micromolar (µM) concentration is required to have postsynaptic effects on 5HTR-containing neurons (Cornelisse *et al.*, 2007; Marinelli *et al.*, 2004). This discrepancy in $5HT_{ext}$ levels possibly represents the difference in microenvironments between the cerebrospinal fluid and synapses in the extracellular space. Conceivably, most 5HT molecules released from synapses are rapidly taken back by transporters while a small portion is diffused away from the synapses. If this is the case, synaptic 5HT concentration must be much higher than that in the cerebrospinal fluid measured by microdialysis. In addition, $5HT_{ext}$ released from synapses is not evenly distributed *in vivo* across postsynaptic cell membranes. The amount of distribution depends on the distance from the release site. The concept of concentration differences in microenvironments is also supported by other studies. For instance, utilizing fast-scan cyclic voltammetry at a time resolution of 100 msec, it has been observed that $5HT_{ext}$ at the extrasynaptic site was rapidly (~5 s) increased from nM to µM in response to electrical stimulation (Bunin *et al.*, 1998). In contrast, comparable stimulations could only induce a 1-2 fold increase in $5HT_{ext}$ in the cerebrospinal fluid determined by microdialysis at intervals of 20 min (McQuade *et al.*, 1997). This suggests that the $5HT_{ext}$ concentration in the extrasynapse compartment is much higher than in the cerebrospinal fluid, supporting the view of microenvironmental variation. Note that such high concentration of $5HT_{ext}$ at synapses and/or extrasynaptic sites may not cause a neural disorder because: 1) the time duration of the effect is usually only a few milliseconds; 2) numbers and types of synaptic neurons involved in the action are highly localized; 3) the neurotransmitter can be rapidly removed by 5HT reuptake transporters. However, it could be problematic if there are prolonged and widespread effects on many 5HTR-containing neurons.

Except for MDMA, few 5HT-promoting drugs acting alone could elevate 5HT to high levels for a long period. This may occur mainly when two or more 5HT metabolic pathways are simultaneously disrupted during medication switching or polypharmacy, resulting in "excessive" $5HT_{ext}$ that exerts an adverse effect on mental health. For instance, Shioda et al. investigated the interaction between MAOIs and SSRIs for understanding changes in $5HT_{ext}$ relevant to serotonin syndrome (Shioda *et al.*, 2004). In their investigation, male Wistar rats received co-injection of tranylcypromine (3.5 mg/kg), a nonselective MAOI, and fluoxetine (10 mg/kg), an SSRI. Hypothalamic $5HT_{ext}$ was determined by microdialysis while serotonin syndrome was estimated by measuring body-core temperature and neuromuscular activity. The environmental temperature was set at 23 ±1 °C. Under this condition, drug interaction caused an increase in $5HT_{ext}$ by 40-fold above the pre-drug level, lasting at least 6 hours. There were obvious signs of serotonin syndrome including hyperthermia, head shakes and tremor, suggesting that $5HT_{ext}$ at this level causes both physical and behavioral problems. Since each drug alone could only evoke a 2-3 fold increase, it appears that their combined effect is not simply additive, but synergistic. Other neurotransmissions, for instance, dopaminergic and glutamatergic systems that usually are not affected by single drug treatments are also elevated. This suggests that neural circuits consisting of several neuronal systems are involved in the syndrome, which is beyond the scope of this analysis and will not be discussed further here.

It is critical to determine the threshold level of $5HT_{ext}$ responsible for evoking serotonin syndrome. Baumann et al examined the harmful potential of MDMA by scoring behavior signs of serotonin syndrome (e.g., flat-body posture and forepaw treading) in correlation with elevated $5HT_{ext}$ in the frontal cortex and nucleus accumbens of male Sprague-Dawley rats (Baumann *et al.*, 2008a; Baumann *et al.*, 2008b). In their behavioral and neurochemical studies, animals received a first injection of MDMA at the dose of 1 mg/kg followed by a second injection of 3 mg/kg 60 min later. The ambient temperature in which animals were examined was 22±2 °C. As a result, $5HT_{ext}$ was increased by 5- to 8-fold in response to 1 mg/kg and 18- to 33-fold following 3 mg/kg. While MDMA at 1 mg/kg had no effect on animal behavior, 3 mg/kg of MDMA were able to induce symptoms of the serotonin syndrome, suggesting that the $5HT_{ext}$ threshold to evoke the serotonin syndrome may be in a range between 9 to 18-fold.

Compared to other 5HT-promoting drugs, a single injection of MDMA at clinically relevant doses has a transient effect on 5HT, lasting only 15-30 min in terms of peak effect. To obtain a significant response, many investigators employ multiple injections to mimic the binge use of MDMA in humans. Theoretically, multiple injections could produce several 5HT peak responses which would complicate the elucidation of causal relationships between 5HT and behavioral effects. To simplify the analysis, 5HT precursors (e.g., 5HTP) in combination with MAOI (e.g., clorgyline) are more applicable for elucidating the 5HT threshold. Many studies have demonstrated that a single injection of 5HTP combined with clorgyline was sufficient to produce a dose-dependent increase in $5HT_{ext}$ in the brain, causing serotonin syndrome (Ma *et al.*, 2008; Nisijima *et al.*, 2004; Nisijima *et al.*, 2000; Nisijima *et al.*, 2001; Shioda *et al.*, 2004). To obtain the threshold level of $5HT_{ext}$, we designed a 5HTP dosing regimen in clorgyline-pretreated rats. Male Sprague–Dawley rats were examined under controlled ambient temperature (22 ± 1°C) and humidity (70%) in test chambers (Ma *et al.*, 2008; Zhang *et al.*, 2009). Animals received 2 mg/kg clorgyline, 2 hours before injection of 5HTP at the dose range of 1 to 25 mg/kg. $5HT_{ext}$ was determined in the frontal cortex and hypothalamus while behavioral responses were recorded with a 4-level scale by scoring the severity of tremor, head shakes, forepaw treading, hindlimb abduction, myotonia and Straub tail.

Specifically, $5HT_{ext}$ was potentially elevated to as high as 100-fold above baseline following a single injection of 5HTP into clorgyline-pretreated rats, consistent with the results of previous studies (Shioda *et al.*, 2004). In fact, this level of $5HT_{ext}$ is already well over the threshold for inducing a serotonin syndrome. Our data showed that a 55-fold increase in $5HT_{ext}$ caused the severe syndrome manifested by high hyperthermia and all other typical signs described in animals (Jacobs *et al.*, 1975) or severe symptoms resembling those described in humans (Mills, 1995). An open question is whether such a high $5HT_{ext}$ level is really evoked in the human brain despite drug interactions.

Our studies further demonstrated that a 10-fold increase in $5HT_{ext}$ was sufficient to cause a mild syndrome, showing hypothermia, head shakes and myoclonus but no signs indicative of advanced or severe syndrome (e.g., tremor, forepaw treading or hindlimb abduction) (Ma *et al.*, 2008; Zhang *et al.*, 2009). This suggests that the brain is most likely intolerant of the double digit increase in $5HT_{ext}$ in terms of fold-change in the cerebrospinal fluid. There was no syndrome-related behavioral changes in response to a 5- to 9-fold increase although head shakes were still apparent, supporting the conclusion that a single digit increase is highly tolerable. The head shaking behavior disappeared when the increased $5HT_{ext}$ decreased to 5-fold and less, consistent with suggestion that a less than 5-fold increase in $5HT_{ext}$ in the brain is essentially safe. Clearly, more investigation into this important area of research is needed. It is crucial to relate changes in behavior to changes in $5HT_{ext}$ level, particularly after long-term interactions between 5HT-promoting drugs.

3. Involvement of 5HT receptors

There are at least 14 subtypes of 5HT receptors (5HTRs) in the brain (Green, 2006). One may wonder what the role is an individual subtype in the development of the serotonin syndrome. It should be kept in mind that 5HT is the natural agonist for these subtypes in the brain; and involvement of individual subtypes depends on not only the availability of the agonist but also the binding affinity to the agonist. It has been recognized for many years that the strength of 5HT affinity can be widely different. For instance, a μM concentration may be required for binding 50% of low affinity $5HT_{2A}Rs$ whereas a nM concentration for high affinity $5HT_{1A}Rs$ (Dalpiaz *et al.*, 1995; Peroutka *et al.*, 1981; Peroutka *et al.*, 1983). Thus, the role of each subtype in the syndrome is complicated, and depends on multiple factors.

For better comparison between subtypes, pK_d or pK_i values are commonly used to indicate the strength or the selectivity of ligands to a given subtype. It is worthy pointing out that their values are not always reliable or consistent between tests due to variables in experimental conditions. Despite this, they provide trends for estimating affinity between receptor and ligand. It is well known that affinity values are strongly associated with functional activity of receptors in the brain (Clemett *et al.*, 1999; Knight *et al.*, 2004; Rossi *et al.*, 2008). Although few studies are available to completely map the affinity between 5HT and its receptor subtypes under the same condition, a valuable reference is found from the database provided by the International Union of Basic and Clinical Pharmacology or IUPHAR (Table 1). Based on their database, the 14 subtypes can be ranked in the order of affinity strength from high (sub-nM; or 10^{-9} M) to low activity (μM or 10^{-6} M). Although so many subtypes are available in the brain, it appears that they are not functioning simultaneously. Their neurological function is selectively elicited by 5HT in a concentration-dependent manner. For instance, 5HT at the concentration of approximately 10^{-9} M could competitively displace 50% of competitors at the $5HT_{1A}R$ site, implying that a sub-nM concentration is able to activate $5HT_{1A}Rs$. In support of this, as demonstrated *in vitro* in dorsal raphe slices, the inward current indicative of $5HT_{1A}R$ activity can be elicited by 5HT at concentration less than 1 nM with the EC_{50} at 30 nM (Penington *et al.*, 1993), closely in line with results obtained by radioligand binding assays. By contrast, other subtypes such as $5HT_{2A}Rs$ are unlikely to be affected by 5HT at such a low concentration. It should be noted that $5HT_{1A}Rs$ are densely located on the somatodendritic sites of serotonergic neurons in the raphe (Li *et al.*, 1997) and their high binding affinity to 5HT is physiologically important for their role in negative feedback regulation.

$5HT_{1A}Rs$ are also distributed on many types of postsynaptic neurons (Bert *et al.*, 2006), particularly glutamatergic and GABAergic neurons in the cortices (de Almeida *et al.*, 2008; Martin-Ruiz *et al.*, 2001). It has often been shown that these postsynaptic $5HT_{1A}Rs$ can be activated by 5HT at a relatively high concentration between 1-10 μM (Goodfellow *et al.*, 2009; Schmitz *et al.*, 1998), controlling the functional balance between glutamate and GABA transmissions.

Thus, since $5HT_{ext}$ in response to most psychoactive drugs is normally in a nanomolar range (Gardier *et al.*, 2003), these postsynaptic receptors are unlikely to be able to be activated. In the case of $5HT_{ext}$ exceeding maximum tolerable limits, there is a global activation of postsynaptic $5HT_{1A}Rs$, causing fluctuation of GABA and glutamate, two functionally opposite neurotransmitters. It is likely that this fluctuation in the CNS is responsible for neuromuscular hyperactivity (Paterson *et al.*, 2009), but verification of this hypothesis related to the serotonin syndrome involving head weaving, tremor and forepaw treading awaits for further investigation.

$5HT_{1A}R$	9.1 - 9.7
$5HT_7R$	8.1 - 9.6
$5HT_{1D}R$	8.0 - 9.0
$5ht_{1e}R$	8.0 - 8.2
$5HT_{2B}R$	7.9 - 8.4
$5HT_{1F}R$	7.7 - 8.0
$5HT_{1B}R$	7.4 - 9.0
$5HT_{2C}R$	6.8 - 8.6
$5HT_6R$	6.8 - 7.5
$5ht_{5a}R$	6.7 - 6.9
$5HT_{3A}R$	6.5 - 6.9
$5HT_{2A}R$	6.0 - 8.4
$5HT_{3AB}R$	6.0
$5HT_4R$	5.9 - 7.0

Data from the database provided by the International Union of Basic and Clinical Pharmacology or IUPHAR (http://www.iuphar-db.org/DATABASE/ObjectDisplayForward?familyId=1&objectId=1)

Table 1. 5HT binding affinity (pK_i)

Based on $5HT_{ext}$ concentration, we suggest that the 5HT subtypes can be hypothetically classified into 5 groups: sub-nM, nM, high nM, sub-µM and µM. As elucidated in Table 2, more and more subtypes in the brain are involved when the concentration is increased. It is likely that, as $5HT_{ext}$ exceeds its upper tolerable level, lower affinity subtypes are activated along with higher affinity ones.

$5HT_{ext}$ levels	Sub-nM	nM	High nM	Sub-µM	µM
Affected receptors	$5HT_{1A}R$ (presynaptic) $5HT_7R$	$5HT_{1A}R$ (presynaptic) $5HT_7R$ $5HT_{2B}R$ $5ht_{1e}R$ $5HT_{1D}R$	$5HT_{1A}R$ (presynaptic) $5HT_7R$ $5HT_{2B}R$ $5ht_{1e}R$ $5HT_{1D}R$ $5HT_{1B}R$ $5HT_{1F}R$	$5HT_{1A}R$ (pre-postsynaptic) $5HT_7R$ $5HT_{2B}R$ $5ht_{1e}R$ $5HT_{1D}R$ $5HT_{1B}R$ $5HT_{1F}R$ $5HT_{2C}R$ $5HT_6R$ $5ht_{5a}R$ $5HT_{3A}R$	$5HT_{1A}R$ (pre-postsynaptic) HT_7R $5HT_{2B}R$ $5ht_{1e}R$ $5HT_{1D}R$ $5HT_{1B}R$ $5HT_{1F}R$ $5HT_{2C}R$ $5HT_6R$ $5ht_{5a}R$ $5HT_{3A}R$ $5HT_{2A}R$ $5HT_4R$ $5HT_{3AB}R$

Table 2. Neurological relationship between subtype activation and increased $5HT_{ext}$. More and more subtypes are activated as $5HT_{ext}$ levels are increased from sub-nM to µM

Using a radioligand receptor binding assay, studies have shown that $5HT_{2A}Rs$ have low affinity and are not easily bound to 5HT (Peroutka *et al.*, 1981; Peroutka *et al.*, 1983). The significance of this finding has not been recognized simply because the threshold level of $5HT_{ext}$ concentration in maintaining mental health is not fully appreciated. $5HT_{2A}Rs$ are mainly (but not exclusively) distributed on glutamatergic neurons in the cortices (de Almeida *et al.*, 2007) and are critical for regulating the interconnection between the cortex and raphe. It has been widely documented in *in vitro* studies that $5HT_{2A}Rs$ cannot be activated until the 5HT concentration reaches 20-50 μM (Aghajanian *et al.*, 1997; Zhou *et al.*, 1999), suggesting that their activation threshold is much higher than $5HT_{1A}Rs$. Importantly, the answer to whether $5HT_{2A}Rs$ are involved in the syndrome is a life-or-death issue. This is mainly because $5HT_{2A}Rs$ are the major receptor strongly associated with hyperthermia and other autonomic hyperactivity (Mazzola-Pomietto *et al.*, 1995; Zhang *et al.*, 2011). Hyperthermia is believed to be the major cause for severe brain injury as demonstrated by MDMA studies (Malberg *et al.*, 1998). Collectively, the serotonin syndrome resulting from activation of $5HT_{1A}Rs$ and $5HT_{2A}Rs$ is associated with excessive $5HT_{ext}$ up to a single to double-digit μM concentration at synapses. This suggests that $5HT_{ext}$ in the synapse has arisen by 1000 times while it usually remains at a low nanomolar level. The effect is likely to correspond to a double-digit fold increase in the cerebrospinal fluid measured by *in vivo* microdialysis (Zhang *et al.*, 2009).

Most recent data have revealed that the functional activity of some 5HTR subtypes can be altered by environmental factors (Krishnamoorthy *et al.*, 2010; Nicholas *et al.*, 2003; Zhang *et al.*, 2011). For instance, Nicholas et al demonstrated that, compared to a normal experimental condition, $5HT_{1A}R$ activity was markedly reduced in animals examined in warm ambient temperatures. On the other hand, the responsivity of $5HT_{2A}Rs$ could be markedly enhanced in warmer environments (Zhang *et al.*, 2011). Similarly, it has been found in an *in vitro* binding kinetic assay that 5HT affinity to $5HT_{2A}Rs$ was strongly increased in a temperature-dependent manner (Dalpiaz *et al.*, 1995). Taken together, $5HT_{ext}$ concentration required for activation of 5HTRs may vary. Indeed, it has been observed that serotonin syndrome evoked by MDMA and other 5HT-promoting drugs is severely augmented in hot and crowded conditions at raves and dance clubs (Parrott, 2002) but ameliorated in a cooling environment (Krishnamoorthy *et al.*, 2010).

4. Summary

The aim of this review is to elucidate possible ranges of brain $5HT_{ext}$ in association with therapeutic benefit, tolerable side effects and the development of serotonin syndrome. $5HT_{ext}$ is derived from either spontaneous release or active stimulation, which initially presents at synapses and subsequently diffuses into the cerebrospinal fluid. In contrast to 5HT at synapses, $5HT_{ext}$ in the cerebrospinal fluid can be easily determined by microdialysis. Physiologically, a low nanomalor concentration (nM) is crucial for normal function of maintaining constant activation of postsynaptic 5HTRs. In the case of depression, $5HT_{ext}$ is at a lower level, resulting in reduced activity of 5HTRs that can affect mood and behaviors. An appropriate elevation of $5HT_{ext}$ promotes a positive mood and an energized sense of physical well-being. Therapeutically, $5HT_{ext}$ elevated by 5HT-promoting drugs such as SRIs or MAOIs up to 5-fold in the cerebrospinal fluid is sufficient to improve the mood.

Unfortunately, MDMA ('Ecstasy") hijacks such positive aspects of the serotonergic effect, causing drug abuse. Although generally safe, 5HT-promoting drugs could evoke an increase in $5HT_{ext}$ in the cerebrospinal fluid exceeding 10-fold. When it occurs, the $5HT_{ext}$ level in the synapses most likely reaches a single to double-digit µM concentration, which would globally activate $5HT_{1A}$Rs and/or $5HT_{2A}$Rs on glutamatergic and GABAergic neurons. Such global activation involved in both glutamate and GABA transmission leads to neurological disorders, which are manifested as serotonin syndrome. Importantly, the actions of $5HT_{ext}$ on 5HTRs are highly variable, and strongly depend on behavioral states as well as external environments. Thus, the severity of the syndrome can vary from mild, to moderate, to life-threatening (Krishnamoorthy *et al.*, 2010; Parrott, 2002).

5. Acknowledgements

The authors are grateful to Professor Howard Prentice for editorial assistance to the final manuscript. This research was supported by USPHS DA029863.

6. References

Adell, A. & Artigas, F. (1991a). Differential effects of clomipramine given locally or systemically on extracellular 5-hydroxytryptamine in raphe nuclei and frontal cortex. An in vivo brain microdialysis study. *Naunyn Schmiedebergs Arch Pharmacol,* 343, 237-44.

Adell, A., Carceller, A. & Artigas, F. (1991b). Regional distribution of extracellular 5-hydroxytryptamine and 5-hydroxyindoleacetic acid in the brain of freely moving rats. *J Neurochem*, 56, 709-12.

Aghajanian, G.K. & Marek, G.J. (1997). Serotonin induces excitatory postsynaptic potentials in apical dendrites of neocortical pyramidal cells. *Neuropharmacology*, 36, 589-99.

Alvarez, J.C., Gluck, N., Arnulf, I., Quintin, P., Leboyer, M., Pecquery, R., Launay, J.M., Perez-Diaz, F. & Spreux-Varoquaux, O. (1999). Decreased platelet serotonin transporter sites and increased platelet inositol triphosphate levels in patients with unipolar depression: effects of clomipramine and fluoxetine. *Clin Pharmacol Ther*, 66, 617-24.

Baumann, M.H., Clark, R.D., Franken, F.H., Rutter, J.J. & Rothman, R.B. (2008a). Tolerance to 3,4-methylenedioxymethamphetamine in rats exposed to single high-dose binges. *Neuroscience*, 152, 773-84.

Baumann, M.H., Clark, R.D. & Rothman, R.B. (2008b). Locomotor stimulation produced by 3,4-methylenedioxymethamphetamine (MDMA) is correlated with dialysate levels of serotonin and dopamine in rat brain. *Pharmacol Biochem Behav*, 90, 208-17.

Baumann, M.H., Wang, X. & Rothman, R.B. (2007). 3,4-Methylenedioxymethamphetamine (MDMA) neurotoxicity in rats: a reappraisal of past and present findings. *Psychopharmacology (Berl)*, 189, 407-424.

Becquet, D., Faudon, M. & Hery, F. (1990). The role of serotonin release and autoreceptors in the dorsalis raphe nucleus in the control of serotonin release in the cat caudate nucleus. *Neuroscience*, 39, 639-47.

Bert, B., Fink, H., Hortnagl, H., Veh, R.W., Davies, B., Theuring, F. & Kusserow, H. (2006). Mice over-expressing the 5-HT_{1A} receptor in cortex and dentate gyrus display exaggerated locomotor and hypothermic response to 8-OH-DPAT. *Behav Brain Res*, 167, 328-41.

Beyer, C.E. & Cremers, T.I. (2008). Do selective serotonin reuptake inhibitors acutely increase frontal cortex levels of serotonin? *Eur J Pharmacol*, 580, 350-4.

Bhide, N.S., Lipton, J.W., Cunningham, J.I., Yamamoto, B.K. & Gudelsky, G.A. (2009). Repeated exposure to MDMA provides neuroprotection against subsequent MDMA-induced serotonin depletion in brain. *Brain Res*, 1286, 32-41.

Bianchi, M., Moser, C., Lazzarini, C., Vecchiato, E. & Crespi, F. (2002). Forced swimming test and fluoxetine treatment: in vivo evidence that peripheral 5-HT in rat platelet-rich plasma mirrors cerebral extracellular 5-HT levels, whilst 5-HT in isolated platelets mirrors neuronal 5-HT changes. *Exp Brain Res*, 143, 191-7.

Bigos, K.L., Pollock, B.G., Aizenstein, H.J., Fisher, P.M., Bies, R.R. & Hariri, A.R. (2008). Acute 5-HT reuptake blockade potentiates human amygdala reactivity. *Neuropsychopharmacology*, 33, 3221-5.

Birkett, M.A., Shinday, N.M., Kessler, E.J., Meyer, J.S., Ritchie, S. & Rowlett, J.K. (2011). Acute anxiogenic-like effects of selective serotonin reuptake inhibitors are attenuated by the benzodiazepine diazepam in BALB/c mice. *Pharmacol Biochem Behav*, 98, 544-51.

Blier, P. (2001). Pharmacology of rapid-onset antidepressant treatment strategies. *J Clin Psychiatry*, 62 Suppl 15, 12-7.

Boyer, E.W. & Shannon, M. (2005). The serotonin syndrome. *N Engl J Med*, 352, 1112-20.

Browning, M., Reid, C., Cowen, P.J., Goodwin, G.M. & Harmer, C.J. (2007). A single dose of citalopram increases fear recognition in healthy subjects. *J Psychopharmacol*, 21, 684-90.

Bruns, D. & Jahn, R. (1995). Real-time measurement of transmitter release from single synaptic vesicles. *Nature*, 377, 62-5.

Bruns, D., Riedel, D., Klingauf, J. & Jahn, R. (2000). Quantal release of serotonin. *Neuron*, 28, 205-20.

Bunin, M.A. & Wightman, R.M. (1998). Quantitative evaluation of 5-hydroxytryptamine (serotonin) neuronal release and uptake: an investigation of extrasynaptic transmission. *J Neurosci*, 18, 4854-60.

Calcagno, E., Canetta, A., Guzzetti, S., Cervo, L. & Invernizzi, R.W. (2007). Strain differences in basal and post-citalopram extracellular 5-HT in the mouse medial prefrontal cortex and dorsal hippocampus: relation with tryptophan hydroxylase-2 activity. *J Neurochem*, 103, 1111-20.

Celada, P. & Artigas, F. (1993). Plasma 5-hydroxyindoleacetic acid as an indicator of monoamine oxidase-A inhibition in rat brain and peripheral tissues. *J Neurochem*, 61, 2191-8.

Clemett, D.A., Kendall, D.A., Cockett, M.I., Marsden, C.A. & Fone, K.C. (1999). Pindolol-insensitive [^{3}H]-5-hydroxytryptamine binding in the rat hypothalamus; identity with 5-hydroxytryptamine7 receptors. *Br J Pharmacol*, 127, 236-42.

Cornelisse, L.N., Van der Harst, J.E., Lodder, J.C., Baarendse, P.J., Timmerman, A.J., Mansvelder, H.D., Spruijt, B.M. & Brussaard, A.B. (2007). Reduced 5-HT_{1A}- and $GABA_B$ receptor function in dorsal raphe neurons upon chronic fluoxetine treatment of socially stressed rats. *J Neurophysiol*, 98, 196-204.

Cremers, T.I., Rea, K., Bosker, F.J., Wikstrom, H.V., Hogg, S., Mork, A. & Westerink, B.H. (2007). Augmentation of SSRI effects on serotonin by 5-HT_{2C} antagonists: mechanistic studies. *Neuropsychopharmacology*, 32, 1550-7.

Curcio, J.J., Kim, L.S., Wollner, D. & Pockaj, B.A. (2005). The potential of 5-hydryoxytryptophan for hot flash reduction: a hypothesis. *Altern Med Rev*, 10, 216-21.

Dalpiaz, A., Gessi, S., Borea, P.A. & Gilli, G. (1995). Binding thermodynamics of serotonin to rat-brain 5-HT_{1A}, 5-HT_{2A} and 5-HT_3 receptors. *Life Sci*, 57, PL141-6.

David, D.J., Bourin, M., Jego, G., Przybylski, C., Jolliet, P. & Gardier, A.M. (2003). Effects of acute treatment with paroxetine, citalopram and venlafaxine in vivo on noradrenaline and serotonin outflow: a microdialysis study in Swiss mice. *Br J Pharmacol*, 140, 1128-36.

Dawson, L.A., Nguyen, H.Q., Smith, D.I. & Schechter, L.E. (2000). Effects of chronic fluoxetine treatment in the presence and absence of (±)pindolol: a microdialysis study. *Br J Pharmacol*, 130, 797-804.

de Almeida, J. & Mengod, G. (2007). Quantitative analysis of glutamatergic and GABAergic neurons expressing 5-HT_{2A} receptors in human and monkey prefrontal cortex. *J Neurochem*, 103, 475-86.

de Almeida, J. & Mengod, G. (2008). Serotonin 1A receptors in human and monkey prefrontal cortex are mainly expressed in pyramidal neurons and in a GABAergic interneuron subpopulation: implications for schizophrenia and its treatment. *J Neurochem*, 107, 488-96.

El Mansari, M. & Blier, P. (2005). Responsiveness of 5-HT_{1A} and 5-HT_2 receptors in the rat orbitofrontal cortex after long-term serotonin reuptake inhibition. *J Psychiatry Neurosci*, 30, 268-74.

Evrard, A., Malagie, I., Laporte, A.M., Boni, C., Hanoun, N., Trillat, A.C., Seif, I., De Maeyer, E., Gardier, A., Hamon, M. & Adrien, J. (2002). Altered regulation of the 5-HT system in the brain of MAO-A knock-out mice. *Eur J Neurosci*, 15, 841-51.

Ferrer, A. & Artigas, F. (1994). Effects of single and chronic treatment with tranylcypromine on extracellular serotonin in rat brain. *Eur J Pharmacol*, 263, 227-34.

Gardier, A.M., David, D.J., Jego, G., Przybylski, C., Jacquot, C., Durier, S., Gruwez, B., Douvier, E., Beauverie, P., Poisson, N., Hen, R. & Bourin, M. (2003). Effects of chronic paroxetine treatment on dialysate serotonin in 5-HT_{1B} receptor knockout mice. *J Neurochem*, 86, 13-24.

Gartside, S.E., Cowen, P.J. & Sharp, T. (1992). Effect of 5-hydroxy-L-tryptophan on the release of 5-HT in rat hypothalamus in vivo as measured by microdialysis. *Neuropharmacology*, 31, 9-14.

Gobbi, M., Frittoli, E., Uslenghi, A. & Mennini, T. (1993). Evidence of an exocytotic-like release of [^{3}H]5-hydroxytryptamine induced by d-fenfluramine in rat hippocampal synaptosomes. *Eur J Pharmacol*, 238, 9-17.

Goodfellow, N.M., Benekareddy, M., Vaidya, V.A. & Lambe, E.K. (2009). Layer II/III of the prefrontal cortex: Inhibition by the serotonin 5-HT1A receptor in development and stress. *J Neurosci*, 29, 10094-103.

Grahame-Smith, D.G. & Green, A.R. (1974). The role of brain 5-hydroxytryptamine in the hyperactivity produced in rats by lithium and monoamine oxidase inhibition. *Br J Pharmacol*, 52, 19-26.

Green, R.A. (2006). Neuropharmacology of 5-hydroxytryptamine. *Br J Pharmacol*, 147 Suppl 1, S145-52.

Greenwood, B.N., Strong, P.V., Brooks, L. & Fleshner, M. (2008). Anxiety-like behaviors produced by acute fluoxetine administration in male Fischer 344 rats are prevented by prior exercise. *Psychopharmacology (Berl)*, 199, 209-22.

Hervas, I., Queiroz, C.M., Adell, A. & Artigas, F. (2000). Role of uptake inhibition and autoreceptor activation in the control of 5-HT release in the frontal cortex and dorsal hippocampus of the rat. *Br J Pharmacol*, 130, 160-6.

Jacobs, B.L. & Klemfuss, H. (1975). Brain stem and spinal cord mediation of a serotonergic behavioral syndrome. *Brain Res*, 100, 450-7.

Jongsma, M.E., Bosker, F.J., Cremers, T.I., Westerink, B.H. & den Boer, J.A. (2005). The effect of chronic selective serotonin reuptake inhibitor treatment on serotonin 1B receptor sensitivity and HPA axis activity. *Prog Neuropsychopharmacol Biol Psychiatry*, 29, 738-44.

Karunatilake, H. & Buckley, N.A. (2006). Serotonin syndrome induced by fluvoxamine and oxycodone. *Ann Pharmacother*, 40, 155-7.

Kim, D.K., Tolliver, T.J., Huang, S.J., Martin, B.J., Andrews, A.M., Wichems, C., Holmes, A., Lesch, K.P. & Murphy, D.L. (2005). Altered serotonin synthesis, turnover and dynamic regulation in multiple brain regions of mice lacking the serotonin transporter. *Neuropharmacology*, 49, 798-810.

Knight, A.R., Misra, A., Quirk, K., Benwell, K., Revell, D., Kennett, G. & Bickerdike, M. (2004). Pharmacological characterisation of the agonist radioligand binding site of 5-HT_{2A}, 5-HT_{2B} and 5-HT_{2C} receptors. *Naunyn Schmiedebergs Arch Pharmacol*, 370, 114-23.

Krishnamoorthy, S., Ma, Z., Zhang, G., Wei, J., Auerbach, S.B. & Tao, R. (2010). Involvement of 5-HT_{2A} Receptors in the Serotonin (5-HT) Syndrome caused by Excessive 5-HT Efflux in Rat Brain. *Basic Clin Pharmacol Toxicol*, 107, 830-41.

Lanteri, C., Hernandez Vallejo, S.J., Salomon, L., Doucet, E.L., Godeheu, G., Torrens, Y., Houades, V. & Tassin, J.P. (2009). Inhibition of monoamine oxidases desensitizes 5-HT_{1A} autoreceptors and allows nicotine to induce a neurochemical and behavioral sensitization. *J Neurosci*, 29, 987-97.

Li, Q., Battaglia, G. & Van de Kar, L.D. (1997). Autoradiographic evidence for differential G-protein coupling of 5-HT_{1A} receptors in rat brain: lack of effect of repeated injections of fluoxetine. *Brain Res*, 769, 141-51.

Ma, Z., Zhang, G., Jenney, C., Krishnamoorthy, S. & Tao, R. (2008). Characterization of serotonin-toxicity syndrome (toxidrome) elicited by 5-hydroxy-l-tryptophan in clorgyline-pretreated rats. *Eur J Pharmacol*, 588, 198-206.

Malberg, J.E. & Seiden, L.S. (1998). Small changes in ambient temperature cause large changes in 3,4-methylenedioxymethamphetamine (MDMA)-induced serotonin neurotoxicity and core body temperature in the rat. *J Neurosci*, 18, 5086-94.

Marinelli, S., Schnell, S.A., Hack, S.P., Christie, M.J., Wessendorf, M.W. & Vaughan, C.W. (2004). Serotonergic and nonserotonergic dorsal raphe neurons are pharmacologically and electrophysiologically heterogeneous. *J Neurophysiol*, 92, 3532-7.

Martin-Ruiz, R., Puig, M.V., Celada, P., Shapiro, D.A., Roth, B.L., Mengod, G. & Artigas, F. (2001). Control of serotonergic function in medial prefrontal cortex by serotonin-2A receptors through a glutamate-dependent mechanism. *J Neurosci*, 21, 9856-66.

Mathews, T.A., Fedele, D.E., Coppelli, F.M., Avila, A.M., Murphy, D.L. & Andrews, A.M. (2004). Gene dose-dependent alterations in extraneuronal serotonin but not dopamine in mice with reduced serotonin transporter expression. *J Neurosci Methods*, 140, 169-81.

Mazzola-Pomietto, P., Aulakh, C.S., Wozniak, K.M., Hill, J.L. & Murphy, D.L. (1995). Evidence that 1-(2,5-dimethoxy-4-iodophenyl)-2-aminopropane (DOI)-induced hyperthermia in rats is mediated by stimulation of 5-HT_{2A} receptors. *Psychopharmacology (Berl)*, 117, 193-9.

McQuade, R. & Sharp, T. (1997). Functional mapping of dorsal and median raphe 5-hydroxytryptamine pathways in forebrain of the rat using microdialysis. *J Neurochem*, 69, 791-6.

Mills, K.C. (1995). Serotonin syndrome. *Am Fam Physician*, 52, 1475-82.

Moret, C. & Briley, M. (1992). Effect of antidepressant drugs on monoamine synthesis in brain in vivo. *Neuropharmacology*, 31, 679-84.

Nakatani, Y., Sato-Suzuki, I., Tsujino, N., Nakasato, A., Seki, Y., Fumoto, M. & Arita, H. (2008). Augmented brain 5-HT crosses the blood-brain barrier through the 5-HT transporter in rat. *Eur J Neurosci*, 27, 2466-72.

Nicholas, A.C. & Seiden, L.S. (2003). Ambient temperature influences core body temperature response in rat lines bred for differences in sensitivity to 8-hydroxy-dipropylaminotetralin. *J Pharmacol Exp Ther*, 305, 368-74.

Nisijima, K., Shioda, K., Yoshino, T., Takano, K. & Kato, S. (2004). Memantine, an NMDA antagonist, prevents the development of hyperthermia in an animal model for serotonin syndrome. *Pharmacopsychiatry*, 37, 57-62.

Nisijima, K., Yoshino, T. & Ishiguro, T. (2000). Risperidone counteracts lethality in an animal model of the serotonin syndrome. *Psychopharmacology (Berl)*, 150, 9-14.

Nisijima, K., Yoshino, T., Yui, K. & Katoh, S. (2001). Potent serotonin (5-HT_{2A}) receptor antagonists completely prevent the development of hyperthermia in an animal model of the 5-HT syndrome. *Brain Res*, 890, 23-31.

Parker, G. & Brotchie, H. (2011). Mood effects of the amino acids tryptophan and tyrosine: 'Food for Thought' III. *Acta Psychiatr Scand.*

Parrott, A.C. (2002). Recreational Ecstasy/MDMA, the serotonin syndrome, and serotonergic neurotoxicity. *Pharmacol Biochem Behav*, 71, 837-44.

Parsons, B., Roxas, A., Jr., Huang, Y.Y., Dwork, A. & Stanley, M. (1992). Regional studies of serotonin and dopamine metabolism and quantification of serotonin uptake sites in human cerebral cortex. *J Neural Transm Gen Sect*, 87, 63-75.

Paruchuri, P., Godkar, D., Anandacoomarswamy, D., Sheth, K. & Niranjan, S. (2006). Rare case of serotonin syndrome with therapeutic doses of paroxetine. *Am J Ther*, 13, 550-2.

Paterson, N.E., Malekiani, S.A., Foreman, M.M., Olivier, B. & Hanania, T. (2009). Pharmacological characterization of harmaline-induced tremor activity in mice. *Eur J Pharmacol*, 616, 73-80.

Penington, N.J., Kelly, J.S. & Fox, A.P. (1993). Whole-cell recordings of inwardly rectifying K^+ currents activated by 5-HT_{1A} receptors on dorsal raphe neurones of the adult rat. *J Physiol*, 469, 387-405.

Peroutka, S.J., Lebovitz, R.M. & Snyder, S.H. (1981). Two distinct central serotonin receptors with different physiological functions. *Science*, 212, 827-9.

Peroutka, S.J. & Snyder, S.H. (1983). Multiple serotonin receptors and their physiological significance. *Fed Proc*, 42, 213-7.

Perry, K.W. & Fuller, R.W. (1993). Extracellular 5-hydroxytryptamine concentration in rat hypothalamus after administration of fluoxetine plus L-5-hydroxytryptophan. *J Pharm Pharmacol*, 45, 759-61.

Popa, D., Cerdan, J., Reperant, C., Guiard, B.P., Guilloux, J.P., David, D.J. & Gardier, A.M. (2010). A longitudinal study of 5-HT outflow during chronic fluoxetine treatment using a new technique of chronic microdialysis in a highly emotional mouse strain. *Eur J Pharmacol*, 628, 83-90.

Pothakos, K., Robinson, J.K., Gravanis, I., Marsteller, D.A., Dewey, S.L. & Tsirka, S.E. (2010). Decreased serotonin levels associated with behavioral disinhibition in tissue plasminogen activator deficient (tPA-/-) mice. *Brain Res*, 1326, 135-42.

Rollema, H., Wilson, G.G., Lee, T.C., Folgering, J.H. & Flik, G. (2011). Effect of co-administration of varenicline and antidepressants on extracellular monoamine concentrations in rat prefrontal cortex. *Neurochem Int*, 58, 78-84.

Rossi, D.V., Burke, T.F., McCasland, M. & Hensler, J.G. (2008). Serotonin-1A receptor function in the dorsal raphe nucleus following chronic administration of the selective serotonin reuptake inhibitor sertraline. *J Neurochem*, 105, 1091-9.

Rutter, J.J. & Auerbach, S.B. (1993). Acute uptake inhibition increases extracellular serotonin in the rat forebrain. *J Pharmacol Exp Ther*, 265, 1319-24.

Rutter, J.J., Gundlah, C. & Auerbach, S.B. (1995). Systemic uptake inhibition decreases serotonin release via somatodendritic autoreceptor activation. *Synapse*, 20, 225-33.

Schaefer, T.L., Skelton, M.R., Herring, N.R., Gudelsky, G.A., Vorhees, C.V. & Williams, M.T. (2008). Short- and long-term effects of (+)-methamphetamine and (±)-3,4-methylenedioxymethamphetamine on monoamine and corticosterone levels in the neonatal rat following multiple days of treatment. *J Neurochem*, 104, 1674-85.

Schmitz, D., Gloveli, T., Empson, R.M., Draguhn, A. & Heinemann, U. (1998). Serotonin reduces synaptic excitation in the superficial medial entorhinal cortex of the rat via a presynaptic mechanism. *J Physiol*, 508 (Pt 1), 119-29.

Seidl, R., Kaehler, S.T., Prast, H., Singewald, N., Cairns, N., Gratzer, M. & Lubec, G. (1999). Serotonin (5-HT) in brains of adult patients with Down syndrome. *J Neural Transm Suppl*, 57, 221-32.

Sharp, T., Boothman, L., Raley, J. & Queree, P. (2007). Important messages in the 'post': recent discoveries in 5-HT neurone feedback control. *Trends Pharmacol Sci*, 28, 629-36.

Sharp, T. & Cowen, P.J. (2011). 5-HT and depression: is the glass half-full? *Curr Opin Pharmacol*, 11, 45-51.

Shaw, K., Turner, J. & Del Mar, C. (2002). Are tryptophan and 5-hydroxytryptophan effective treatments for depression? A meta-analysis. *Aust N Z J Psychiatry*, 36, 488-91.

Shioda, K., Nisijima, K., Yoshino, T. & Kato, S. (2004). Extracellular serotonin, dopamine and glutamate levels are elevated in the hypothalamus in a serotonin syndrome animal model induced by tranylcypromine and fluoxetine. *Prog Neuropsychopharmacol Biol Psychiatry*, 28, 633-40.

Silva, J.H., Gomez, M.V., Guatimosim, C. & Gomez, R.S. (2008). Halothane induces vesicular and carrier-mediated release of [3H]serotonin from rat brain cortical slices. *Neurochem Int*, 52, 1240-6.

Spanos, L.J. & Yamamoto, B.K. (1989). Acute and subchronic effects of methylenedioxymethamphetamine [(±)MDMA] on locomotion and serotonin syndrome behavior in the rat. *Pharmacol Biochem Behav*, 32, 835-40.

Stenfors, C., Yu, H. & Ross, S.B. (2001). Pharmacological characterisation of the decrease in 5-HT synthesis in the mouse brain evoked by the selective serotonin re-uptake inhibitor citalopram. *Naunyn Schmiedebergs Arch Pharmacol*, 363, 222-32.

Sternbach, H. (1991). The serotonin syndrome. *Am J Psychiatry*, 148, 705-13.

Tao, R. & Auerbach, S.B. (1994). Increased extracellular serotonin in rat brain after systemic or intraraphe administration of morphine. *J Neurochem*, 63, 517-24.

Tao, R., Ma, Z. & Auerbach, S.B. (2000). Differential effect of local infusion of serotonin reuptake inhibitors in the raphe versus forebrain and the role of depolarization-induced release in increased extracellular serotonin. *J Pharmacol Exp Ther*, 294, 571-9.

Turner, E.H., Loftis, J.M. & Blackwell, A.D. (2006). Serotonin a la carte: supplementation with the serotonin precursor 5-hydroxytryptophan. *Pharmacol Ther*, 109, 325-38.

Wimbiscus, M., Kostenko, O. & Malone, D. (2010). MAO inhibitors: risks, benefits, and lore. *Cleve Clin J Med*, 77, 859-82.

Wolf, W.A. & Kuhn, D.M. (1986). Uptake and release of tryptophan and serotonin: an HPLC method to study the flux of endogenous 5-hydroxyindoles through synaptosomes. *J Neurochem*, 46, 61-7.

Zhang, G., Krishnamoorthy, S., Ma, Z., Vukovich, N.P., Huang, X. & Tao, R. (2009). Assessment of 5-hydroxytryptamine efflux in rat brain during a mild, moderate and severe serotonin-toxicity syndrome. *Eur J Pharmacol*, 615, 66-75.

Zhang, G. & Tao, R. (2011). Enhanced responsivity of 5-HT_{2A} receptors at warm ambient temperatures is responsible for the augmentation of the 1-(2,5-dimethoxy-4-iodophenyl)-2-aminopropane (DOI)-induced hyperthermia. *Neurosci Lett*, 490, 68-71.

Zhou, F.M. & Hablitz, J.J. (1999). Activation of serotonin receptors modulates synaptic transmission in rat cerebral cortex. *J Neurophysiol*, 82, 2989-99.

4

Brain Commissural Anomalies

Behpour Yousefi
Semnan University of Medical Sciences,
Iran

1. Introduction

The human brain commissures include the corpus callosum (neocortical), the anterior commissure (paleocortical), the fornix (archicortical) [the hippocampal commissure (also called commissure of psalterium Davidi or David's lyre in the older literature)] (Raybaud, 2010) and the posterior commisure (Keene, 1938). The largest of the commissures in advanced mammals is the corpus callosum that holds its name from its compactness (Raybaud, 2010) which develops embryologically in intimate relationship to the hippocampal formation, fornix, septum pellucidum, and cingulate gyrus (Swayze et al., 1990). It has already been accepted that the commissural fibers are important for transfer of complex cognitive information between the brain hemispheres (Zaidel, 1994; van der Knaap & van der Ham, 2011) and coordinated transfer of information is essential for the cerebral functions (Moldrich et al., 2010). In normal condition, commissural fibers must be actively guided across the midline to reach their targets in the contralateral hemisphere. When the underlying mechanisms regulating the guidance of commissural fibers fail, pathological dysgenesis of one or more commissures ensues. It is suggested that a complex set of cellular and molecular mechanisms regulate commissural development (Ren et al., 2006). Malformation of the corpus callosum is a various condition, which can be observed either as isolated form or as one manifestation in the context of congenital syndromes (Schell-Apacik et al., 2008). Based on survey of 596 network families, the most frequently clinical findings reported about agenesis of the corpus callosum are developmental delay, visual problems, language delay, seizures, and muscle- tone issues (Schilmoeller & Schilmoeller, 2000). Furthermore, agenesis of the corpus callosum results in disabilities in social cognition that appears to be secondary to deficits in complex cognitive operations such as reasoning, concept formation, and problem solving (Doherty et al., 2006). Also, there is no evidence that individuals with partial agenesis of the corpus callosum have better outcomes than individuals with complete agenesis the corpus callosum (Moes et al., 2009). Although, the embryology, anatomy, functions, anomalies and molecular mechanisms of the human brain commissures have been extensively studied over the past years. However a need to an overall and new collection on the basis of the other recent studies was seriously felt. Therefore this chapter is to provide a collection of the fundamental principles of the embryogenesis, organization, congenital malformations of brain commissures. The chapter presents new information about prevalence, the brain disorders associated with commissural anomalies, the etiology and the pathogenetic mechanisms that have been understood in recent years in this issue in the neurosciences.

2. Embryology of the brain commissures

2.1 Embryology of the corpus callosum

The corpus callosum is a new phylogenetic acquisition of the placental mammals (Raybaud, 2010) that develops from anterior to posterior pattern **(**Richards et al., 2004**)** through a process of interhemispheric midline fusion with groups of specialized midline glial guiding the callosal fibers to the other side (Raybaud, 2010). The corpus callosum begins to differentiate as a commissural plate (Rakic & Yakovlev, 1968).within the dorsal third of the lamina terminalis at about 39th embryonic day (Sarnat, 2007). The primitive lamina terminalis corresponds to the closing point of the anterior neuropore. Its dorsal part grows and forms the lamina reunions (6-8 intra uterine weeks). From ventral to dorsal, the lamina reunions **(Fig. 1)** gives rise to the area praecommissuralis (origin of the anterior commissure), to the primordium hippocampi (10 intra uterine weeks,fornix), and to the massa commissuralis (10 intra uterine weeks., corpus callosum) (Destrieux et al., 1998).The plate acts as a passive bed for axonal passage and provides a preformed glial pathway to guide decussating growth cones of commissural axons (Silver et al., 1982).In the human embryo the genu of corpus callosum begins to develop around 8th week after conception (Giedd et al., 1996) and inter-hemispheric crossing fibres begin to transverse the massa commissuralis in this region at 11 to 12 weeks post-conceptional age(Griffiths et al., 2009) and progress caudally, forming the body (corpus) and the splenium (Rakic & Yakovlev, 1968), so that at 18 weeks' gestation the genu and body are detected cleanly; but the splenium is thin and not fully developed (Malinger & Zakut ,1993).The last part of the corpus callosum that to form at the weeks 18-20 post-conceptional ages is the rostrum (Griffiths et al., 2009; Destrieux et al., 1998). It is reported that the adult morphology of corpus callosum is achieved by 16.4 weeks (115 days) (Loser & Alvord, 1968) so that it is clearly identified. The studies have shown that the linear association are between the corpus callosum length, thickness and width, with age before **(** Achiron & Achiron, 2001) and after birth **(**children and young people aged 4 to 18 years) **(** Giedd et al., 1996; Pujol et al., 1993). The length of the corpus callosum increases a 10-fold during gestation and rapid growth of thickness increases during the period between 16 and 20 weeks' gestation. Additionally, the maximal growth of the corpus callosum width and thickness was observed between 19 and 21 weeks' gestation, while the growth of its length appeared to be constant. Further growth is accelerated until 21–22 weeks and then remains stable throughout the rest of gestation. This rapid development of the fetal corpus callosum depends on the first phase of neuronal migration (Achiron, 2001) and follows the expansion of the hemispheres, in a rostro-caudal and then dorso-ventral circular movement (Destrieux et al., 1998).The studies have been shown that, Although the basic structure of the corpus callosum is completed by 18–20 weeks' gestation, but continues to increase in size over the third trimester (Malinger & Zakut, 1993) and grows dramatically during the first 2 postnatal years (Keshavan et al., 2002).The results of these evolutions correspond to axonal elimination and myelination and progressively changing pattern of callosal connections of the newborn and infant into the adult pattern. In spite of the development of the corpus callosum from anterior to posterior pattern ,the preoligodendrocytes are thought to appear first in the genu and splenium (Huppi et al., 1998) and attains the adult levels by the age of 10 years (Yakovlev & Lecours, 1967).Of course ,the Magnetic resonance imaging studies indicate that the maturation in the corpus callosum may be more protracted.Differences in the size and form of the corpus callosum in adults have been shown to relate differences in hemispheric representation of cognitive abilities (Witelson, 1989).

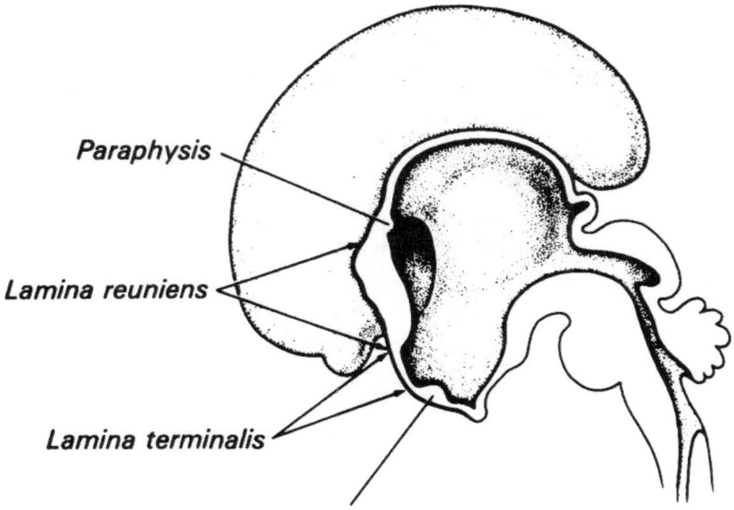

Fig. 1. Rostral midline telencephalon at approximately 7 week's gestational age. Thickening of dorsal aspect of thin rostral wall of telencephalon (primitive lamina terminalis) represents lamina reuniens of His, which will eventually form precursors of corpus callosum and anterior commissure (Barkovich & Norman, 1988)

2.2 Embryology of the anterior commissure

The anterior commissure contains the paleopallial (Lamantia & Rakic, 1990) and the neocortical parts (Guénot, 1998). It is phylogenetically, the oldest of the great forebrain commissures (Raybaud, 2010; Griffiths et al., 2009). At the 8th week of development ,the early fibres of the anterior commissure appear laterally, gradually ,these fibers come nearer to the midline at the week 9, cross the midline at the week 9 (Bayer & Altman, 2006) or week 10 (Rakic & Yakovlev, 1968; Griffiths et al., 2009). Crossing fibres can be detected in the area praecommissuralis by 10 weeks post-conceptional age (12 weeks post-last menstrual period) in the lamina reunions **(Fig. 1)** (Griffiths et al., 2009) and at 13th gestational week both the anterior commissure and optic chiasm are well developed (Ren et al., 2006). The progression of these fibers is facilitated by commissural cellular glial tunnels that provide axonal guidance cues for them along their path (Katz et al., 1983; Lent et al., 2005). During these processes the anterior commissure is surrounded by a glial fibrillary acidic protein $^{+}$/ vimentin $^{+}$ glial tunnel and a tunnel of GFAP- and VN-positive glial cells with TN between the cell bodies from 12th week post conceptional age until at least 17th week postovulation (Lent et al., 2005).

2.3 Embryology of the fornix

The fornix begins as two fiber bundles arising in the area of praecommissuralis and passes dorsally into the hippocampal primordium of the lamina reunions **(Fig. 1)**. The fornices pass towards the medial wall of their epsilateral hemispheric vesicle (Griffiths et al., 2009) as early as week 8 (Bayer & Altman, 2006) or week 9 (Rakic & Yakovlev, 1968) and diverge as they do so (Griffiths et al., 2009). By weeks 10–11(13 weeks post-last menstrual period) some of the fornical fibers cross the midline and form the early hippocampal commissure (Griffiths et al., 2009). Glial fibrillary acidic protein -expressing glial cells were seen surrounding the fornix as of 16–17 weeks postovulation (Lent et al., 2005). The early fornix is a short and slightly curved bundle that contains more hippocampal-septal or septohippocampal fibers that connecting the hippocampus with the septal area (Vasung et al., 2010). Myelination of the fornix is first evident near term, with strong myelin basic protein immunoreactivity that presents in the angular bundle, alveus, and fimbria and relatively scant immunoreactivity in the nascent perforant pathway.Myelination in the hippocampus increases in childhood until adolescence, after which the pattern to stay in the same condition (Arnold & Trojanowski, 1996).

2.4 Embryology of the posterior commissure

The posterior commissure can be seen in stage 12 mm embryos as a large of fibers and until stage 25-37 mm (about 7-8 weeks), it is presented as a very well-developed commissure. During these stages the attachment of the fibers to the cells of the subcommissural organ is still continuing. Some of the fibers connect with the subcommissural organ, others with the thalamus and tegmental region of the embryos of 25-37 mm. Myelination of the posterior commissure begins about the 14th week of development and proceeds to develop in the various fibres in the following order: (1) at the 14th week a few myelinated fibres are found in the ventral part of the commissure, and also in the nucleus chiefly connected with this group of fibres, the nucleus of the posterior commissure. (2) at about the 24th week myelination is found in the fibres connecting with the subcommissural organ and the medial longitudinal fasciculus (Keene, 1938). Recent researches have revealed that development of both the posterior commissure and the underlying subcommissural organ are tightly related to one another and that these structures are under the control of regulatory genes such as *Pax2*, *Pax5*, *Pax6* and *Msx1* (Estivill-Torrús et al., 2001).

3. Anatomical organization of the brain commissures

3.1 Anatomy of the corpus callosum

The corpus callosum holds its name from its compactness; it is the largest of the commissures in advanced mammals (Raybaud, 2010). Although, this commissure has proven to be an important structure in the human brain, it is possible to live without this white matter structure (van der Knaap & van der Ham, 2011). The corpus callosum consists mainly of myelinated axons of various sizes (Griffiths et al., 2009), a certain amount of non-myelinated fibers (Tomasch, 1954), neuralgial cells and a certain number of blood vessels that connect the homologous regions of cerebral cortex of both hemispheres (Griffiths et al., 2009) from the anterior commissure anteriorly to the hippocampal commissure posteriorly (Raybaud, 2010). The callosal commissural neurons are located predominantly in intermediate cortical layers (Richards, 2004). The corpus callosum can be subdivided into several functionally and morphologically distinct sub regions which are arranged according to the topographical organization of cortical areas (Witelson, 1989); the small comma-shaped rostrum tucked under the genu (Griffiths et al., 2009) genu, truncus or midbody and splenium in sequential order from anterior to posterior (Witelson, 1989; Griffiths et al., 2009). The isthmus usually appears as a mild focal narrowing found where the fornix joins the corpus callosum (Velut et al., 1998; Hofer & Frahm, 2006) which contains, connecting fibers of motor, somatosensory and primary auditory areas (Aralasmak et al., 2006; Raybaud, 2010; Aboitiz & Montiel, 2003; Aboitiz et al., 1992; Buklina, 2005; Fabri et al., 2005). The upper surface of the corpus callosum is lined with the indusium griseum (gray velum) (Jea, 2008). The rostrum of the corpus callosum extends anteriorly from the anterior commissure to the posterior inferior aspect of the genu and commonly assumed to be the last callosal segment to develop (Kier et al., 1997); its fibers are likely to connect the fronto-basal cortex (Velut et al., 1998; Hofer & Frahm, 2006). The genu (knee) is a thickened part of the corpus callosum, so named because of the sudden alteration in orientation; it is located between the rostralis and the callosal body. It forms the anterior boundary of the septum pellucidum and its fibers take part the formation of the forceps minor that connect the prefrontal cortex, the anterior cingulate area (Hofer & Frahm, 2006) and higher order sensory areas (Aralasmak et al., 2006; Raybaud, 2010; Aboitiz & Montiel, 2003; Buklina, 2005; Fabri et al., 2005). Its ventral part contains the fibers of the ventro-medial prefrontal cortex; its dorsal part includes the fibers of the dorso-lateral prefrontal cortex (Velut et al., 1998). The callosal body is the horizontal portion that extends from the genu to the point where the fornix abuts the undersurface of the corpus callosum. It borders the septum pellucidum superiorly. The fibers of the callosal body extends laterally between the cingular bundle superiorly and the occipito-frontal fascicle inferiorly and across the anterior radiations of the thalamus and forming the roofs of the lateral ventricular bodies. They connect the precentral cortex (premotor area, supplementary motor area), the adjacent portion of the insula, and the overlying cingulate gyrus mostly (Velut et al., 1998; Hofer & Frahm, 2006).The commissural fibers of the isthmus connect the pre- and postcentral gyri (motor and somatoisensory strips) (Velut et al., 1998; Hofer & Frahm, 2006) and the primary auditory area (Aboitiz, 2003; Aboitiz, 1992). The splenium is the thickest portion of the corpus callosum. It protrudes in the ambient cistern and overhangs the tectal plate, while the vein of Galen sweeps around it. Its morphology is extremely variable, from rounded to flat. It should be located above or just at the line drawn along the third ventricular floor (Widjaja et al., 2008). Fibers of the splenium form the forceps major and participate in the tapetum, or sagittal stratum, in the lateral wall of the posterior cornu of the lateral ventricle. It contains the commissural fibers for the posterior parietal cortex, the medial occipital cortex (Aboitiz, 2003; Velut et al., 1998; Hofer &Frahm, 2006), which connects visual areas in the occipital lobe (Aralasmak et al., 2006; Raybaud, 2010; Aboitiz & Montiel, 2005; Aboitiz et al., 1992; Buklina, 2005; Fabri et al., 2005) and the medial temporal cortex (Aboitiz, 2003; Velut et al., 1998; Hofer &Frahm, 2006). In regard to callosal size and width, most of the articles have shown that callosal size to be directly related to the number of interhemispheric connections (Bloom & Hynd, 2005) and vary between individuals and between sexes (Luders et al., 2010; Aboitiz et al., 1992; Junle et al., 2008; Clarke & Zaidel, 1994; Hasan et al., 2008). Additionally, Age related thinning of the corpus callosum is often reported (van der Knaap & van der Ham, 2011), however, these findings are still controversial.

3.2 Anatomy of the anterior commissure

The anterior commissure **(Fig. 2. A, B)** in humans is classically composed of two distinct tracts, the anterior (olfactive limb) and posterior limbs (temporal limb) (Patel et al., 2010; Mitchell et al., 2002; Peltier et al., 2011). The anterior limb forms an open "U" and those of

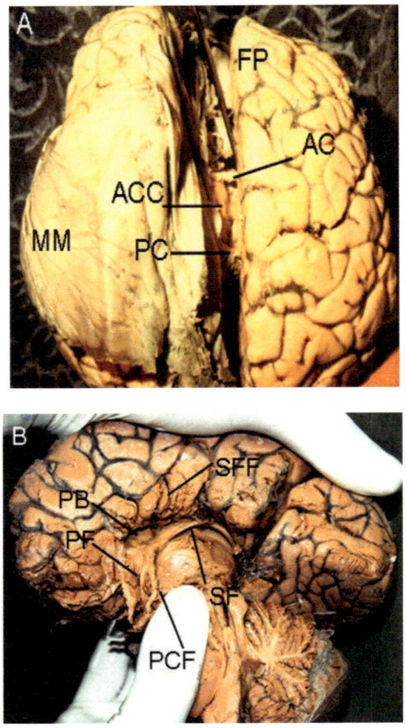

Fig. 2. Photographs of the brain in the superior view (A) and midsagittal plane (B) showing an absence of the corpus callosum (ACC), septum pellucidum, cingulum sulcus, interthalamic adhesion and hippocampal commissure.The anterior (AC) and posterior commissure (PC) are seen within the hemispheres. Other abbreviations in this figure: FP, frontal pole; MM, meningeal membrane; SF, separated fornix; PF, precommissural fornix; PCF, postcommissural fornix; PB, Probst bundle.

the posterior limb make an apposed flattened "M" shapes when viewed in the axial plane (Mitchell et al., 2002). The anterior limb is much smaller and varies considerably in size between subjects, it contains the small bundles of fibers that leave the main bulk of the commissure at the level of the anterior perforated substantia and connects the olfactory bulbs (Di Virgilio et al., 1999), their nuclei (Di Virgilio et al., 1999; Mitchell et al., 2002) and the inferior posterior orbital gyri (Patel et al., 2010; Di Virgilio et al., 1999). In addition, the small numbers of axons are detected in the anterior limb that crossing the midsagittal plane which is believed to convey fibers to territories others than the temporal cortex. Fibers of the posterior limb, which form the major and neocortical portion of the anterior commissure (Di Virgilio et al., 1999), travels within the basal parts of the putamen, the caudate nucleus and below the anterior border of the globus pallidus (Di Virgilio et al., 1999; Turner, 1979; Peltier et al., 2011) into the temporal cortex and projects to the amygdale (Turner, 1979; Di Virgilio et al., 1999; Jellison et al., 2004; Patel et al., 2010), (basolateral nucleus) (Martínez-Lorenzana et al., 2004), temporal pole (Jellison et al., 2004; Patel et al., 2010) parahippocampal, inferior temporal and fusiform gyri (Di Virgilio et al., 1999; Johnston et al., 2008; Demeter et al., 1985). The remaining fibers (a few fibers) of the posterior limb travel into the occipital lobe (Di Virgilio et al., 1999) and intermingle with other fasciculi in various directions to form a dense 3D network (Peltier et al., 2011). Also, additional afferent fibers from the occipital cortex (Patel et al., 2010; Di Virgilio et al., 1999) precentral gyrus and central fissure have been detected through the posterior limb (Di Virgilio et al., 1999).

3.3 Anatomy of the fornix

The fornix **(Fig. 2. B)** provides bidirectional connectivity between the hippocampus and subcortical structures (Swanson & Cowan, 1977; Cassel et al., 1997). It contains the main efferent bundle (Carpenter, 1991) of large fibers connecting the hippocampal formation to the mamillary body (Atlas et al., 1986) and anterior thalamic nuclei. It is also has afferent cholinergic tracts from the septal nuclei and a smaller amount pathways from other basal forebrain to the hippocampus and entorhinal cortex, respectively (Gaffan et al., 2001; Mesulam et al., 1983; Ridley et al., 1996; Selden et al., 1998). The most of the fibres in the fornix begin from the subicular cortex and the pyramidal cells of the hippocampus. Those fibres converge into a discrete bundle as the fimbria at the medial surface of the alveus of the head of the hippocampus (Standring, 2005). The fimbria on the anterosuperior curvature of the hippocampus (Chance et al., 1999) lies to posterior end of thalamus, then arcs posterosuperiorly and medially to form the crus of the fornix (Atlas et al., 1986). Beneath the splenium (Chance et al., 1999), about 20% of the fibres (Lamantia & Rakic, 1990) cross the midline between the fornical crura at a point known as the commissure of the fornix (Chance et al., 1999; Lamantia & Rakic, 1990). Anteriorly, upon reaching the septum pellucidum on the midline and under the corpus callosum the crura meet to form the body of the fornix (Atlas et al., 1986; Lamantia & Rakic, 1990). Most text-books state that the two fornices merge but evidence from MR imaging indicates it is more accurate to say that they join but always maintain an obvious, separate identity (Griffiths et al., 2009).There, they course in the lower margin of the pellucidal leaves until they reach the superioranterior edge of the foramen of Monro (Lamantia & Rakic ,1990). As they descend, above the interventricular foramina the body of the fornix diverges into right and left fascicles which split into a precommissural fornix and a posterior commissural fornix (the columns of the fornix) near the anterior commissure (Williams et al., 1989). In each side the column or posterior commissural fornix (Carpenter, 1991; Meibach & Siegel, 1977) [hippocampo-mammillary tract (Lamantia & Rakic, 1990)] which contains the majority of the fornical fibres (Carpenter, 1991; Meibach & Siegel, 1977) and the fibres from the subicular area (Lamantia & Rakic, 1990) bend ventrally in front of the interventricular foramina and caudal to the anterior commissure, to join the anterior thalamus and hypothalamus (Atlas et al., 1986), predominantly to mamillary body. The precommissural (hippocampo-septal tract) contains the remaining portion of the fornical fibres which arising from the cornu ammonis (Lamantia & Rakic, 1990; Meibach & Siegel, 1977) and the subiculum (some of the fibres) and terminate exclusively in the (Meibach & Siegel, 1977) septum area (Chance, 1999) and septal nuclei (Meibach & Siegel, 1977; Lamantia & Rakic, 1990). The distribution of neurons contributing to the fornix in rhesus monkeys (Macaca mulatta) have been shown that the medial fornix originates from cells in the caudal half of the subiculum, the lamina principalis interna of the caudal half of the presubiculum, and from the perirhinal cortex (area 35). The intermediate portion of the fornix originates from cells in the rostral half of the subiculum and prosubiculum, the anterior presubiculum (only from the lamina principalis externa), the caudal presubiculum (primarily from lamina principalis interna), the rostral half of CA3, the EC (primarily 28I and 28M), and the perirhinal cortex (area 35). The lateral parts of the fornix arise from the rostral EC (28L only) and the most rostral portion of CA3.

Subcortically, the medial septum, nucleus of the diagonal band, supramammillary nucleus, lateral hypothalamus, dorsal raphe nucleus, and the thalamic nucleus reuniens all send projections through the fornix, which presumably terminate in the hippocampus and adjacent parahippocampal region (Saunders & Aggleton, 2007). In conclusion, it is apparent that schizophrenia and to some extent gender have an influence on the neuroanatomy of the fornix (Church et al., 1999).

3.4 Anatomy of the posterior commissure

The posterior commissure **(Fig. 2. A)** extends from the region of the pineal recess to the tectal commissure. Its caudal end corresponds with the position of the orifice of the mesocoelic recess. It contains the coarse and fine fibres. The coarse fibres lie close to the ventricular roof and also skirt the mesocoelic recess, whereas the fine ones occupy a position nearer to the exterior, and are continued into the tectal commissure. thus the cephalic part of the commissure consists of ventral coarse fibres and fine dorsal ones, and the caudal part has a more complicated arrangement of fibres, due to the forward folding of the roof of the mid-brain in that region. The following connexions for the posterior commissure are reported: a) the coarse fibres directly connect with the nucleus of the posterior commissure and also indirectly through the nucleus of the posterior commissure or interstitial nucleus with the ipsilateral medial longitudinal bundle, b) Other fibres, chiefly coarse ones, connect with the regions of the tegmentum and the capsule of the red nucleus, c) fine fibres situated in the dorsal part of the commissure connect with the thalamus, d) the commissure consisting of horizontal fibres which may be traced in a lateral direction, it is thought that this connexion may be striatal,or possibly cortical, e) a small connexion with the habenular ganglia, and the habenulo-peduncular tracts, h) a fine connection with the pineal gland is also established (Keene, 1938). Also, studies in rat have demonstrated that the activity of the subcommissural organ depends on serotoninergic fibers originated in the raphe nuclei, some of which reach the subcommissural organ through the posterior commissure (Mikkelsen et al., 1997). In the chick brain, the tract of the posterior commissure emerges in the caudal pretectum as the first transversal tract. It is formed by dorsally projecting axons from neurones located in the ventral pretectum, and by ventrally projecting axons from neurones located in the dorsal pretectum (Ware & Schubert, 2011).

4. The vessels of the brain commissures

4.1 The arteries of the brain commissures

4.1.1 The arteries of the corpus callosum

The blood supply to the corpus callosum originates from both of the arterial systems of the brain; the carotid system and the vertebral-basilar system.

4.1.1.1 The carotid system

The carotid system contributes mainly to this supply via the pericallosal artery (Kakou et al., 1998; Wolfram-Gabel et al., 1989; Türe et al., 1996) which is the main artery of the corpus callosum (Wolfram-Gabel et al., 1989; Türe et al., 1996). It curves around the genu and continues posteriorly along the dorsal surface of the corpus callosum (Yasargil, 1984). Its posterior extension followed a cork-screw-like tortuousity, anastomosed with the posterior pericallosal artery in the splenial region, and formed the dense portion of the pericallosal pial plexus within the callosal sulcus. Usually, some of the branches arising from this network circle around the splenium and joint with branches of the medial posterior choroidal arteries in the tela choroidea of the third ventricle. In addition to pericallosal artery, the anterior communicating artery accessorily contributes to it by an inconstant artery called median artery of the corpus callosum (Wolfram-Gabel et al., 1989; Türe et al., 1996). The pericallosal artery gives rise to four types of branches that supply the corpus callosum, these are the callosal artery; cingulocallosal artery; long callosal artery; and recurrent cingulocallosal artery. **The callosal arteries** are thin branches which directly supply the indusium griseum and the superficial surface of the corpus callosum in the midline (Kahilogullari et al., 2008; Türe et al., 1996**). The cingulocallosal arteries** bring the chief supply to the corpus callosum. These arise from the inferolateral aspect of the pericallosal artery and run laterally into the callosal sulcus, where they are divided into three arterial subgroups (Türe et al., 1996) which supply the corpus callosum, the cingulate gyrus and the radiation of the corpus callosum.The cingulocallosal arteries anastomosing with each other and with branches arising from the subcallosal, median callosal and long callosal arteries to form the pericallosal pial plexus. **The long callosal artery** is found almost in half of the hemispheres, it is an another branch arising from the pericallosal artery, courses parallel with it in the callosal sulcus and has multiple branches that contributed to the pericallosal pial plexus (Kahilogullari et al., 2008; Türe et al., 1996) .The artery ends in the body of the corpus callosum or in the medial longitudinal striae at the splenium and anastomosis with the posterior pericallosal artery of the same hemisphere or is crossed the midline and anastomosed with the posterior pericallosal artery of the opposite hemisphere, both within the callosal sulcus in the splenial region. **The recurrent cingulocallosal artery** is a thin branch, arises from major cortical branches of the pericallosal artery: It courses on the medial surface of the cingulate gyrus toward the callosal sulcus, present in 45% of the subjects (Türe et al., 1996) and contributed to the pericallosal pial plexus (Kahilogullari et al., 2008; Türe et al., 1996). In addition to the pericallosal artery, the perforating branches of **the anterior communicating artery** participate in providing blood supply to the corpus callosum.The hypothalamic artery (which do not supply the corpus callosum); subcallosal artery; and median callosal artery spring from these branches. In 80% of the specimens, either the subcallosal artery or the median callosal artery are present and contributed to blood supply of the corpus callosum, especially to the anterior portion. **The subcallosal** artery is a major contributor to the blood supply of the medial portions of the rostrum and genu of the corpus callosum.**The median callosal artery** is present in 30% of the specimens an anatomical variations. This artery followed the same course as that of the subcallosal artery and supplies the same structures, except that its distal extension reached the body and frequently even the splenium of the corpus callosum (Kahilogullari et al., 2008; Türe et al., 1996; Kakou et al., 1998).

4.1.1.2 The vertebral-basilar system

The vertebral-basilar system contributes to the blood supply of the corpus callosum by the terminal and choroidal branches of the posterior cerebral artery (Wolfram-Gabel et al., 1989; Türe et al., 1996). The posterior cerebral artery is divided into four segments: the end segment of which comprises the posterior extension of the posterior cerebral artery that runs along or inside both the parieto-occipital sulcus and the distal part of the calcarine fissure and gives the parieto-occipital and calcarine arteries (Párraga et al., 2010). The posterior cerebral artery contributes in providing blood supply to the corpus callosum by the posterior pericallosal artery (also known as the splenial artery), in particular the splenial portion, in all hemispheres. It arises from the main trunk of the parieto-occipital artery or its precuneal branch (52%) of the third segment of the posterior cerebri artery (32%), the calcarine artery (7%), the temporo-occipital artery (7%), or the second segment of the posterior cerebri artery (2%). In addition to the posterior pericallosal artery, a very fine artery that contributed to the blood supply of the splenium is observed in 25% of the hemispheres.It origins from the precuneal branch of the parieto-occipital artery, the hippocampal artery, the medial posterior choroidal artery, or the lateral posterior choroidal artery. It has been named this artery the "accessory posterior pericallosal artery (Türe et al., 1996).

4.1.2 The arteries of the anterior commissure and the fornix

The medial portion of the anterior commissure and the column of the fornix, are supplied by the small perforating branches of the hypothalamic arteries(Türe et al., 1996; Dunker & Harris , 1976) and the remaining anterior cerebral artery proximal to the anterior communicating artery (Dunker & Harris , 1976). The hypothalamic arteries arise from the posteroinferior aspect of the anterior communicating artery (Türe et al., 1996). Also, the inferior branch of the posterior pericallosal (Türe et al., 1996) and lateral posterior choroidal arteries supply the crus of the fornix.

4.2 The veins of the brain commissures

The venous drainage of the corpus callosum is essentially via callosal and callosocingulate veins empty into the deep venous system of the brain (Kakou et al., 1998; Wolfram-Gabel & Maillot , 1992). Most of these veins pass caudally and anastomoses together at the central level of the corpus callosum and form the subependymal veins and are collected by the septal and the medial atrial veins. All these veins are tributaries of the internal cerebral veins (Wolfram-Gabel & Maillot, 1992).

5. Functional correlation of the brain commissures

It has long been accepted that the commissural fibres are important for transfer of complex cognitive information (Zaidel, 1994; van der Knaap & van der Ham, 2011). In this issue the corpus callosum has an important role than other commissures. The corpus callosum involves in lower-level processes (Schulte & Müller-Oehring, 2010), transferring sensory information (Banich, 1998), interhemispheric visuomotor integration (Banich, 1998; Schulte & Müller-Oehring, 2010; Mordkoff & Yantis, 1993), hemispheric specialty (Doron & Gazzaniga, 2008) and contribution in development of higher-order cognitive functions (Gazzaniga, 2000; Doron & Gazzaniga, 2008). So, the corpus callosum is needed to maintain an integrated sense of self with regards to body awareness and planning of actions. (Uddin , 2011), as in regard to visuomotor integration, the integration of perception and action by the corpus callosum promoting a unified experience of the way that we perceive the visual world and prepare our actions (Bloom & Hynd, 2005). It appears that the corpus callosum employs a differentiated role with callosal areas transmitting different types of information depending on the cortical destination of connecting fibers (Bloom & Hynd, 2005). The anterior corpus callosum is necessary for awareness of initiation of goal-directed movements and subjective feelings of agency (Uddin, 2011) and associate with inhibitory functions in situations of semantic competition (Stroop) and local-global interference (Bloom

& Hynd, 2005); in addition to intact fronto-parietal cortical functioning (Uddin, 2011) . The posterior corpus callosum integrity seems proved for maintaining a sense of limb ownership, as this region interconnects parietal areas involved in self-body representation (Uddin, 2011). Also, it connects temporo-parietal and occipital cortical regions in related with facilitation functions from redundant targets and local-global features. It is reported that an intact (posterior) corpus callosum and interaction between ipsilateral and contralateral hemispheres are required for coordination of the hand movements (Eliassen et al., 1999). Additionally, it is suggested that the posterior callosal area associated with the superior colliculi connect visual extrastriate areas as the key structures for interhemispheric neural coactivation explaining visuomotor integration between hemispheres (Iacoboni et al., 2000). A study has shown that lesions of the posterior or mid-body corpus callosum or complete commissurotomy conflict intermanual coordination; injuries of the posterior corpus callosum and parietal cortical areas cause the alien hand sign; and lesions of the frontal lobe or anterior corpus callosum results the anarchic hand (Aboitiz et al., 2003).Studies in acallosal and split brain patients have revealed that the absence or loss of the corpus callosum integrity contributes to impairment in sensory and cognitive integration (Fabri et al., 2001; Yamauchi et al., 1997) and large individual differences in interhemispheric transfer among split-brain patients (Zaidel et al., 2003). In split-brain patients, however, several investigators have noted that transfer of some types of visual information is usually spared (Eviatar & Zaidel, 1994; Uddin et al., 2008).The condition that cortical commissures are no longer available some information can be transferred between the hemispheres through subcortical pathways (Funnell et al., 2000) by the subcortical coordination of cortical networks (Uddin et al., 2008). In regard to involvement of the corpus callosum in lower-level visuomotor functions, split-brain research indicates that the corpus callosum acts in an inhibitory fashion within a subcortico-cortical network (Corballis et al., 2002; Roser & Corballis, 2003), while recent research on callosal degradation without disconnection have shown cooperative role for the corpus callosum (Schulte & Müller-Oehring, 2010) in conscious perception (Marzi et al., 1996; Müller-Oehring et al., 2009). In addition to mentioned functions, recently the enhanced redundancy gain (co-activation model) (Bucur et al., 2005; Schulte et al., 2006; Turatto et al., 2004) and mediate interhemispheric processing advantages (Corballis et al., 2003; Iacoboni et al., 2000; Roser & Corballis, 2003) a possible role for the corpus callosum are reported. In regard to this question, how the corpus callosum mediates this transfer? There are, two contrasting theories of interhemispheric interaction in the literature, excitatory and inhibitory messages, although there is more evidence to support the notion that the corpus callosum plays an excitatory function in interhemispheric communication rather than an inhibitory function, there is some evidence that inhibition occurs. The nature of functions may occur at different times depending on the task or may even occur simultaneously to achieve an interhemispheric balance between component brain functions (Bloom & Hynd, 2005). How the corpus callosum regulates this transfer of information between cortical areas seems uncertain (van der Knaap & van der Ham, 2011).

6. Interhemispheric transfer time

Consumed time of transfer time of information between hemispheres is shorter and more equal for women than men (Moes et al., 2007) and is faster from right-to-left than from left-to right (Barnett & Corballis, 2005; Iwabuchi & Kirk, 2009). The causes of these differences may be; faster axonal conduction in the right hemisphere relative to the left (Barnett & Corballis, 2005) or the degree of hemispheric specialization (Nowicka et al., 1996; Rugg &

Beaumont, 1978) more gray matter relative to white matter in the left hemisphere than in the right (Gur et al., 1980); other anatomical differences between both hemispheres. It appears that the ratio of gray and white matter may be underlying functional asymmetry (Schulte & Müller-Oehring, 2010). A correlation between callosal connectivity and prolonged interhemispheric transfer time have been reported in split-brain patients and in acallosal patients (Iacoboni et al., 2000; Mooshagian et al., 2009; Paul et al., 2007; Reuter-Lorenz et al., 1995; Roser & Corballis, 2002).

7. Brain commissural anomalies

7.1 Malformations of the corpus callosum

It is observed in a variety of conditions that disrupt early cerebral development, including chromosomal and metabolic disorders, as well as intrauterine exposure to teratogens and infection (Paul et al., 2007). Callosal agenesis can be detected prenatally by routine sonography, for which the important signs include absence of the cavum septum pellucidum, colpocephaly, high-riding third ventricle, and widening of the interhemispheric fissure (Tang et al., 2009). On the basis of the known embryology of the corpus callosum, two primary or "true" types and two secondary types of callosal abnormalities have been documented. The two types of true agenesis of the corpus callosum include (1) defects in which axons form but are unable to cross the midline because of absence of the massa commissuralis and leave large aberrant longitudinal fiber bundles known as Probst bundles **(Fig. 3. A, B)**, along the medial hemispheric walls; and (2) defects which the commissural axons or their parent cell bodies fail to form in the cerebral cortex (Sidman & Rakic, 1982). The former, probably the most common type of agenesis of the corpus callosum, occurs in BALB mice and all agenesis of the corpus callosum syndromes, in which Probst bundles are

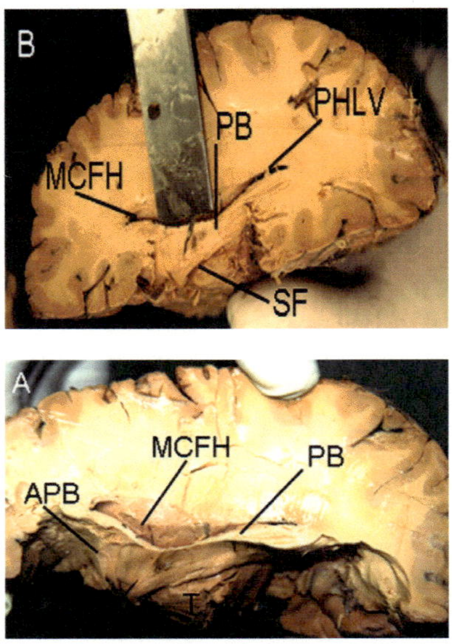

Fig. 3. Photographs of transverse sections of the cerebral hemisphere (A, B) showing the Probst bundle (PB) and medially concave frontal horn (MCFH). Other abbreviation in this figure (B): APB, anterior part of the Probst bundle; T, thalamus; SF, separated fornix; PHLV, posterior horn of the lateral ventricle.

seen. The latter occurs in Walker- Warburg syndrome and other types of lissencephaly, in which Probst bundles are generally not seen.The two types of secondary of callosal abnormalities include (1) absence of the corpus callosum associated with major malformations of the embryonic forebrain prior to formation of the anlage of the corpus callosum; and (2) degeneration or atrophy of the corpus callosum, which results in striking thinning that may again be mistaken for true agenesis of the corpus callosum (Dobyns, 1996). Its incidence in the general population is 3-7 per 1000 birth; in children with developmental disabilities is 2-3 per 100 (Grogono, 1968; Jeret et al., 1985; Glass et al., 2008), among patients undergoing cranial magnetic resonance imaging at a tertiary care referral institution was determined to be 0.25% (Hetts et al .,2006). Also, a population-based survey indicates that the combined prevalence of agenesis and hypoplasia of the corpus callosum before age 1 year is only 1.8 per 10,000 live births (Glass et al., 2008). In addition to mentioned incidences, a epidemiologic study in Hungary has been shown that the overall birth prevalence of total or partial agenesis and hypoplasia involved 2.05 per 10,000 live births, including 2.73 per 10,000 among boys, and 1.33 per 10,000 among girls. The birth prevalence of total and partial agenesis of the corpus callosum involved 1.02 per 10,000 live births, with 1.36 per 10,000 among boys and 0.66 per 10,000 among girls. The birth prevalence of hypoplasia of the corpus callosum involved 1.02 per 10,000 live births, with 1.36 per 10,000 among boys, and 0.66 per 10,000 among girls. The male/female sex ratio was 2.2 for both total or partial agenesis and hypoplasia of the corpus callosum (Szabó , 2011).The morphological anomalies of the corpus callosum may be **agenesis (complete and partial), dysgenesis (Fig. 4), hypoplasia** and **hyperplasia** (Raybaud, 2010; Yousefi & Kokhei, 2009; Hetts et al .,2006; Hanna , 2011).

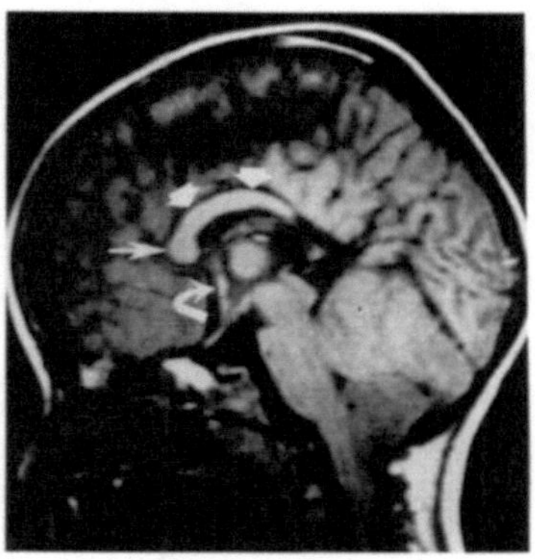

Fig. 4. Corpus callosum is markedly dysgenic; the genu (long Stright arrow) and body (short straight arrows) are present, but splenium and rostrum are absent. Anterior cornmissure (curved arrow) is present and of normal size (Barkovich & Norman, 1988).

7.1.1 Complete agenesis of the corpus callosum (Figs. 2. A, B & 5. A)

The agenesis (complete or partial) is one of the most commonly observed features in the malformations of the brain (Chiappedi & Bejor, 2010), a part of many syndromes (Chiappedi & Bejor, 2010; Penny, 2006) and/or somatic anomalies (Barkovich, 2005; Barkovich & Norman, 1988; Marszal et al., 2000; Hetts, 2006). Primary complete agenesis usually occurs earlier in embryologic development, while partial agenesis occurs in later gestation (Penny, 2006). Complete agenesis of the corpus callosum, in which patients do not develop a callosal structure. It is rarely limited to the callosal structure (Raybaud, 2010) and usually sporadic (Chouchane et al., 1999). This form of anomaly is often associated with defects or absence of the other forebrain commissures (Raybaud, 2010). Most of the patients with complete agenesis and without telencephalic dysgenesis or syndromic features typically have the Probst's bundles (Szriha, 2005). During embryogenesis, the fibres are thought to arrive at the midplane, where they are hindered in their further migration across the midline and then change their direction of growth (Rosenthal-Wisskirchen, 1967) into an anteroposterior direction which leads to the formation of bundles in each hemisphere (Rosenthal-Wisskirchen, 1967; Lee et al., 2004; Hetts et al., 2006; Meyer & Röricht, 1998).The formation of bundles upon the lateral ventricular lumen, giving it a crescentic and a bull's head appearance to the section of the lateral and third ventricles on the coronal view. This bundle is called the bundle of Probst in the literature (Raybaud ,2010) .The Probst bundle may be intermingled with upper border of the separated fornix (Hetts et al., 2006; Meyer & Röricht, 1998; Yousefi & Kokhei, 2009). In this condition, frontally, it has comma (Meyer & Röricht, 1998), a U-turn shape (Ozaki et al., 1987) and the lower area of the anterior portion (radiated fibres) of the Probst bundle is attached to the ventral branches of the precommissure fornix. Posteriorly it forms a thin layer on the upper medial wall of the lateral ventricles (Meyer and Röricht, 1998) and accumulates as an anomalous fascicle below the cingulum (Ozaki et al., 1987) or attaches to the crus of fornix at the beginning of the fimbria (Yousefi & Kokhei, 2009). The etiology of Agenesis of the corpus callosum is heterogeneous, including cytogenetic abnormalities, metabolic disorders and genetic syndromes (Dobyns, 1996). At the molecular level, the process of development of the corpus callosum is complex. These processes rely on intricate cell-to-cell signaling mechanisms. Disruption or desynchronization of these mechanisms could lead to partial or complete callosal agenesis (Prasad et al., 2007).

7.1.2 Partial agenesis of the corpus callosum

Partial agenesis of the corpus callosum **(Fig. 4)** results from an arrest of growth between 12 and 18 weeks of gestation and usually involves the dorsal part or splenium (Kier & Truwit, 1996). It is suggested that a deviation in the normal course of the pericallosal arteries may be the sign of corpus callosal partial agenesis. In such cases the arteries closely follow the contour of the corpus callosum at its anterior part (the genu and the body), but take an upward direction at the level of the missing splenium (Volpe et al., 2006). Additionally, an insult to the developing corpus callosum may inhibit the complete formation of this large commissural bundle and lead to partial agenesis (hypogenesis) of the corpus callosum, when only the early formed portions appear (Szriha, 2005).

7.1.3 Callosal hypoplasia

Callosal hypoplasia is a developmental disorder that may be induced by teratogens (radiation, alcohol) or compression (e.g. intracranial masses, obstructive hydrocephalus (Davila-Gutuerrez, 2002; Paupe et al., 2002) rather than a primary malformative abnormality.Thus, callosal hypoplasia more likely depends upon an external factor affecting the number and size of callosal axons. This is apparently confirmed by a experience since callosal hypoplasia was often associated with additional brain anomalies (Ghi et al., 2010). Hypoplasia and partial agenesis of the corpus callosum may occur in isolation form, in these conditions, neurological outcome is reported by some to be similar to that in cases with absent corpus callosum1(Moutard et al. ,2003; Mordefroid et al., 2004; Ghi et al., 2010). Callosal hypoplasia include a significant size reduction of the anterior genu, posterior genu (Walterfang, 2008, Walterfang, 2009a; Vidal et al., 2006; Just et al., 2007; Walterfang et al., 2009b; 2009c), isthmus (Walterfang, 2008; Walterfang, 2009a; Vidal et al., 2006; Just et al., 2007; Cao et al., 2010)and the posterior midbody (Cao et al., 2010), a smaller splenium width (Bersani, 2010; Vidal et al., 2006; Just et al., 2007; Hutchinson et al ., 2008), a cyst in the splenium (Bamiou et al., 2007) and a smaller anterior midbody. The anterior midbody is known to increase in size until the late twenties (Bersani, 2010). Other abnormal shapes of the corpus callosum are reported such as global shape due different bending degrees of the callosal body (He et al ., 2010) ; a slit-like left paracallosal lesion extending from the genu towards the splenium (Faber et al., 2010) with an additionally smaller anterior corpus callosum for boys (Hutchinson et al ., 2008). The studies have shown a correlation between illness duration and callosal shape in patients with bipolar disorder. Therefore, the corpus callosum degeneration and axonal loss is repeatedly described in some of the psychiatric disorders (Evangelou et al., 2000; Manson et al., 2006; Warlop et al., 2008; Gadea et al., 2009). Callosal thinning by defective myelination or decreased fiber density, can manifest itself in pathology specific symptoms. Also, a lot of variations are seen in patients in related to age, sex and type of symptoms (van der Knaap, 2011).

7.1.4 Callosal hypertrophy

Hypertrophy of the corpus callosum is a classical marker of neurofibromatosis type 1. It has also been recently identified as a characteristic of a macrocephaly syndrome with polymicrogyria and developmental delay (Pierson et al., 2008). Finally, investigations have shown that the corpus callosum is particularly vulnerable to closed head trauma (Peru et al., 2003). There are evidences that the chromosomes of 8, 11, 13-15 and 18 involvement in abnormal corpus callosum morphogenesis and it can occur as an X-linked (Jeret et al., 1987; Davila-Gutierrez, 2002) or autosomal-recessive condition, or can present as an incidental finding during imaging in apparently normal patients (Davila-Gutierrez, 2002).

7.2 Malformations of the anterior commissure

In the classic commissural agenesis, in about 50% of the cases, the anterior commissure is either absent or too thin to be recognized (Raybaud & Girard, 1998), or apparent but hypoplastic (Raybaud & Girard, 1998; Griffiths, 2009), probably due to the absence of its neocortical component. It is classically mentioned that in some of the cases it may be enlarged, as if compensating for the missing corpus callosum (Raybaud & Girard, 1998; Probst, 1973; Barr Melodie & Corballis Michael, 2002) whereas in other studies, it is reported that this commissure was small (Bamiou et al., 2007; Barkovich & Norman, 1988; Atlas, 1986) in the patients who had complete agenesis of the corpus callosum associated with cranial abnormalities and some of syndromes and small but had a normal configuration in the patient with isolated callosal agenesis (Barkovich & Norman, 1988; Atlas, 1986).The anterior commissure is dislocated in more than a third of the cases (38%), low on the lamina terminalis, halfway between the foramen of Monro and the optic chiasm (Raybaud, 2010). In addition to above anomalies, the unilateral anterior commissure run posterior to the columna fornicis in the brain of a 20 year-old man was reported (Hori, 1997). Association of the callosal agenesis with absent or hypoplastic of the anterior commissure is most likely the result of either an anomaly of the primitive lamina tenminalis, either of these situations would inhibit the formation of the beds of tissue into which both the commissural and callosal fibers are induced to grow. The presence of normal anterior commissures in those patients with a partially formed corpus callosum suggests that the insult to the brain that disrupts callosal formation occurs after the bed for ingrowth of the anterior commissure is formed (Barkovich & Norman, 1988). Investigations have shown that in adult humans, *Pax6* mutations are associated with cerebral malformations and structural abnormalities of the interhemispheric pathway, with an absent or hypoplastic anterior commissure (Sisodiya et al., 2001). Also, the anterior commissure is reduced in *Pax6cKO* mutants (Abouzeid et al., 2009; Sisodiya et al., 2001). Of interest, there is circumstantial evidence that a hypertrophied anterior commissuer may reflect compensation for the lack of the corpus callosum in terms of interhemispheric transfer function (Fischer et al., 1992).

7.3 Malformations of the fornix

A focus on the fornix abnormalities and their association with hippocampal anomalis may figure importantly in our understanding of the pathophysiology of schizophrenia (Kuroki et al., 2006).

7.3.1 Anomalies of the fornix

The fornical defects associates with other commissural agenesis such as missing of the hippocampal commissure. The accumulation of the fornical fibers in the lower edge of the medial telencephalic medullary velum, the separated fornix **(Fig. 5. A)** (Meyer & Rorich, 1998; Yousefi & Kokhei, 2009), variation in the pattern of distribution (Griffiths, 2009; Yousefi & Kokhei, 2009) of the precommissural fornix to more than three branches **(Fig. 5. A)** on the medial surface of the frontal lobe, thickening one of **(Fig. 5. B)** precommissural fornix branches and continued to curve inferior posteriorly parallel with the posterior commissural fornix, so that is visualized without dissection (Yousefi & Kokhei, 2009), entrapped some fibres of the genu of the corpus callosum with the fornical fibers bundle (Hori, 1997), enter the fornix to the basal forebrain without the normal division, a bulky connection between the anterior parts of the fornices producing a very prominent hippocampal commissure (Griffiths, 2009; Barkovich, 1990) are repoted in literature as anomalies of the fornix. Also, a recent postmortem and a in-vivo studies confirming decreased axonal density (Ozdogmus et al., 2009, Concha et al., 2010) and a near complete absence of unmyelinated axons of the fimbria-fornix bilaterally in the temporal lobe epilepsy and unilateral mesial temporal sclerosis patients (Ozdogmus et al., 2009; Concha et al., 2010) due to the intriguing possibility that a specific subset of projection fibers may be lost in temporal lobe epilepsy (Concha et al ., 2010). Reduced fractional anisotropy and cross-sectional area simultaneous increase mean diffusivity in the fornix in the schizophrenia patients which indicate that fornix abnormalities may be due to either immaturity or degeneration of the fiber tract. These abnormalities may reflect decreased axonal density, axonal damage, or decreased degree of myelination. Atrophy of the fornix is the other condition that is detected in 86% of the temporal lobe epilepsy patients with unilateral hippocampal atrophy and in almost all patients with bilateral symmetrical hippocampal atrophy. This finding suggests that hippocampal atrophy may cause secondary fornix

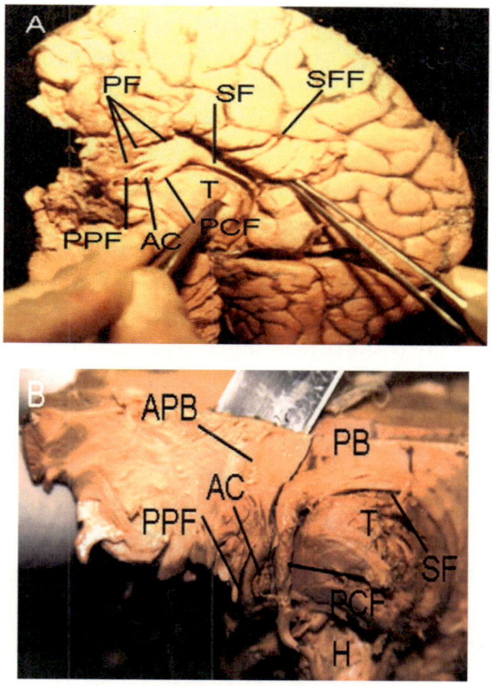

Fig. 5. The photograph of the brain in midsagittal plane (A), exposed of medial aspect of the frontal lobe (B) showing the separated fornix (SF) and associated branches. The precommissural fornix (PF) is abnormal. Other abbreviations in this figure: PCF, postcommissural fornix; AC, anterior commissure; PB, Probst bundle; SFF, sulci as fan-like fashion. APB, anterior part of the Probst bundle; T, thalamus; H, hypothalamus.

atrophy (Concha et al., 2010). In addition to the schizophrenic patients, the fornical anomalies are common in patients with myelomeningocele and Chiari II malformation. These fornical anomalies include; intact but thin, thin and right greater than left, atrophic and left greater than right, left intact and right crus deficient, thin body and crura,, thin crura, defects in crura, bilaterally deficient crus and body and frank defects in the fornices associated with atresia or hypoplasia of crura and body of fornices. In these patients such defects are associated with memory and learning deficits (Vachha et al., 2006). Beyond mechanical stretching of periventricular axons, chronic hydrocephalus has been shown to be associated with microvascular changes in the cerebral white matter, which include capillary compression and calcium-mediated proteolysis that may account for the defects within the limbic fibres (Del Bigio, 2001).

7.3.2 Asymmetry of the fornix

In addition to patients (Baldwin et al., 1994), significant differences in the fornical volume were seen between the right and left sides of the fornix in healthy individual (Zahajszkyet al., 2001).This asymmetry is present in the position of the two columns of the fornix in relation to the septum pellucidum. This difference was seen in most of the subjects caudal located of the left fornical column to the right (Supprian &, Hofmann, 1997). In patients asymmetric volume loss in the fornix is detected on the same side as the abnormal hippocampus and hippocampal sclerosis, the correlation may be related to the anatomy of this white matter tract. Because some of axons of the fonnix originate in the pyramidal cells of the hippocampus, it is suggested that hippocampal neuronal loss may result in wallenian degeneration and subsequent atrophy of the ipsilatenal fornix (Baldwin et al., 1994). The fornix asymmetry is more likely to be of developmental origin as opposed to secondary alterations (Supprian &, Hofmann, 1997) and the degree of asymmetry between the fomices varied from 41% to 82% (mean, 68%).It is apparent that men have a lower density of fibers in the fornix than women, the density of fibers on the left in men is significantly greater for patients with schizophrenia than those whom total fibre number is not significantly affected by gender or diagnosis (Chance et al., 1999). In regard to the precise course of the fornix in commissural agenesis associated with meningeal dysplasia cases, it is highly variable and posteriorly appears to be influenced by the anatomy of the interhemispheric cysts to a major degree. The fornix maintains a high-riding path as it courses cephalad and does not appear to give a postcommissural branch; instead, the fornix passes more anteriorly than usual before passing posteriorly to enter the basal forebrain (Griffiths, 2009). In children with complete commissural agenesis the path of fornix appears to be remarkably constant, although the fornix travels more laterally than usual away from its partner. Some of the cases may show the Shift of the fornix into a rostroventral direction (Boretius et al., 2009).

7.4 Malformations of the posterior commissure

The hypoplasia (Abouzeid et al., 2009) and absence of the posterior commissure may be associated with other forebrain commissures anomalies (Meyer & Rorich, 1998; Abouzeid et al., 2009) in the patients with *Pax6* (p.R159fs47) mutations (Abouzeid et al., 2009). Experimental studies in the null mutant mice with lacking subcommissural organs or with subcommissural organs alterations have shown that a normal posterior commissure fails to form (Louvi & Wassef, 2000; Estivill-Torrus et al., 2001; Fernandez-Llebrez et al., 2004; Ramos et al., 2004) due to lack of the homeobox gene Msx1 (Fernandez-Llebrez et al., 2004; Ramos et al., 2004). Additionally, in mutant mice lacking the transcription factor PAX6, the posterior commissure fails to develop (Estivill-Torrus et al., 2001).Also, in WEXPZ.En1 transgenic mice in which engrailed-1 is expressed ectopically in the dorsal midline of the diencephalon (Danielian & McMahon, 1996) the posterior commissure development is delayed and frequent errors in axonal pathfinding happen (Louvi & Wassef, 2000). In addition to mentioned anomalies, relocation of the posterior commissure on the subcommissural organs is reported in the one-eyed pinhead mutants of the zebrafish Dario rerio due to slightly displaced of the subcommissural organs from its normal midline position (Hoyo-Becerra et al., 2010).

8. Brain malformations associated with commissural disorders

It has been proposed that the corpus callosum is useful as an indicator of both congenital and degenerative brain disorders in children, since the corpus callosum is formed contemporaneously with many other major telencephalic structures (Barkovich & Norman, 1988; Hetts et al., 2006). Callosal dysgenesis is frequently associated with other central nervous system malformations and /or somatic anomalies (Barkovich & Norman, 1988; Marszal et al., 2000; Hetts et al., 2006) and its defects are rarely isolated (Barkovich & Norman, 1988; Hetts et al., 2006). Since formation of the corpus callosum is complex and this characteristic may explain why most cases of callosal agenesis are not isolated (Tang et al., 2009). The type, number, and severity of related anomalies, however, are deferent. Brain anomalies associated with commissural disorders can be arranged based on the morphology of the cerebral commissures and associated malformations of the midline, of cortical development, of white matter, and of the diencephalon and rhombencephalon (Hetts et al., 2006).

8.1 Midline anomalies

Midline malformations association with agenesis or dysgenesis of the corpus callosum include **interhemispheric cysts (Figs. 6, 9), lipomas** (Hetts et al., 2006; Byrd, 1990; Johnston, 1934; Probst, 1973, Barkovich et al., 2001; Raybaud, 2010; Truwit, 1990) and **craniocerebral midline defects** (Raybaud, 2010) which can be confirmed by imaging studies (Davila-Gutierrez, 2002). Agenesis of the commissures with interhemispheric cysts is felt to have different causes, possibly related to a meningeal rather than neural disorder (Raybaud, 2010). There are two broad classes of interhemispheric cysts (**Fig. 6**), communicating and non-communicating (Johnston, 1934; Probst, 1973; Barkovich et al., 2001). The communicating cysts are expansions of the ventricular tela choroidea and the non-communicating cyst is multiloculated meningeal cystic dysplasia (Raybaud, 2010; Davila-Gutierrez, 2002). Additionally, another classification of the callosal agenesis with cysts has been advised: type 1, in which there is one single cystic cavity that communicates with the ventricles and subdivides in three subgroups on the bases of being a) with macrocephaly and hydrocephalus, b) with macrocephaly and hydrocephalus associated with a developmental ventricular obstruction (thalamic fusion, hamartoma), and c) with microcephaly (Barkovich et al., 2001). Type 2 refers to the cases where the interhemispheric cysts are multiloculated (Davila-Gutierrez, 2002; Barkovich et al., 2001) and independent from the ventricles; it is subdivided into three subgroups on the bases of being a) hydrocephalus and an essentially normal brain, b) affects girls and is made of multiple cysts different from cerebro spinal fluid with frontoparietal polymicrogyria and periventricular nodular heterotopias and one or two dilated ventricles (Barkovich et al., 2001) and c) with multiloculated cysts, large subcortical heterotopia, and dysmorphic head and brain. However, it needs to be confirmed (Raybaud, 2010). In the form of a single ventricular diverticulation cyst, the commissural agenesis is usually not associated with significant hemispheric dysplasia or malformations of cortical development. The main feature is the markedly expanded tela choroidea, the septum pellucidum, fornices, and bundles of Probst are missing (Raybaud, 2010). Such conditions have been previously described as "septo-optic dysplasia": with total absence of the corpus callosum (Sener, 1993) or agenesis of the corpus callosum with dehiscent fornices (De León et al., 1995). In commissural agenesis with multilocular cysts, most of the cases have cerebral dysplasia. The CT density and the MR signals of some of these cysts commonly are different from those of the cerebro spinal fluid, histological peculiarity to explain protein content different from that of the cerebro spinal fluid, children usually are born with hydrocephalus and the size of the cysts usually increased during gestation (Raybaud, 2010). An association of agenesis or dysgenesis of the corpus callosum with subarachnoidal cysts also have been recognized for example, reported two sisters that presented corpus callosal agenesis, neuralsensory deafness, and subarachnoideal cysts with hydrocephalus, the cysts being located in the pineal region and obstruct the cerebral aqueduct, as an autosomal-recessive trait (Hendriks et al., 1999). Interhemispheric meningeal lipomas are the second meningeal dysplasia which commonly

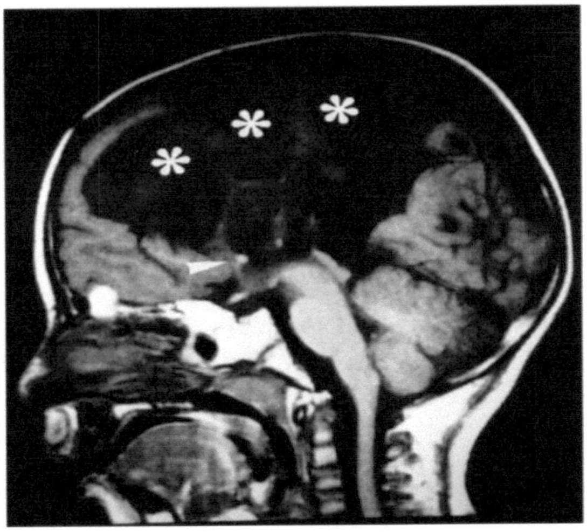

Fig. 6. Midsagittal T1-weighted image shows a complete callosal defect with interhemispheric cyst and cortical dysplasia in a 3-year-old boy, histologically verified as glioependymal (asterisks) (Utsunomiya et al., 1997).

associated with a malformation of the commissures (Raybaud, 2010). The most common location being the depth of the interhemispheric fissure where the lipoma often extend toward into the choroid plexuses (Truwit, 1990). The mechanism of the malformative association is not really known. It has been known for some time that, depending on the appearance (tubulonodular or curvilinear) and location (ventral or dorsal) of the lipoma, the dysplasia of the corpus callosum was different, while that the commissural defect does not correlate with the size or shape of the lipoma (dorsal tubulonodular lipoma can be observed with normal callosal morphology). A study, depending on the location of the lipoma, has classified it into four topographic groups: anterior, transitional (or global: covering the callosum from the front to the back), posterior, and inferior (below the hippocampal commissure). The anterior lipoma (15%) is associated with major commissural hypogenesis, the more posterior transitional lipoma (24%) with a complete but hypoplastic commissural plate, the posterior ones (48%) with minor shortening or tapering of the splenium, and the inferior ones (12%), with minor commissural abnormalities only. Craniocerebral midline defects: along the neural tube, commissuration is primarily a basal process, and in cases of commissural agenesis other commissuration defects may be observed anywhere along the ventral cord and brainstem (Raybaud, 2010). In the basal forebrain, other commonly associated defects involve the anterior optic pathway (Raybaud & Girard, 1998) and the hypothalamo-pituitary axis (Raybaud & Girard, 1998). Because the development of the corpus callosum itself is associated with the dorsalization of the hemispheres, other disorders of the dorsalization may be observed, primarily at the level of the cerebellum: a Dandy–Walker malformation (or related defect) is commonly associated with an agenesis of the corpus callosum (Johnston, 1943; Raybaud, 1982). The rare rhombencephalon synapsis is often found in association with septal defects/ septo-optic dysplasia (Michaud et al., 1982; Jellinger, 2002) and obviously the midline skull defects commonly include commissural agenesis or dysgenesis, especially the frontonasal dysplasia (Guion-Almeida et al., 1996; Wu et al., 2007) and the basal, notably sphenoidal cephaloceles (Koenig et al., 1982).

8.2 Malformations of cortical development

All abnormalities of cortical development may be associated with anomalies of the commissures. Migration disorders are probably the most typical (Raybaud, 2010). Periventricular nodular heterotopias (Volpe et al., 2006; Tang et al.,2009; Raybaud, 2010) are commonly found (Raybaud, 2010) and dysplastic-appearing deep gray nuclei characterized by small size, abnormal shape with periventricular nodular heterotopias (Volpe et al., 2006; Tang et al.,2009) see only in delayed sulcation (Tang et al., 2009). Abnormal sulcation associated with commissural anomalies are reported as the most common malformation of cortical development (Hetts et al., 2006; Byrd, 1990; Barkovich & Norman 1988; Tang et al., 2009). Major hemispheric dysplasia with large subcortical heterotopia and cortical dysplasia **(Figs. 7, 9)** are characteristic as well (Raybaud, 2010). Cortical dysplasia (Donmez et al., 2009; Volpe et al., 2006) may be associated with an interhemispheric glioependymal cyst and porencephaly (Utsunomiy et al., 1997), and small porencephalic cysts (Volpe et al., 2006). In addition to above malformations of cortical development, the gyral abnormalities have been described previously in relation to anomalies of the cerebral commissures and abnormal gyral patterns which are characterized either by abnormal, too numerous infoldings or by absent sulcation (Tang et al., 2009). These abnormalities include polymicrogyria (Utsunomiy et al., 1997 ; Hetts et al., 2006; Tang et al., 2009) classic lissencephaly (Hetts et al., 2006: Tang et al., 2009; Volpe et al., 2006, Donmez et al ., 2009), cobblestone lissencephalies (Hetts et al., 2006), schizencephaly (Hetts et al., 2006; Tang et al., 2009), schizencephaly with bilateral frontoparietal holohemispheric clefts (Utsunomiy et al., 1997), pachygyria (Tang et al., 2009), heterotopia pachygyria (Hetts et al., 2006) or diffuse pachygyria (Utsunomiy et al., 1997) and other nonclassified abnormalities (Tang et al., 2009).

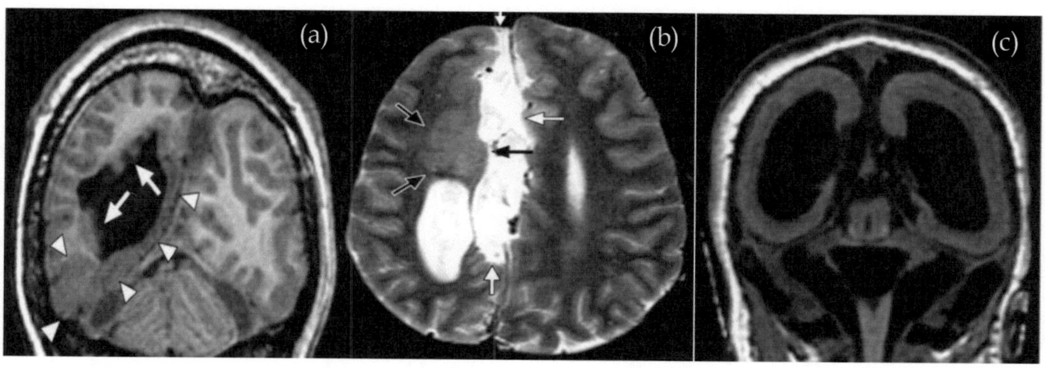

Fig. 7. Anomalies of cortical development of varying extent and severity were found in patients with callosal hypogenesis or agenesis. Coronal (a) T1-weighted image in 4-year-old boy shows periventricular nodular heterotopia (arrows) and dysplastic occipital cortex (arrowheads) in addition to dysplastic cerebellum. Axial (b) T2-weighted image in 17-year-old boy shows dysplastic frontal and cingulate cortex (black arrows) adjacent to interhemispheric cyst (white arrows). Coronal (c) T1-weighted image in 6-year-old girl shows lissencephaly with four-layer (Hetts et al., 2006).

8.3 Brain white matter anomalies

Definition of commissural anomalies is an abnormality of the white matter (Van Essen., 1997).The white matter has been postulated to contribute to normal sulcation. Abnormalities of sulcation may be possibly associated with a decreased volume of white matter, as reported in literature (Hetts et al., 2006). The following sulcal abnormalities have been descripted in the brains with commissural defects: The sulci of the medial surface of the hemisphere radiated in a fan-like fashion (Meyer & Röricht, 1998; Sztriha, 2005; Yousefi & Kokhei, 2009) towards the lateral **(Fig. 5. A)** wall of the third ventricle (Yousefi & Kokhei, 2009; Sztriha, 2005) without a visible callosomarginal (Meyer & Röricht, 1998; Yousefi & Kokhei, 2009) and cingulate sulcution (Yousefi & Kokhei, 2009; Sztriha, 2005). The parieto-occipital and calcarine sulci cross in the medial surface and enter toward the lateral ventricle and a lack of a well-defined cingulum (Atlas et al., 1986; Yousefi & Kokhei, 2009). Beyond the expected eversion of the cingulum and radial orientation of paramedian gyri that routinely accompany callosal agenesis (Hetts et al., 2006; Yousefi & Kokhei, 2009). Abnormalities of sulcation ranged from overly shallow olfactory sulci to frank hemispheric dysplasia (Hetts et al., 2006). The basis of one theory (Van Essen, 1997) it is possible that the absence of normal connections between hemispheres and formation of aberrant connections within the same hemisphere can delay the formation of primary sulci and perhaps even contribute to the abnormal sulcal morphology seen in so many of the cases. Reductions in extracallosal white matter volume and the presence of moderately or severely reduced extracallosal white matter volume in patients with agenesis of the corpus callosum and patients with hypogenesis of the corpus callosum may represent a primary dysplasia or hypogenesis, with fewer axons forming during development, or a secondary regression, possibly due to retraction of axons that do not find their way across midline to synapse with their homologues and thereby gain the neurotrophic support necessary for survival (Hetts, 1998). The thickness of the mid-body of the corpus callosum positively correlates with volume of cerebral white matter in children with cerebral palsy and developmental delay. Assessment of the thickness of the corpus callosum might help in estimating the extent of the loss of volume of cerebral white matter in children with a broad spectrum of periventricular white matter injury (Panigrahy et al., 2005).

8.4 Abnormal morphology of the lateral ventricle

Anomalies of the lateral ventricle are always seen in association with abnormal sulcal morphology and constantly influence at least the frontal horn and occurrs on the side with the abnormal cortical infoldings. The morphological abnormalities of the lateral ventricle include enlargement of the ventricular atria (Tang et al., 2009), widening, colpocephaly, disproportionate dilatation of the trigones and occipital horns (Bekiesiñska et al., 2004; Atlas et al., 1986; Utsunomiy et al., 1997), keyhole dilatation of the temporal horns which is thought to result from deficient hippocampal formation (Atlas et al., 1986; Utsunomiy et al., 1997), narrow frontal horns (Bekiesiñska et al., 2004), abnormal curvature of the anterior horn **(Fig . 3. A, B)** (Yousefi & Kokhei, 2009) which is explained as secondary deformity of anterior horn (Atlas et al., 1986) and irregularity of the ventricular wall due to periventricular nodular heterotopia, choroid plexus cysts, abnormal brain stem, germinal matrix and intraventricular hemorrhage (Tang et al., 2009). Also, some of the cases with callosal hypoplasia show abnormal cerebrospinal fluid spaces (Bekiesiñska et al., 2004).

Among above abnormal morphology of the lateral ventricle, colpocephaly, a selective ventriculomegaly (**Fig. 8. b**) of the occipital horns more than the frontal or temporal horns of the lateral ventricles, is a common finding in agenesis of the corpus callosum, and it appears that callosal agenesis is probably the second most frequent cause of colpocepbaly after periventricular leukomalacia (Sarnat, 1992). The deficiency of white matter around the occipital horns due to absence of the posterior fornix of the corpus callosum is the reason (Davila-Gutierrez, 2002). Holoprosencephaly (Raybaud, 2010), some subtypes of microcephaly (significantly associated) (Vermeulen et al., 2010) and DCX (doublecortin) related lissencephaly may be associated with callosal agenesis in humans (Kappeler et al, 2007). Of course the agenesis is a defining feature of the ARX (Xp22.13) related lissencephaly with callosal agenesis (Kitamura et al., 2002).

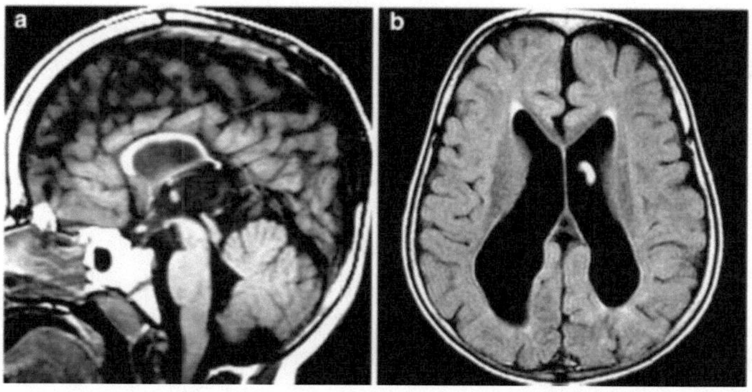

Fig. 8. Diffusely hypoplastic commissural plate with ventriculomegaly, Midline (a) sagittal T1WI.The commissural plate is complete but thin with a tiny splenium, Axial FLAIR (b). Diffuse ventriculomegaly without real evidence of leukomalacia: this points to a global white matter disorder that may be developmental or acquired, not a commissural disorder (Raybaud, 2010).

8.5 The diencephalon and rhombencephalon abnormalities

Abnormalities of the cerebellum (hemispheres, vermis), brainstem, orbits, pituitary, state of white matter myelination, and olfactory (apparatus and sulci which are more frequent in agenesis of the corpus callosum patients) have been reported in many patients with callosal anomalies (Hetts et al., 2006). Additional findings about the cerebellum include cerebellar hypoplasia with lissencephaly (Miyata et al., 2004), small or absent of the vermis, an asymmetric appearance of the fourth ventricle, small or absent of the cerebellum, the small cerebellum with abnormal orientation of the folia and posterior fossa cyst or hydrocephalus. An abnormal brain stem appears as dysplastic, small, compressed; also, dysgenesis of the corpus callosum can occur in association with dysgenesis of the frontal, parietal, and occipital lobes (Kawamura et al., 2002).

9. Syndromes that include commissural dysgenesis as a defining feature

OMIM (Online Mendelian Inheritance in Man, Johns Hopkins University, March 16, 2010) lists 189 specific syndromes in which a commissural agenesis is or may be present (Raybaud, 2010). Also, agenesis/dysgenesis of the corpus callosum has been described with congenital metabolic diseases (Dobyns, 1989; Kiratli, 1999, Kolodny, 1989). The syndromes of the Aicardi **(Fig. 9),**Acrocallosal, Andermann and Shapiro are characterized by agenesis of the corpus callosum while others are only sporadically associated (Jeret et al., 1987). The CRASH syndrome with clinical features of callosal agenesis, retardation, adducted thumbs, spasticity, hydrocephalus (Yamasaki et al., 1997; Sztriha et al., 2000; Fransen et al., 1997; Weller & Gärtner, 2001) aphasia (Fransen et al., 1997; Weller & Gärtner, 2001) is related to a mutation of *L1* gene at Xq28 (Yamasaki et al., 1997; Sztriha et al., 2000). This gene involves in encoding a cell adhesion molecule which is involved in the fasciculation of the axons, as well as synaptic targeting and cellular migration (Schmid et al., 2008). The Miller-Dieker syndrome and Walker-Warburg syndrome are defined with partial to complete agenesis of the corpus callosum (Davila-Gutierrez, 2002). Walker-Warburg syndrome is the most severe phenotype of the group of the "cobblestone brains" that also includes the Fukuyama and the muscle-eye-brain syndromes characterized by congenital muscular dystrophy and neuronal migration disorder in which there is overmigration of the neurons beyond the pia limiting membrane. The neurons overmigrate and form abnormal arrangements in the cortical and meningeal layers. This disorganization of the tissular pattern and the abnormal extracellular matrix signals in turn results in failure of the white matter to form perfectly. Of the three phenotypes, the Walker–Warburg syndrome is the most severe, irregular (cobblestone) cortical surface, disorganized cortex, thin cerebral mantle with lack of white matter and ventriculomegaly, absence of the commissures; the underdeveloped brainstem often has a Z shape; and the cerebellum is hypoplastic with correspondingly huge posterior fossa cisterns (Raybaud, 2010). Syndromic craniosynostoses (Apert, Crouzon, Pfeiffer mostly) the typical occurrence of corpus callosal dysgenesis and/or septum pellucidum defects may well be intrinsically part of the syndromes (Raybaud & Di Rocco, 2007), All syndromic craniosynostoses result from a defect of one of the FGFR genes (FGFR2 on 10q25- q26 for Apert, Crouzon and Pfeiffer; FGFR1 on 8q11.22- p12 for Pfeiffer alsoes (Doherty & Wlash, 1996; Kamiguchi & Lemmon, 1997).

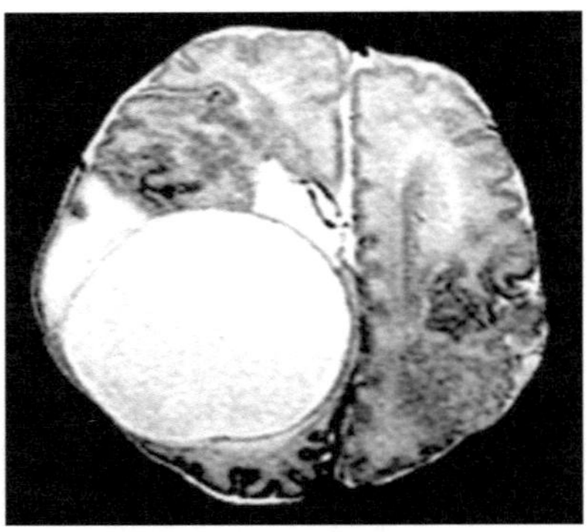

Fig. 9. Aicardi syndrome, newborn girl. Huge right-sided choroid plexus cyst with adjacent parenchymal damage. Note the multiple subcortical heterotopias and cortical dysplasia in the adjacent right frontal lobe and in the left parietal lobe (Raybaud, 2010).

10. Clinical and paraclinical features

Children with isolated form of agenesis or dysgenesis of the corpus callosum are asymptomatic or presented a mild hypotonia (Francesco et al., 2006). The intelligence

(Davila-Gutierrez, 2002) and electroencephalographic (Francesco et al., 2006), usually are normal (Davila-Gutierrez, 2002; Francesco et al., 2006), although these children have a lower capacity for processing somatosensory information (Friefeld et al., 2000). Many of the cases have shown a hypertelorism as mild facial dysmorphism which may be a clue to the neuroanatomic anomaly and further justify neuroimaging studies (Davila-Gutierrez, 2002), a more thorough neurological examination reveals defects in transfer of information. Additionally, the mental retardation can be exist (Serur et al., 1988) and neurodevelopmental outcome reported to be poor in 15–28% of cases (Moutard et al., 2003; Pilu et al., 1993). While in children with associated brain malformations (Francesco et al., 2006), the neurological features relate to the severity and variety of the accompanying cerebral defects (Davila-Gutierrez, 2002) and include epilepsy (Davila-Gutierrez, 2002; Francesco et al., 2006) mental retardation, hydrocephalus, and morphologic and growth abnormalities that vary from hypotonia to severe spasticity, ataxia, autistic behavior, learning disabilities, and behavioral disorders (Davila-Gutierrez, 2002).

10.1 Electroencephalographic features

The most characteristic feature is the continued asynchrony of sleep spindles after 18 months of age. However, this asynchrony is not an overall asymmetry, and the morphology and number of spindles in the two hemispheres are relatively equal over an extended period of stage 2 sleep (Sarnat, 1992).

10.2 Prenatal diagnosis

Laterally displacement of the lateral ventricles and atrium, upward movement of the third ventricle (Comstock et a1., 1985), absence or alteration of the cavum septum pellucidum (Meizner et a1, 1987), high-riding third ventricle, and widening of the interhemispheric fissure (Tang et al., 2009) and posterior ventriculomegaly or colpocephaly have been described as an imaging sign of the corpus callosal malformations (Lockwood et a1., 1988). Also, it is found that when complete corpus callosal dysgenesis exists, the frontal region is small and the cavum septi pellucidi is not evident (Tepper & Zale1, 1996). Male fetuses are more likely to have an isolated agenesis/dysgenesis of the corpus callosum that is considered benign in its clinical expression (Davila-Gutierrez, 2002).

11. Conclusion

It is highly likely that agenesis of the brain commissural might have been developed as a result of an early embryological abnormally growth and development .The commissures formation is a complex process and involved many commissuration factors, so that a isolated commissural agenesis is uncommon, and the abnormality is usually associated with other cerebral or craniocerebral or syndromic defects. A full understanding of the embryological, anatomical and functional of the commissures could us to the diagnosis and handling of these abnormalities.

12. Acknowledgment

I would wish to offer my best thanks to Dr.Farhad Azizzadeh because of his diligent endeavors to edit this chapter.

13. References

Aboitiz, F.; Carrasco, X. Schröter, C. Zaidel, D. Zaidel, E. & Lavados, M. (2003). The alien hand syndrome: classification of forms reported and discussion of a new condition. *Neurological Sciences,* Vol. 24, No. 4, (November 2003), pp. 252-7, ISSN 1590-1874

Aboitiz, F. & Montiel, J. (2003).One hundred million years of interhemispheric communication: the history of the corpus callosum. *Brazilian Journal of Medical and Biological Research,* Vol. 36, No. 4, (April 2006), pp. 409-20, ISSN 0100-879X

Aboitiz, F.; Scheibel, AB. Fisher, RS. & Zaidel, E. (1992). Fiber composition of the human corpus callosum. *Brain Research,* Vol. 589, No. 1-2, (December 1992), pp. 143-53, ISSN 0006-8993

Abouzeid, H.; Youssef, MA. ElShakankiri, N. Hauser, P. Munier, FL. & Schorderet, DF. (2009). Pax6 aniridia and interhemispheric brain anomalies. *Molecular vision,* Vol. 15, (October 2009), pp. 2074-83, ISSN 1090-0535

Achiron, R. & Achiron, A. (2001). Development of the human fetal corpus callosum: a high-resolution, cross-sectional sonographic study. *Ultrasound in Obstetrics & Gynecology,* Vol. 18, No. 4, (October 1968), pp. 343-7, ISSN 0960-7692

Aralasmak, A.; Ulmer, JL. Kocak, M. Salvan, CV. Hillis, AE. & Yousem, DM. (2006). Association, commissural, and projection pathways and their functional deficit reported in literature. *Journal of Computer Assisted Tomography,* Vol. 30, No. 5, (September-October 2006), pp. 695-715, ISSN 0363-8715

Arnold, SE. & Trojanowski, JQ. (1996). Human fetal hippocampal development: I. Cytoarchitecture, myeloarchitecture, and neuronal morphologic features. *The Journal of Comparative Neurology,* Vol. 367, No. 2, (April 1996), pp. 274-92, ISSN 0021-9967

Atlas, SW.; Zimmerman, RA. Bilaniuk, LT. Rorke, L. Hackney, DB. Goldberg, HI. & Grossman, RI. (1986). Corpus callosum and limbic system: neuroanatomical MR evaluation of developmental anomalies. *Radiology,* Vol. 160, No. 2, (August 1986), pp. 355-362, ISSN 0033-8419

Baldwin, GN.; Tsuruda; JS. Maravilla, KR. Hamill, GS. & Hayes, CE. (1994). The fornix in patients with seizures caused by unilateral hippocampal sclerosis: detection of unilateral volume loss on MR images. A J R. *American Journal of Roentgenology,* Vol. 162, No. 5, (May 1994), pp. 1185-9, ISSN 0361-803X

Bamiou, DE.; Free, SL. Sisodiya, SM. Chong, WK. Musiek, F. Williamson, KA. Van Heyningen, V. Moore, AT. Gadian, D. & Luxon, LM. (2007). Auditory interhemispheric transfer deficits, hearing difficulties, and brain magnetic resonance imaging abnormalities in children with congenital aniridia due to PAX6 mutations. *Archives of Pediatrics & Adolescent Medicine,* Vol. 161, No. 5, (May 2007), pp. 463-9, ISSN 1072-4710

Banich, MT. (1998). The missing link: the role of interhemispheric interaction in attentional processing. *Brain and Cognition,* Vol. 36, No. 2, (March 1998), pp. 128-57, ISSN 0278-2626

Barkovich, AJ. (1990). Apparent a typical callosal dysgenesis. *A J N R. American Journal of Neuroradiology,* Vol. 11, No. 2, (March-April 1990), pp. 333-40, ISSN 0195-6108

Barkovich, AJ. (2005). Congenital Malformations of the Brain and Skull. *Pediatric Neuroimaging,* 4th ed, pp. 296–304, Lippincott Williams & Wilkins, ISBN 0-7817-5766-5. USA.

Barkovich, AJ. & Norman D. (1988). Anomalies of the corpus callosum: correlation with further anomalies of the brain. *A J R. American Journal of Roentgenology,* Vol. 151, No. 1, (July 1988), pp. 171-9, ISSN 0361-803X

Barkovich, AJ.; Simon, EM. & Walsh CA. (2001). Callosal agenesis with cyst. A better understanding and new classification. *Neurology,* Vol. 56, No. 2, (January 2001), pp. 220–227, ISSN 0028-3878

Barnett, KJ . & Corballis, MC. (2005). Speeded right-to-left information transfer: the result of speeded transmission in right-hemisphere axons? *Neuroscience letters,* Vol. 380, No. 1-2, (May 2005), pp. 88-92, ISSN 0304-3940

Barr Melodie, S. & Corballis Michael, C.(2002). The role of the anterior commissure in callosal agenesis. *Neuropsychology,* Vol. 16, No. 4, (October 2002), pp. 459-71, ISSN 0028-3932

Bayer, SA. & Altman, J. (2006).The human brain during the late first trimester. *Atlas of Central Nervous System Development.Vol. 4,* PP. 576, Boca Raton CRC Press (Taylor France Group), ISBN 0849314232, Florida in the *USA.*

Bekiesińska-Figatowska, M.; Chrzanowska, KH. Jurkiewicz, E. Wakulińska, A. Rysiewskis, H. Gładkowska-Dura, M. & Walecki, J. (2004). Magnetic resonance imaging of brain abnormalities in patients with the Nijmegen breakage syndrome. *Acta Neurobiologiae Experimentalis,* Vol. 64, No. 4, (2004), pp. 503-509, ISSN 0065-1400

Bersani, G.; Quartini, A. Iannitelli, A., Paolemili, M. Ratti, F. Di Biasi, C. & Gualdi, G. (2010). Corpus callosum abnormalities and potential age effect in men with schizophrenia: an MRI comparative study. *Psychiatry Research,* Vol. 183, No .2, (August 2010), pp. 119-25, ISSN 0165-1781

Bloom, JS. & Hynd, JW. (2005). The role of the corpus callosum in interhemispheric transfer of information: excitation or inhibition? *Neuropsychology Review,* Vol. 15, No. 2, (June 2005), pp. 59-71, ISSN *1040-7308*

Boretius, S.; Michaelis, T.Tammer, R. Ashery-Padan, R. Frahm, J. & Stoykova, A. (2009). In vivo MRI of altered brain anatomy and fiber connectivity in adult pax6 deficient mice. *Cerebral Cortex,* Vol. 19, No. 12, (December 2009), pp. 2838-47, ISSN 1047-3211

Bucur, B.; Madden, DJ. & Allen, PA. (2005). Age-related differences in the processing of redundant visual dimensions. *Psychology and Aging,* Vol. 20, No. 3, (September 2005), pp. 435-46, ISSN 0882-7974

Buklina, SB. (2005). The corpus callosum. Interhemispheric interactions and the function of the right hemisphere of the brain. *Neuroscience and Behavioural Physiology,* Vol. 35, pp. 473-80, ISSN 0882-7974

Byrd, S.; Radkowski, M. Falnnery, A. & McLone, D. (1990). The clinical and radiologic evaluation of absence of the corpus callosum. *European Journal of Radiology,* Vol. 10, No. 1, (January-February 1990), pp. 65–73, ISSN 0720-048X

Carpenter, MB. (1991). Olfactory pathways, hippocampal formation and the amygdale. In: Carpenter MB (ed) *Core Text of Neuroanatomy,* 4th ed. pp 361–384, Williams and Wilkins, Baltimore, ISBN 0683014579. U S A

Cao, Q.; Suna, L. Gong, G. Lv, J. Cao, X. Shuai, L. Zhu, C. Zang, Y. & Wang, Y. (2010). The macrostructural and microstructural abnormalities of corpus callosum in children with attention deficit/hyperactivity disorder: a combined morphometric and diffusion tensor MRI study. *Brain Research*, Vol. 1310, (January 2010), pp. 172-80, ISSN 0006-8993

Chance, SA.; Highley, JR. Esiri, MM. & Crow, TJ. (1999). Fiber content of the fornix in schizophrenia: lack of evidence for a primary limbic encephalopathy.*The American Journal of psychiatry*, Vol. 156, No. 11, (November 1999), pp. 1720-4, ISSN 0002-953X

Chiappedi, M. & Bejor, M. (2010). Corpus callosum agenesis and rehabilitative treatment. *Italian Journal of Pediatrics [Electronic Resource]*, Vol. 17, No. 36, (September 2010), pp. 64, ISSN 1824-7288

Chouchane, M.; Benouachkou-Debuche, V. Giroud, M. Durand, C. & Gouyon. JB. (1999). Agenesis of the corpus callosum: etiological and clinical aspects, diagnostic methods and prognosis. *Archives de Pédiatrie*, Vol. 6, No. 12, (December 1999), pp. 1306-11, ISSN 929-693X

Church, MW. & Gerkin, K. (1988). Hearing disorders in children with fetal alcohol syndrome: Findings from case reports. *Pediatrics*, Vol. 82, No. 2, (August 1988), pp. 147-154, ISSN 0031-4005

Clarke, JM. & Zaidel, E. (1994). Anatomical-behavioral relationships: corpus callosum morphometry and hemispheric specialization. *Behavioural Brain Research*, Vol. 64, No. 1-2, (October 1994), pp. 185-202, ISSN *0166-4328*

Comstock, CH.; Culp, D. Gonzalez, J. & Boal, DB. (1985). Agenesis of the corpus callosum in the fetus: its evolution and significance. *Journal of Ultrasound in Medicine*, Vol. 4, No. 11, (November 1985), pp. 613-16, ISSN 0278-4297

Concha, L. ; Livy, DJ. Beaulieu, C. Wheatley, BM. & Gross, DW. (2010). In vivo diffusion tensor imaging and histopathology of the fimbria-fornix in temporal lobe epilepsy. *The Journal of Neuroscience*, Vol. 30, No. 3, (January 2010), pp. 996-1002, ISSN 0270-6474

Corballis, MC.; Corballis, PM. & Fabri, M. (2003). Redundancy gain in simple reaction time following partial and complete callosotomy. *Neuropsychologia*, Vol. 42, No. 1, pp. 71-81, ISSN 0028-3932

Corballis, MC .; Hamm, JP. Barnett, KJ . & Corballis, PM . (2002). Paradoxical interhemispheric summation in the split brain. *Journal of Cognitive Neuroscience*, Vol. 14, No. 8, (November 2002), pp. 1151-7, ISSN 0898-929X

Danielian, PS. & McMahon, AP. (1996). Engrailed-1 as a target of the Wnt-1 signalling pathway in vertebrate midbrain development. *Nature*, Vol. 383, No. 6598, (September 1996), pp. 332–334, ISSN 0028-0836

Davila-Gutuerrez, G. (2002). Agenesis and dysgenesis of the corpus callosum. *Seminars in Pediatricneurology*, Vol. 9, No. 4, (December 2002), pp. 292-301, ISSN1071-9091

Del Bigio, MR. (2001). Pathophysiologic consequences of hydrocephalus. *Neurosurgery Clinics of North America*, Vol. 12, No. 4, (October 2001), pp. 639–649, ISSN 1042-3680

De León, GA.; Radkowski, MA. & Gutierrez, FA. (1995). Single forebrain ventricle without prosencephaly: agenesis of the corpus callosum with dehiscent fornices. *Acta Neuropathologica*, Vol. 89, No. 5, pp. 454-8, ISSN 0001-6322

Destrieux, C.; Velut, S. & Kakou, M. (1998). Development of the corpus callosum. *Neurochirurgi,* Vol. 44, No.1 suppl, (May 1998), pp. 11-6, ISSN 0028-3770

Di Virgilio, G.; Clarke, S. Pizzolato, G. & Schaffner, T. (1999). Cortical regions contributing to the anterior commissure in man. *Experimental Brain Research,* Vol. 124, No. 1, (January 1999), pp. 1-7, ISSN 0014-4819

Dobyns, WB. (1996). Absence makes the search grow longer.*The American Jounal of Human Genetics,* Vol. 58, No. 1, (January 1996), pp. 7-16, ISSN 0002-9297

Doherty, D.; Tu, S. Schilmoeller, K. & Schilmoeller, G. (2006). Health-related issues in individuals with agenesis of the corpus callosum. *Child: Care, Health and Development,* Vol. 32, No. 3, (May 2006), PP. 333-42, ISSN 0305-1862

Doherty, P. & Wlash, F. (1996). CAM-FGF receptor interaction: a model for axonal growth. *Molecular and Cellular Neurosciences,* Vol. 8, No. 2-3, (August 1996), pp. 99–111, ISSN 1044-7431

Donmez, FY.; Yildirim, M. Erkek, N. Demir Karacan, C. & Coskun, M. (2009). Hippocampal abnormalities associated with various congenital malformations. *Child's Nervous System,* Vol. 25, No. 8, (August 2009), pp. 933-9, ISSN 0256-7040

Doron, KW. & Gazzaniga, MS. (2008). Neuroimaging techniques offer new perspectives on callosal transfer and interhemispheric communication. *Cortex,* Vol. 44, No. 8, (September 2008), pp. 1023-9, ISSN 0010-9452

Dunker, RO. & Harris, AB. (1976). Surgical anatomy of the proximal anterior cerebral artery. *Journal of Neurosurgery,* Vol. 44, No. 3, (March 1976), pp. 359-67, ISSN 0022-3085

Eliassen, JC.; Baynes, K. & Gazzaniga, MS. (1999). Direction information coordinated via the posterior third of the corpus callosum during bimanual movements. *Experimental Brain Research,* Vol. 128, No. 4, (October 1999), pp. 573-7, ISSN 0014-4819

Estivill-Torrús, G.; Vitalis, T. & Fernández-Llebrez, P. & Price, DJ. (2001). The transcription factor Pax6 is required for development of the diencephalic dorsal midline secretory radial glia that form the subcommissural organ. *Mechanism of Development,* Vol. 109, No. 2, (December 2001), pp. 215-24, ISSN 0925-4773

Evangelou, N.; Konz, D. Esiri, MM. Smith, S. Palace, J. & Matthews, PM. (2000). Regional axonal loss in the corpus callosum correlates with cerebral white matter lesion volume and distribution in multiple sclerosis. *Brain,* Vol. 123, No. (Pt 9), (September 2000), pp. 1845-9, ISSN 0006-8950

Eviatar, Z. & Zaidel, E. (1994). Letter matching within and between the disconnected hemispheres. *Brain and Cognition,* Vol. 25, No. 1, (May 1994), pp. 128-37, ISSN 0278-2626

Faber, R.; Azad, A. & Reinsvold, R. (2010). A case of the corpus callosum and alien hand syndrome from a discrete paracallosal lesion. *Neurocase,* Vol. 16, No. 14, (August 2010), pp. 281-5, ISSN 1355-4794

Fabri, M.; Del Pesce, M. Paggi, A. Polonara, G. Bartolini. M. Salvolini U. & Manzoni, T. (2005). Contribution of posterior corpus callosum to the interhemispheric transfer of tactile information. *Cognitive Brain Research.*Vol. 24, No. 1, (June 2003), pp. 73-80, ISSN 0926-6410

Fernandez-Llebrez, P.; Grondona, JM. Perez, J. Lopez-Aranda, MF. Estivill-Torrus, G. Llebrez-Zayas, PF. Soriano, E. Ramos, C. Lallemand, Y. Bach, A. & Robert, B. (2004). Msx1-deficient mice fail to form prosomere 1 derivatives, subcommissural organ, and posterior commissure and develop hydrocephalus. *Journal of Neuropathology and Experimental Neurology*, Vol. 63, No. 6, (June 2004), pp. 574–86, ISSN 0022-3069

Fischer, M.; Ryan, SB. & Dobyns, WB. (1992). Mechanisms of interhemispheric transfer and patterns of cognitive function in acallosal patients of normalintelligence. *Archives of Neurology*, Vol. 49, No. 3, (March 1992), pp. 271-77, ISSN 0003-9942

Francesco, P.; Maria-Edgarda, B. Giovanni, P. Dandolo, G. & Giulio, B. (2006). prenatal diagnosis of agenesis of corpus callosum: what is the neurodevelopmental outcome? Pediatrics international, Vol. 48, No. 3, (June 2006), pp. 298-304, ISSN 1328-8067

Fransen, E.; Van Camp, G. Vits, L. Willems, PJ. (1997). L1- associated diseases: clinical geneticists divide, molecular geneticists unite. *Human Molecular Genetics*, Vol. 6, No. 10, (1997), pp. 1625–32, ISSN 0964-6906

Friefeld, S.; MacGregor, D. Chuang, S. & Saint-Cyr, J. (2000). Comparative study of inter- and intrahemispheric somatosensory functions in children with partial and complete agenesis of the corpus callosum. *Developmental Medicine and Child Neurology*, Vol. 42, No. 12, (December 2000), pp. 831-8, ISSN 831-8.0012-1622

Funnell, MG.; Corballis, PM. & Gazzaniga, MS. (2000). Insights into the functional specificity of the human corpus callosum. *Brain*, Vol. 123, No. Pt5, (May 2008), pp. 920-6, ISSN 0006-8950

Gadea, M.; Marti-Bonmatí, L. Arana, E. Espert, R. Salvador, A. & Casanova, B. (2009) . Corpus callosum function in verbal dichotic listening: inferences from a longitudinal follow-up of relapsing-remitting multiple sclerosis patients. *Brain & Language*, Vol. 110, No. 2, (August 2009), pp. 101-5, ISSN 0093-934X

Gaffan, D.; Parker, A. & Easton, A. (2001). Dense amnesia in the monkey after transection of fornix, amygdala and anterior temporalstem, *Neuropsychologia*, Vol. 39, No. 1, (2001), pp. 51-70, ISSN 0028-3932

Gazzaniga, MS. (2000).Cerebral specialization and interhemispheric communication. Does the corpus callosum enable the human condition? *Brain*, Vol. 123, No. Pt 7, (July 2000), pp. 1293-326. ISSN 0006-8950

Ghi, T.; Carletti, A. Contro, E. Cera, E. Falco, P. Tagliavini, G. Michelacci, L. Tani, G. Youssef, A. Bonasoni, P. Rizzo, N. Pelusi, G. & Pilu, G.(2010). Prenatal diagnosis and outcome of partial agenesis and hypoplasia of the corpus callosum. *Ultrasound in Obstetrics & Gynecology*, Vol. 35, No. 1, (January 2010), pp. 35-41, ISSN 0960-7692

Giedd, JN.; Rumsey, JM. Castellanos, FX. Rajapakse, JC. Kaysen, D. Vaituzis, AC. Vauss, YC. Hamburger, SD. & Rapoport, JL. (1996). A quantitative MRI study of the corpus callosum in children and adolescents. *Brain Research Developmental Brain Research*, Vol. 91, No. 2, pp. 274-80, ISSN 0165-3806

Glass, HC.; Shaw, GM. Ma, C. & Sherr, EH. (2008). Agenesis of the corpus callosum in California 1983-2003: a population-based study. *American Journal of Medical Genetics. Part A*, Vol. 146A, No. 19, (October 2008), pp. 2495–500, ISSN 1552-4825

Griffiths, PD.; Batty, R. Reeves, MJ. & Connolly, DJ. (2009). Imaging the corpus callosum, septum pellucidum and fornixin children: normal anatomy and variations of normality. *Neuroradiology,* Vol. 51, No. 5, (May 2009), pp. 337-45, ISSN 0028-3940

Grogono, JL. (1968). Children with agenesis of the corpus callosum. *Developmental Medicine and Child Neurology,* Vol. 10, No. 5, (January 1968), pp. 613-16, ISSN 0012-1622

Guion-Almeida, ML; Richieri-Costa, A. Saavedra, D. & Cohen, MM Jr. (1996). Frontonasal dysplasia: analysis of 21 cases and literature review. *International Journal of Oral and Maxillofacial Surgery,* Vol. 25, No. 2, (April 1996), pp. 91-7, ISSN 0901-5027

Guénot, M. (1998). Interhemispheric transfer and agenesis of the corpus callosum. Capacities and limitations of the anterior commissure. *Neurochirurgie (Paris),* Vol. 44, No. suppl 1, (May 1998), pp. 113-115, ISSN 0028-3770

Gur, RC.; Packer, IK. Hungerbuhler, JP. Reivich, M. Obrist, WD. Amarnek, WS. & Sackeim, HA. (1980). Differences in the distribution of gray and white matter in human cerebral hemispheres. *Science,* Vol. 207, No. 4436, (March 1980), pp. 1226-8, ISSN 0036-8075

Hasan, KM.; Kamali, A. Kramer, LA. Papnicolaou, AC. Fletcher, JM. & Ewing-Cobbs, L. (2008). Diffusion tensor quantification of the human midsagittal corpus callosum subdivisions across the lifespan. *Brain Research,* Vol. 1227, (August 2008), pp. 52-67, ISSN 0006-8993

Hanna, RM.; Marsh, SE. Swistun, D. Al-Gazali, L. Zaki, MS. Abdel-Salam, GM. Al-Tawari, A. Bastaki, L. Kayserili, H. Rajab, A. Boglárka, B. Dietrich, RB. Dobyns, WB. Truwit, CL. Sattar, S. Chuang, NA. Sherr, EH. & Gleeson, JG. (2011). Distinguishing 3 classes of corpus callosal abnormalities in consanguineous families. *Neurology,* Vol. 76, No. 4, (January 2011), pp. 373-82, ISSN 0028-3878

Hendriks, YM.; Laan, L. Vielvoye, G. & Van Haeringen, A. (1999). Bilateral sensorioneural deafness, partial agenesis of the corpus callosum, and arachnoid cysts in two sisters. *American Journal of Medical Genetics,* Vol. 86, No. 2, (September 1999), pp. 183-6, ISSN 0148-7299

He, Q.; Duan, Y. Karsch, K. & Miles, J. (2010). Detecting corpus callosum abnormalities in autism based on anatomical landmarks. *Psychiatry Research,* Vol. 183, No. 2, (August 2010), pp. 126–32, ISSN 0165-1781

Hetts, SW.; Sherr, EH. Chao, S. Gobuty, S. & Barkovicg H. (2006). Anomalies of the corpus callosum: an MR analysis of the phenotypic spectrum of associated malformations. *A J R. American Journal of Roentgenology,* Vol. 187, No. 5, (November 2006), pp. 1343-48, ISSN 361-803X

Hofer, S. & Frahm, J. (2006). Topography of the human corpus callosum revisited—comprehensive fiber tractography using diffusion tensor magnetic resonance imaging. *Neuroimage,* Vol. 32, No. 3, (Setember 2006), pp. 989-94, ISSN 1053-8119

Hori, A. (1997). Anatomical variants of brain structure: confused spatial relationship of the fornix to the corpus callosum and anterior commissure. *Annals of Anatomy,* Vol. 179, No. 6, (December 1997), pp. 545-7, ISSN 0940-9602

Hoyo-Becerra, C.; López-Avalos, MD. Cifuentes, M. Visser, R. Fernández-Llebrez, P. & Grondona, JM. (2010). The subcommissural organ and the development of the posterior commissure in chick embryos Cell. *Cell and Tissue Research,* Vol. 339, No. 2, (February 2010), pp. 383–95, ISSN 0302-766X

Huppi, PS.; Maier, SE. Peled, S. Zientara, GP. Barnes, PD. Jolesz, FA. & Volpe, JJ. (1998). Microstructural development of human newborn cerebral white matter assessed in vivo by diffusion tensor magnetic resonance imaging. *Pediatric Research,* Vol. 44, No. 4, (October 1993), pp. 584-90, ISSN 0031-3998

Hutchinson, AD.; Mathias, JL. & Banich, MT. (2008). Corpus callosum morphology in children and adolescents with attention deficit hyperactivity disorder: a meta-analytic review. *Neuropsychology,* Vol. 22, No. 3, (May 2008), pp. 341–9, ISSN 0894-4105

Iacoboni, M.; Ptito, A. Weekes, NY. & Zaidel, E. (2000). Parallel visuomotor processing in the split brain: cortico- subcortical interactions. *Brain,* Vol. 123, No. Pt4, (April 2011), pp. 759-69, ISSN 0006-8950

Iwabuchi, SJ. & Kirk, IJ. (2009). Atypical interhemispheric communication in left-handed individuals. *Neuroreport,* Vol. 20, No. 2, (January 2009), pp. 166-9, ISSN 0959-4965

Jea, A.; Vachhrajani, S. Widjaja, E. Nilsson, D. Raybaud, C. Shroff, M. & Rutka, JT. (2008). Corpus callosotomy in children and the disconnection syndromes: a review. *Child's Nervous System,* Vol. 24, No. 6, (June 2008), pp. 685-92, ISSN 0256-7040

Jellinger, KA. (2002). Rhombencephalosynapsis. *Acta Neuropathologica,* Vol. 103, No. 3, (March 2002), pp. 305–6, ISSN 0001-6322

Jeret, JS.; Serur, D. Wisniewski, K. & Fisch, C. (1985). Frequency of agenesis of the corpus callosum in the developmentally disabled population as determined by computerized tomography. *Pediatric Neuroscience,* Vol. 12, No .2, pp. 101-3, ISSN 0255-7975

Johnston, TB. (1934). A note on the peduncle of the flocculus and the posterior medullary velum. *Journal of Anatatomy,* Vol. 68, No. 4, (July 1934), pp. 471–479, ISSN 0021-8782

Junle, Y.; Youmin, G. Yanjun, G. Mingyue, M. Qiujuan, Z. & Min, X. (2008). A MRI quantitative study of corpus callosum in normal adults. *Journal of Medical Colleges of PLA,* Vol. 23, pp. 346-51, ISSN 1000-1948

Just, MA.; Cherkassky, VL. Keller, TA. Kana, RK. & Minshew, NJ. (2007). Functional and anatomical cortical underconnectivity in autism: evidence from an fMRI study of an executive function task and corpus callosum morphometry.*Cerebral Cortex,* Vol. 17, No .4, (April 2007), pp. 951-61, ISSN 1047-3211

Kahilogullari, G.; Comert, A. Arslan, M. Esmer, AF. Tuccar, E. Elhan, A. Tubbs, RS. & Ugur HC. (2008). Callosal branches of the anterior cerebral artery: an anatomical report. *Clinical Anatomy,* Vol. 21, No. 5, (July 2008), pp. 383-8, ISSN 0897-3806

Kakou, M.; Velut, S. & Destrieux, C. (1998). Arterial and venous vascularization of the corpus callosum. *Neurochirurgie,* Vol. 44, No. (Suppl. 1), (May 1998), pp. 31-7, ISSN 0028-3770

Kamiguchi, H. & Lemmon, V. (1997). Neural cell adhesion molecule L1: signaling pathways and growth cone motility. *Journal of Nneuroscience Research,* Vol. 49, No. 1, (July 1997), pp. 1–8, ISSN 0360-4012

Kappeler, C.; Dhenain, M. Phan Dinh Tuy, F. Saillour, Y. Marty, S. Fallet-Bianco, C. Souville, I. Souil, E. Pinard, JM. Meyer, G. Encha-Razavi, F. Volk, A. Beldjord, C.

Chelly, J. & Francis, F. (2007). Magnetic resonance imaging and histological studies of corpus callosumand hippocampal abnormalities linked to doublecortin deficiency. *The Journal of Comparative Neurology,* Vol. 500, No. 2, (January 2007), pp. 239-54, ISSN 0021-9967

Katz, MJ.; Lasek, RJ. & Silver, J. (1983). Ontophyletics of the nervous system: development of the corpus callosum and evolution of axon tracts. *Proceedings of the National Academy of Sciences of the United States of America,* Vol. 80, No. 19, (Octeber 1983), pp. 5936-40, ISSN 0027-8424

Keene, MF. (1938). The connexions of the posterior commissure: a study of its development and myelination in the human foetus and young infant, of its phylogenetic development, and of degenerative changes resulting from certain experimental lesions. *Journal of Anatomy,* Vol. 72, No. Pt4, (Ju1y 1938), pp. 488-501, ISSN 0021-8782

Keshavan, MS.; Diwadkar, VA. DeBellis, M., Dick, E. Kotwal, R. Rosenberg, DR. Sweeney, JA. Minshew, N. & Pettegrew, JW. (2002). Development of the corpus callosum in childhood, adolescence and early adulthood. *Life Sciences.* Vol. 70, No. 16, (March 1993), pp. 1909-22, ISSN 0024-3205

Kier, L. & Truwit, CL. (1996).The normal and abnormal genu of the corpus callosum: an evolutionary, embryologic, anatomic and MR analysis.A J N R. *American Journal Neuroradiolology,* Vol. 17, No. 9, (October 1996), pp. 1631-41, ISSN 0195-6108

Kitamura, K.; Yanazawa, M. Sugiyama, N. Miura, H. Iizuka-Kogo, A. Kusaka, M. Omichi, K. Suzuki, R. Kato-Fukui, Y. Kamiirisa, K. Matsuo, M. Kamijo, S. Kasahara, M. Yoshioka, H. Ogata, T. Fukuda, T. Kondo, I. Kato, M. Dobyns, WB. Yokoyama, M. & Morohashi, K. (2002). Mutation in ARX causes abnormal development of brain and testes in mice and X-linked lissencephaly with abnormal genitalia in humans. *Nature Genetics,* Vol. 32, No. 3, (November 2002), pp. 359-69, ISSN 1061-4036

Koenig, SB.; Naidich, TP. & Lissner, G. (1982). The morning glory syndrome associated with sphenoidal cephalocele. *Ophtalmology,* Vol. 89, No. 12, (December 1982), pp. 1368–73, ISSN 0161-6420

Kuroki, N.; Kubicki, M. Nestor, PG. Salisbury, DF. Park, HJ. Levitt, JJ. Woolston, S. Frumin, M. Niznikiewicz, M. Westin, CF. Maier, SE. McCarley, RW. & Shenton, ME. (2006). Fornix integrity and hippocampal volume in male schizophrenic patients. *Biological Psychiatry,* Vol. 60, No. 1, (July 2006), pp. 22-31, ISSN 0006-3223

Lamantia, A, & Rakic P. (1990). Cytological and quantitative characteristics of four cerebral commissures in the rhesus monkey. *The Journal of Comparative Neurology,* Vol. 291, No. 4, (January 1990), pp. 520-37, ISSN 0021-9967

Lee, SK.; Mori, S. Kim DJ. Kim, SY. Kim, SY. & Kim, DI. (2004). Diffusion tensor MR imaging visualizes the altered hemispheric fiber connection in callosal dysgenesis. *A J N R. American Journal of Neuroradiolog,* Vol. 25, No. 1, (Janury 2004), pp. 25-8, ISSN 0195-6108

Lent, R.; Uziel, D. Baudrimont, M. & Fallet, C. (2005). Cellular and molecular tunnels surrounding the forebrain commissures of human fetuses. *The Journal of Comparative Neurology,* Vol. 483, No. 4, (March 1983), pp. 375-82, ISSN 0021-9967

Lockwood, CJ.; Ghidini, A. Aggarwal, R. & Hobbins, JC. (1988). Antenatal diagnosis of partial agenesis of the corpus callosum: A benign cause of entriculomegaly. *American Journal of Obstetrics and Gynecology, Vol.* 159, No. 1, (July 1988), pp. 184-6, ISSN 0002-9378

Loser, JD. & Alvord, EC. (1968). Agenesis of the corpus callosum. *Brain,* Vol. 91, No. 3, (September 1968), pp. 553-70, ISSN 0006-8950

Louvi, A. & Wassef, M. (2000). Ectopic engrailed 1 expression in the dorsal midline causes cell death, abnormal differentiation of circumventricular organs and errors in axonal pathfinding. *Development*, Vol. 127, No. 18, (September 2000), pp. 4061–71, ISSN 0950-1991

Luders, E.; Thompson, PM. & Toga, AW. (2010). The development of the corpus callosum in the healthy human brain. *Journal of Neuroscience,* Vol. 30, pp. 10985-90, ISSN 0270-6474

Malinger, G. & Zakut, H. (1993). The corpus callosum: normal fetal development as shown by transvaginal sonography. *A J R. American Journal of Roentgenology.* Vol. 161, No. 5, (November 1968), pp. 1041-3, ISSN 0361-803X

Manson, SC.; Palace, J. Frank, JA. & Matthews, PM. (2006). Loss of interhemispheric inhibition in patients with multiple sclerosis is related to corpus callosum atrophy. *Experimental Brain Research,* Vol. 174, No. 4, (October 2000), pp. 728–33, ISSN 0014-4819

Marszal, E.; Jamroz, E. Pilch, J. Kluczewska, E. Jablecka-Deja, H. & Krawczyk, R. (2000). Agenesis of corpus callosum: clinical description and etiology. *Journal of Child Neurology,* Vol. 15, No. 6, (January 2000), pp. 401-5, ISSN 0883-0738

Marzi, CA.; Smania, N. Martini, MC. Gambina, G. Tomelleri, G. Palamara A. Alessandrini, F. & Prior, M. (1996). Implicit redundant-targets effect in visual extinction. *Neuropsychologia,* Vol. 34, No. 1, (January 2009), pp. 9-22, ISSN 0028-3932

Meibach, RC. & Siegel, A. (1977). Efferent connections of the hippocampal formation in the rat. *Brain Research,* Vol. 124, No. 2, (March 1977), pp. 197-224, ISSN 0006-8993

Meizner, I.; Barki, Y. & Hertzanu, Y. (1987). Prenatal sonographic diagnosis of agenesis of corpus callosum. *Journal of Clinical Ultrasound,* Vol. 15, No. 4, (May 1987), pp. 262-4, ISSN 0091-2751

Mesulam, MM.; Mufson, EJ. Levey, AI. & Wainer, BH. (1983). Cholinergic innervation of cortex by the basal forebrain: cytochemistry and cortical connections of the septal area, diagonal band nuclei, nucleus basalis (substantia innominata), and hypothalamus in the rhesus monkey.*The Journal of Comparative Neurology,* Vol. 214, No. 2, (February 1983), pp. 170-97, ISSN 0021-9967

Meyer, BU. & Roricht, S. (1998). In vivo visualization of the longitudinal callosal fascicle (Probst's bundle) and other abnormalities in an acallosal brain. *Journal of Neurology, Neurosurgery, and Psychiatry,* Vol. 64, No. 1, (January 1998), pp. 138-9, ISSN 022-3050

Mitchell, TN.; Stevens, JM. Free, SL. Sander, JW. Shorvon, SD. & Sisodiya, SM. (2002). Anterior commissure absence without callosal agenesis: a new brain malformation. *Neurology,* Vol. 58, No. 8, (April 2002), pp. 1297-9, ISSN 0028-3878

Mikkelsen, JD.; Hay-Schmidt, A. & Larsen, PJ. (1997). Central innervation of the rat ependyma and subcommissural organ with special reference to ascending serotoninergic projections from the raphe nuclei. *The Journal of Comparative Neurology*, Vol. 384, No. 4, (August 1997), pp. 556-68, ISSN 0021-9967

Miyata, H.; Chute, DJ. Fink, J. Villablanca, P. & Vinters, HV. (2004). Lissencephaly with agenesis of corpus callosum and rudimentary dysplastic cerebellum: a subtype of lissencephaly with cerebellar hypoplasia. *Acta Neuropathologica*, Vol. 107, No. 1, (January 2004), pp. 69-81, ISSN 0001-6322

Moes, P.; Schilmoeller, K. Schilmoeller, G. (2009). Physical, motor, sensory and developmental features associated with agenesis of the corpus callosum. *Child: care, Health and Development*, Vol. 35, No. 5, (September 2009), ISSN 0305-1862

Moes, PE.; Brown, WS. & Minnema, MT. (2007). Individual differences in interhemispheric transfer time (IHTT) as measured by event related potentials. *Neuropsychologia*, Vol. 45, No. 11, (Jun 2007), pp. 2626-30, ISSN 0028-3932

Moldrich, RX.; Gobius, I. Pollak, T. Zhang, J. Ren, T. Brown, L. Mori, S. De Juan Romero, C. Britanova, O. Tarabykin, V. & Richards, LJ. (2010). Molecular regulation of the developing commissural plate. *Journal of Comparative Neurology and Psychology*, (September 2010), Vol. 518, No. 18, pp. 3645-61, ISSN 0021-9967

Mooshagian, E.; Iacoboni, M. & Zaidel, E. (2009). Spatial attention and interhemispheric visuomotor integration in the absence of the corpus callosum. *Neuropsychologia*, Vol. 47, No. 3, (February 2009), pp. 933-7, ISSN 0028-3932

Mordefroid, M.; Grabar, S. Andre, Ch. Merzoug, V. Moutard, ML. & Adamsbaum, C. (2004). Ag´en´esie partielle du corps calleux de l'enfant. *Journal de Radiologie*, Vol. 85, No. 11, (November 2004), pp. 1915-26, ISSN 0221-0363

Mordkoff, JT. & Yantis, S. (1993). Dividing attention between color and shape: Evidence of coactivation. *Percept Psychophys*. Vol. 53, No. 4, (April 1993), pp. 357-66, ISSN 0031-5117

Moutard, ML.; Kieffer, V. Feingold, JV. Kieffer, FV. Lewin, FV. Adamsbaum, CV. G´elot, AV. Campistol I Plana, JV. Van Bogaert, PV. Andr´e, MV. & Ponsot, G. (2003). Agenesis of corpus callosum: prenatal diagnosis and prognosis. *Child's Nervous System*, Vol. 19, No. (7-8), (August 2003), pp. 471-6, ISSN 0256-7040

Müller-Oehring, EM.; Schulte, T. Kasten, E. Poggel, DA. Müller, I. Wüstenberg, T. & Sabel, BA. (2009). Parallel interhemispheric processing in hemineglect: relation to visual field defects. *Neuropsychologia*, Vol. 47, No. 12, (October 2009), pp. 2397-408, ISSN 0028-3932

Nowicka, A.; Grabowska, A. & Fersten, E. (1996). Interhemispheric transmission of information and functional asymmetry of the human brain. *Neuropsychologia*, Vol. 34, No. 2, (February 2009), pp. 147-51, ISSN 0028-3932

Ozaki, HS.; Murakami, TH. Toyoshima, T. & Shimada, M. (1987). The fibers which leave the Probst's longitudinal bundle seen in the brain of an acallosal mouse: a study with the horseradish peroxidase technique. *Brain Research*, Vol. 400, No. 2, (January 1987), pp. 239-46, ISSN 1385-299X

Ozdogmus, O.; Cavdar, S. Ersoy, Y. Ercan, F. & Uzun, I. (2009). A preliminary study, using electron and light-microscopic methods, of axon numbers in the fornix in autopsies of patients with temporal lobe epilepsy. *Anatomical Science International*, Vol. 84, No. (1-2), (April 2009), pp. 2-6, ISSN 1447-6959

Panigrahy, A.; Barnes, PD. Robertson, RL. Sleeper, LA. & Sayre, JW. (2005). Quantitative analysis of the corpus callosum in children with cerebral palsy and developmental delay: correlation with cerebral white matter volume. *Pediatric Radiology,* Vol. 35, No. 12, (December 2005), pp. 1199-207, ISSN 0301-0449

Párraga, RG.; Ribas, GC. Andrade, SE. & De Oliveira, E. (2011). Microsurgical anatomy of the posterior cerebral artery in three-dimensional images. *World Neurosurgery,* vol. 75, No. 2, (February 2011), pp. 233-57, ISSN 1878-8750

Patel, M D.; Toussaint, N. Charles-Edwards, GD. Lin, JP. & Batchelor, PG. (2010). Distribution and fibre field similarity mapping of the human anterior commissure fibres by diffusion tensor imaging. *Magnetic Resonance Materials in Physics, Biology, and Medicine,* Vol. 53, No. 5-6, (December 2005), pp. 399-408, ISSN 0968-5243

Paul, LK.; Brown, WS. Adolphs, R. Tyszka, JM. Richards, LJ. Mukherjee, P. & Sherr, EH. (2007). Agenesis of the corpus callosum: genetic, developmental and functional aspects of connectivity. *Nature Reviews Neuroscience,* Vol. 8, No. 4, (April 2007), pp. 287–299, ISSN 1471-003X

Paupe, A.; Bidat, L. Sonigo, P. Lenclen, R. Molho, M. & Ville, Y. (2002). Prenatal diagnosis of hypoplasia of the corpus callosum in association with non-ketotic hyperglycinemia. *Ultrasound in Obstetrics & Gynecology,* Vol. 20, No. 6, (December 2002), pp. 616-19, ISSN 0960-7692

Peltier, J.; Verclytte, S. Delmaire, C. Pruvo, JP. Havet, E. & LE Gars, D. (2011). Microsurgical anatomy of the anterior commissure: correlations with diffusion tensor imaging fiber-tracking and clinical relevance. *Neurosurgery*. [Epub a head of print], (May 2011), ISSN 0148-396X

Penny, SM. (2006). Agenesis of the corpus callosum: neonatal sonographic detection. *Radiologic Technology,* Vol. 78, No. 1, (Sep-October 2006), pp. 14-8, ISSN 0033-8397

Pierson, TM.; Zimmerman, RA. Tennekoon, GI. & Bönnemann, CG. (2008). Mega-corpus callosum, polymicrogyria, and psychomotor retardation: confirmation of a syndromic entity. *Neuropediatrics,* Vol. 39, No. 2, (April 2009), pp. 123-7, ISSN 0174-304X

Pilu, G.; Sandri, F. Perolo, A. Pittalis, MC. Grisolia, G. Cocchi, G. Foschini, MP. Salvioli, GP. & Bovicelli, L. (1930). Sonography of fetal agenesis of the corpus callosum: a survey of 35 cases.*Ultrasound in Obstetrics & Gynecology,* Vol. 3, No. 5, (September 1993), pp. 318–29, ISSN 0960-7692

Prasad, AN.; Bunzeluk, K. Prasad, C. Chodirker, BN. Magnus, KG. & Greenberg, CR. (2007). Agenesis of the corpus callosum and cerebral anomalies in inborn errors of metabolism. *Congenital Anomalies,* Vol. 47, No. 4, (December 2007), pp. 125-35, ISSN 0914-3505

Probst, FP. (1973). Congenital defects of the corpus callosum.Morphology and encephalographic appearances. *Acta Radiologica. Supplementum,* Vol. 331, pp. 1-152, ISSN 0365-5954

Pujol, J.; Vendrell, P. Junqué, C. Martí-Vilata, JL. & Capdevila, A. (1993). When does human brain development end? evidence of corpus callosum growth up to adulthood. *Annals of Neurology,* Vol. 34, No. 1, (July 1993), pp. 71-5, ISSN 0364-5134

Rakic, P. & Yakovlev, PI. (1968). Development of the corpus callosum and cavum septi in man. *The Journal of Comparative Neurology*, Vol. 132, No. 1, (January 1968), pp. 45-72, ISSN 0021-9967

Ramos, C.; Fernandez-Llebrez, P. Bach, A. Robert, B. & Soriano, E. (2004). Msx1 disruption leads to diencephalon defects and hydrocephalus. *Developmental Dynamics*, Vol. 230, No. 3, (July 2004), pp. 446–60, ISSN 1058-8388

Raybaud, C. (2010). The corpus callosum, the other great forebrain commissures, and the septum pellucidum: anatomy, development, and malformation. *Neuroradiology*, Vol. 52, No. 6, (June 2010), pp. 447-77, ISSN 0028-3940

Raybaud, C. & Di Rocco, C. (2007). Brain malformation in syndromic craniosynostoses, a primary disorder of white matter: a review.*Child's Nervous System*, Vol. 23, No. 12, (December 2007), pp.1379–88, ISSN 0256-7040

Raybaud, C. & Girard, N. (1998). Etude anatomique par IRM des agénésiesand dysplasies commissurales télencéphaliques. Corrélationscliniques et interprétation morphogénétique. *Neurochirurgie*, Vol. 44, No. Suppl. 1, (May 1998), pp. 38-60, ISSN 0028-3770

Ren, T.; Anderson, A. Shen, WB. Huang, H. Plachez, C. Zhang, J. Mori, S. Kinsman, SL. & Richards, LJ. (2006). Imaging, anatomical, and molecular analysis of callosal formation in the developing human fetal brain. *The Anatomical Record, Part A, Discoveries in Molecular, Cellular, and Evolutionary Biology*, Vol. 288, No .2, (February 2006), pp. 191-204, ISSN 1552-4884

Reuter-Lorenz, PA.; Nozawa, G. Gazzaniga, MS. & Hughes, HC. (1995). Fate of neglected targets: a chronometric analysis of redundant target effects in the bisected brain. *Journal of Experimental Psychology, Human Perception and Performance*, Vol. 21, No. 2, (April 1995), pp. 211-30, ISSN 0096-1523

Richards, LJ.; Plachez, C. & Ren, T. (2004). Mechanisms regulating the development of the corpus callosum and its agenesis in mouse and human.*Clinical Genetics*, Vol. 66, No. 4, (October 2009), pp. 276-89, ISSN 0009-9163

Ridley, RM.; Baker, HF. Harder, JA. & Pearson, C. (1996). Effects of lesions of different parts of the septo-hippocampal system in primates on learning and retention of information acquired before or after surgery. *Brain Research Bulletin*, Vol. 40, No. 1, pp. 21-32, ISSN 0361-9230

Rosenthal-Wisskirchen, E. (1967). Pathological-anatomical and clinical observations of the corpus callosum absence with particular consideration of the longitudinal bundle of the corpus callosum.*Dtsch Z Nervenheilkd*, Vol. 192, pp. 1-45, ISSN 0367-004X

Roser, M. & Corballis, MC. (2003). Interhemispheric neural summation in the split brain: effects of stimulus colour and task. *Neuropsychologia*, Vol. 41, No. 7, pp. 830-46, ISSN 0028-3932

Roser, M. & Corballis, MC. (2002). Interhemispheric neural summation in the split brain with symmetrical and asymmetrical displays. *Neuropsychologia*, Vol. 40, No. 8, pp. 1300-12, ISSN 0028-3932

Rugg, MD. & Beaumont, JG. (1978). Interhemispheric asymmetries in the visual evoked response: effects of stimulus lateralisation and task. *Biological Psychology*, Vol. 6, No. 4, (January 1978), pp. 283-92, ISSN 0301-0511

Sarnat, HB. (1992). Agenesis of the Corpus Callosum in Cerebral dysgenesis, in Sarnat HB (ed): *Cerebral Dysgenesis: Embryology and Clinical Expression.* pp. 215-44, New York, Oxford University Press, *ISBN* 9780199238323

Sarnat, HB. (2007). Embryology and malformations of the forebrain commissures. *Handbook of clinical neurology,* Vol. 87, (September 2007), pp. 67-87, ISSN 0072-9752

Saunders, RC. & Aggleton, JP. (2007). Origin and topography of fibers contributing to the fornix in macaque monkeys. *Hippocampus,* Vol. 17, No. 5, pp. 396-411, ISSN 1050-9631

Schell-Apacik, CC.; Wagner, K. Bihler, M. Ertl-Wagner, B. Heinrich, U. Klopocki, E. Kalscheuer, VM. Muenke, M. & Von Voss, H. (2008). Agenesis and dysgenesis of the corpus callosum: clinical, genetic and neuroimaging findings in a series of 41 patients. *American Journal of Medical Genetics,* Vol. 146A, No. 19, (October 2008). pp. 2501-11, ISSN 1552-4825

Schilmoeller, G. & Schilmoeller, K. (2000). Filling a void: facilitating family support through networking for children with a rare disorder. *Family Science Review,* Vol. 13, No. 3-4, (December 2000), pp. 224–33, ISSN 1084-0524

Schmid, RS. & Maness, PF. (2008). L1 and NCAM adhesion molecules as signaling co-receptors in neuronal migration and process outgrowth. *Current Opinion in Neurobiology,* Vol. 18, No. 3, (June 2008), pp. 245–250, ISSN 0959-4388

Schulte, T.; Chen, SH. Müller-Oehring, EM. Adalsteinsson, E. Pfefferbaum, A. & Sullivan, EV. (2006). fMRI evidence for individual differences in premotor modulation of extrastriatal visual-perceptual processing of redundant targets. *Neuroimage,* Vol. 30, No. 3, (April 2006), pp. 973-82, ISSN 1053-8119

Schulte, T. & Müller-Oehring, EM. (2010). Contribution of callosal connections to the interhemispheric integration of visuo motor and cognitive processes. *Neuropsychology Review,* Vol. 20, No. 2, (June 2010), pp. 174-90, ISSN 1040-7308

Selden, NR.; Gitelman, DR. Salamon-Murayama, N. Parrish, TB. & Mesulam, MM. (1998). Trajectories of cholinergic pathways within the cerebral hemispheres of the human brain. *Brain,* Vol. 121, No. (P12), (December 1998), pp. 2249-57, ISSN 0006-8950

Sener, RN. (1993). Septo-optic dysplasia associated with total absence of the corpus callosum: MR and CT features. *European Radiology,* Vol. 3, pp. 551–553, ISSN 0938-7994

Serur, D.; Jeret, JS. & Wisniewski, K. (1988). Agenesis of the corpus callosum: clinical, neuroradiological and cytogenetic studies. *Neuropediatrics,* Vol. 19, No. 2, (May 1988), pp. 87-91, ISSN 0174-304X

Sidman, RL. & Rakic P. (1982). Development of the human central nervous system. In: Haymaker W. Adams RD (eds) *Histology and Histopathology of the Nervous System,* pp. 40-2, Charles C Thomas, Springfield, IL.

Silver, J.; Lorenz, SE. Washten, D. & Coughlim, J. (1982). Axonal guidance during development of the great cerebral commissures: description and experimental studies in vivo on the role of preformed glial pathways. *The Journal of Comparative Neurology.* Vol. 210, No. 1, (September 1998), pp. 10-29, ISSN 0021-9967

Sisodiya, SM.; Free, SL. Williamson, KA. Mitchell, TN. Willis, C. Stevens, JM. Kendall, BE. Shorvon, SD. Hanson, IM. Moore, AT. & Van Heyningen, V. (2001). *PAX6* haploin sufficiency causes cerebral malformation and olfactory dysfunction in humans. *Nature genetics,* Vol. 28, No. 3, (July 2001), pp. 214-16, ISSN 1061-4036

Standring, S. (2005). *Ventricular system and cerebrospinal fluid* (chapter 16). In: Gray's Anatomy 39th edition. S Standring (Ed) Elsevier, PP.287–294, Churchill Livingstone, ISBN 04430 066760

Supprian, T. & Hofmann, E. (1997). The fornix of the human brain: evidence of left/right asymmetry on axial MRI scans.*Surgical and Radiologic Anatomy,* Vol. 19, No. 2, pp. 105-9, ISSN 0930-1038

Swanson. LW. & Cowan, WM. (1977). An autoradiographic study of the organization of the efferent connections of the hippocampal formation in the rat. *The Journal of Comparative Neurology,* Vol. 172, (March 1977), pp. 49-84, ISSN 0021-9967

Swayze, VW2nd.; Andreasen, NC. Ehrhardt, JC. Yuh, WT. Alliger, RJ. & Cohen, GA. (1990). Developmental abnormalities of the corpus callosum in schizophrenia. *Archives of Neurology,* Vol. 47, No. 7, (July 1990), pp. 805-8, ISSN 0003-9942

Szabó, N.; Gergev, G. Kóbor, J. Bereg, E. Túri, S. & Sztriha, L. (2011). Corpus callosum anomalies: birth prevalence and clinical spectrum in Hungary. *Pediatric Neurology,* Vol. 44, No. 6, (June 2011), pp. 420-6, ISSN 0887-8994

Szriha, L. (2005). Spectrum of corpus callosum agenesis. *Pediatric Neurology,* Vol. 32, No. 2, (February 2005), pp. 94-101, ISSN 0887-8994

Sztriha, L.; Frossard, P. Hofstra, R. Verlind, E. & Nork, M. (2000). Novel missense mutation in the LI gene in a child with corpus callosum agenesis, retardation, adducted thumbs, apastic paraparesis, and hydrocephalus. *Journal of Child Neurology,* Vol. 15, No. 4, (April 2000), pp. 239-43, ISSN 0883-0738

Tang, PH.; Bartha, AI. Norton, ME. Barkovich, AJ. Sherr, EH. & Glenn, OA. (2009). Agenesis of the corpus callosum: an MR imaging analysis of associated abnormalities in the fetus. *A J N R. American Journal of Neuroradiology,* Vol. 30, No. 2, (February 2009), pp. 257-63, ISSN 0195-6108

Tepper, R.; Zalel, Y. Gaon, E. Fejgin, M. & Beyth, Y. (1996). Antenatal ultrasonographic findings differentiating complete from partial agenesis of the corpus callosum. *American Journal of Obstetrics and Gynecology,* Vol. 174, No. 3, (March 1996), pp. 877-8, ISSN 0002-9378

Tomasch, J. (1954). Size, distribution and number of fibres in the human corpus callosum. *The Anatomical Record,* Vol. 119, No. 1, (May 1954), pp. 119-35, ISSN 0003-276X

Truwit, CL. & Barkovich, AJ. (1990). Pathogenesis of intracranial lipomas: an MR study in 42 patients. *A J N R. American Journal of Neuroradiology,* Vol. 11, No. 4, (July-August 1990), pp. 665–74, ISSN 0195-6108

Turatto, M.; Mazza, V. Savazzi, S. & Marzi, CA. (2004). The role of the magnocellular and parvocellular systems in the redundant target effect. *Experimental Brain Rresearch,* Vol. 158, No. 2, (September 2004), pp. 141-50, ISSN 0014-4819

Türe, U.; Yaşargil, MG. & Krisht, AF. (1996). The arteries of the corpus callosum: a microsurgical anatomic study. *Neurosurgery,* Vol. 39, No. 6, (December 1996), pp. 1075-84, discussion 1084-5, ISSN 0148-396X

Uddin, LQ . (2011). Brain connectivity and the self: the case of cerebral disconnection. *Consciousness and Cognition,* Vol. 20, No. 1, (March 2011), pp. 94-8, ISSN 1053-8100

Uddin, LQ.; Mooshagian, E. Zaidel, E. Scheres, A. Margulies, DS. Kelly AM. Shehzad, Z. Adelstein, JS. Castellanos, FX. Biswal, BB. & Milham, MP. (2008). Residual functional connectivity in the split-brain revealed with resting-state functional MRI. *Neuroreport*, Vol. 19, No. 7, (May 2008), pp. 703-9, ISSN 0959-4965

Utsunomiya, H.; Ogasawara, T. Hayashi, T. Hashimoto, T. & Okazaki, M. (1997). Dysgenesis of the corpus callosum and associated telencephalic anomalies: MRI. *Neuroradiology,* Vol. 39, No. 4, (April 1997), pp. 302-10, ISSN 0028-3940

Vachha, B.; Adams, RC. & Rollins, NK. (2006). Limbic tract anomalies in pediatric myelomeningocele and Chiari II malformation: anatomic correlations with memory and learning--initial investigation. *Radiology,* Vol. 240, No. 1, (July 2006), pp. 194-202, ISSN 0033-8419

Van der Knaap, LJ. & van der Ham, IJ. (2011). How does the corpus callosum mediate interhemispheric transfer? a review. *Behavioural Brain Research,* Vol. 223, No.1, (September 2011), pp. 211-21, ISSN 0166-4328

Van Essen, DC. (1997). A tension-based theory of morphogenesis and compact wiring in the central nervous system. *Nature,* Vol. 385, No. 6614, (January 1997), pp. 313-18, ISSN 0028-0836

Vasung, L.; Huang, H. Jovanov- Milošević, N. Pletikos, M. Mori, S. & Kostović, I. (2010). Development of axonal pathways in the human fetal fronto-limbic brain: histochemical characterization and diffusion tensor imaging. *Journal of Anatomy,* Vol. 217, No. 4, (October2010), pp. 400-17, ISSN 0021-8782

Velut, S.; Destrieux, C. & Kakou, M. (1998). Morphologic anatomy of the corpus callosum. *Neurochirurgie* (Paris), Vol. 44, No. (Suppl.1), pp. 17-30, ISSN 0028-3770

Vermeulen, RJ.; Wilke, M. Horber, V. & Krägeloh-Mann, I. (2010). Microcephaly with simplified gyral pattern. MRI classification. *Neurology,* Vol. 74, No. 5, (February 2010), pp. 386-91, ISSN 0028-3878

Vidal, CN.; Nicolson, R. DeVito, TJ. Hayashi, KM. Geaga, JA. Drost, DJ. Williamson, PC. Rajakumar, N. Sui, Y. Dutton, RA. Toga, AW. & Thompson, PM . (2006). Mapping corpus callosum deficits in autism: an index of aberrant cortical connectivity. *Biological Psychiatry,* Vol. 60, No. 3, (August 2006), pp. 218-25, ISSN 0006-3223

Volpe, P.; Paladini, D. Resta, M. Stanziano, A. Salvatore, M. Quarantelli, M. De Robertis, V. Buonadonna, AL. Caruso, G. & Gentile, M. (2006). Characteristics, associations and outcome of partial agenesis of the corpus callosum in the fetus. *Ultrasound in Obstetrics & Gynecology,* Vol. 27, No. 5, (May 2006), pp. 509-15, ISSN 0960-7692

Walterfang, M.; Malhi, GS. Wood, AG. Reutens, DC. Chen, J. Barton, S. Yücel, M. Velakoulis, D. & Pantelis, C. (2009b). Corpus callosum size and shape in established bipolar affective disorder. *The Australian and New Zealand Journal of Psychiatry,* Vol. 43, No. 9, (September 2009b), pp. 838-45, ISSN 0004-8674

Walterfang, M.; Wood, AG. Reutens, DC. Wood, SJ. Chen, J. Velakoulis, D. McGorry, PD. & Pantelis, C. (2008). Morphology of the corpus callosum at different stages of schizophrenia: cross-sectional study in first-episode and chronic illness. *The British Journal of Psychiatry,* Vol. 192, No. 9, (July 2008), pp. 429-34, ISSN 0007-1250

Walterfang, M.; Wood, AG. Reutens, DC. Wood, SJ.Chen, J. Velakoulis, D. McGorry, PD. & Pantelis, C. (2009a). Corpus callosum size and shape in first-episode affective and schizophrenia spectrum psychosis. *Psychiatry Research*, Vol. 173, No. 1, (July 2009a), pp. 77-82, ISSN 0165-1781

Walterfang, M.; Wood, AG. Barton, S.Velakoulis, D. Chen, J. Reutens, DC. Kempton, MJ. Haldane, M. Pantelis, C. & Frangou, S. (2009c). Corpus callosum size and shape alterations in individuals with bipolar disorder and their first-degree relatives. *Progress in Neuro-Psychopharmacology& Biological Psychiatry*, Vol. 33, No. 6, (August 2009c), pp. 1050-7, ISSN 0278-5846

Ware, M. & Schubert, FR. (2011). Development of the early axon scaffold in the rostral brain of the chick embryo. *Journal of Anatomy*, Vol. 219, No .2, (August 2011), pp. 203-16, ISSN 0021-8782

Warlop, NP.; Achten, E. Debruyne, J. & Vingerhoets, G. (2008). Diffusion weighted callosal integrity reflects interhemispheric communication efficiency in multiple sclerosis. *Neuropsychologia*.Vol. 46, No. 8, pp. 2258-64, ISSN 0028-3932

Weller, S. & Gärtner, J. (2001). Genetic and clinical aspects of Xlinked hydrocephalus (L1 disease): mutations in the L1CAM gene. *Human Mutation*, Vol. 18, No. 1, pp. 1-12, ISSN 1059-7794

Widjaja, EE.; Nilsson, D. Blaser, S. & Raybaud, C. (2008). White matter abnormalities in children with idiopathic developmental delay. *Acta Radiologica*, Vol. 49, No. 5, (June 2008), pp. 589-95, ISSN 0284-1851.

Williams, PL.; Warwick, R. Dyson, M. & Bannister, LH. (1989). Neurology, *Gray's anatomy*, 37th ed, pp.1039, Churchill Livingstone, New York, ISBN 0443 041776.U S A.

Witelson, SF. (1989). Hand and sex differences in the isthmus and genu of the human corpus callosum. A postmortem morphological study, Brain, Vol. 112, No. pt 3, (June 1989), pp. 799-835, ISSN 0006-8950

Wolfram-Gabel, R.; Maillot, C. & Koritke, JG. (1989). Arterial vascularization of the corpus callosum in man. *Archives d'Anatomie, d'Histologie et d'Embryologie Normales et Expérimentales*, Vol. 72, pp. 43-55, ISSN 0249-5554

Wu, E.; Vargevik, K. & Slavotinek, AM. (2007). Subtypes of frontonasal dysplasia are useful in determining clinical prognosis. *American Journal of Medical Genetics. Part A*, Vol. 143A, No. 24, (December 2007), pp. 3069–78, ISSN 1552-4825.

Yakovlev, PI. & Lecours, AR. (1967). *The myelogenetic cycles of regional maturation of the brain*, A. Minkowski, Editor, Regional development of the brain in early life, pp. 3–70, Blackwell, Oxford.

Yamauchi, H .; Fukuyamam, H. Nagahama, Y. Katsumi, Y. Dong, Y. Konishi. & Kimura, J. (1997). Atrophy of the corpus callosum, cognitive impairment, and cortical hypometabolism in progressive supranuclear palsy. *Annals of Neurology*, Vol. 41, No. 5, (May 1997), pp. 606-14, ISSN 0364-5134

Yamasaki, M.; Thompson, P. & Lemmon, V. (1997). CRASH syndrome mutations in L1CAM correlate with severity of the disease. *Neuropediatrics*, Vol. 28, No. 3, (June 1997), pp. 175-6, ISSN 0174-304X

Yousefi, B. & Kokhei, P. (2009). The Probst bundle associated with anomalies of the precommissural separated fornix in an acallosal brain. *European Journal of Anatomy*, Vol. 13, No. 1, (May 2009), pp. 37-42, ISSN 1136-4890

Zaidel, E. (1994). Interhemispheric transfer in the split brain: long-term status following complete cerebral commissurotomy. In: Davidson RH, Hugdahl K, editors. *Brain Asymmetry*, pp. 491–532, ISBN 0- 262-04144-8 (hardcover). Cambridge: MIT Press.

Zaidel, DW.; Esiri, MM. & Oxbury, JM. (1994). Sex-related asymmetries in the morphology of the left and right hippocampi? a follow-up study on epileptic patients. *Journal of Neurology*, (October 1994); Vol. 241, No. 10, pp. 620-23. , ISSN 0166-4328. 0340-5354.

Zaidal, E.; Iacoboni, M. Zaidel, D. & Bogen, J. (2003). The callosal syndromes.In: Heilman KM, Valenstein E, editors. *Clinical Neuropsychology*. Oxford University Press, pp. 347–403, New York. U S A

Zahajszky, J.; Dickey, CC. McCarley, RW. Fischer, IA. Nestor, P. Kikinis, R. & Shenton, ME. (2001). A quantitative MR measure of the fornix in schizophrenia. *Schizophrenia*, Vol. 47, No. 1, (January 2001), pp. 87-97, ISSN 0920-9964

5

Advances in Neuromodulation: The Orbitofrontal-Striatal Model Of, and Deep Brain Stimulation In, Obsessive-Compulsive Disorder

Robert K. McClure
Department of Psychiatry, School of Medicine, University of North Carolina, Chapel Hill, USA

The chains of habit are too weak to be felt
until they are too strong to be broken
Samuel Johnson

1. Introduction

Obsessive-compulsive disorder is a common chronic neuropsychiatric illness. Estimates of the lifetime prevalence rate of obsessive-compulsive disorder will vary depending on the methods used to gather the epidemiological data and the diagnostic criteria used to define obsessive-compulsive disorder. Estimates of the lifetime prevalence of obsessive-compulsive disorder have been reported to be to be between 1.9%-3.3%, when obsessive compulsive disorder was defined without DSM-III criteria. A slightly lower prevalence of obsessive-compulsive disorder was reported to be between 1.2%-2.4%, when obsessive-compulsive disorder was defined using DSM-III criteria[123]. These estimates of the prevalence of obsessive-compulsive disorder are likely to be accurate because they are based on: a.) population-based data; b.) that was gathered from five US communities; c.) from more than 18,500 outpatients; participating in the NIMH Epidemiologic Catchment Area (ECA) a Study. The lifetime prevalence rates obtained from the NIMH ECA study were 25-60 times higher than previous estimates, which were based on studies of clinical populations. If the true lifetime prevalence of OCD in the United States is 2.5%, then it follows that 6.5 million Americans will be affected by obsessive-compulsive disorder during their lifetime. If the 1-month prevalence rate of OCD in the United States is 1.3 %, then approximately 3.4 million Americans suffer from obsessive compulsive disorder each Month[4]. Regardless of the specific epidemiological and diagnostic methods used to estimate the incidence or prevalence of obsessive-compulsive disorder, literally millions of Americans are affected by the symptoms.

There is strong evidence that obsessive-compulsive disorder impacts the American economy measureably. This premise is supported by the following. First, medical costs, yearly, from obsessive-compulsive disorder have been estimated to be $2.1 billion. Second, indirect costs due to lost productivity have been estimated to be $5.9 billion[5]. Third, health care expenditures in the United States surpassed $2.3 trillion in 2008, were $714 billion spent in 1990, and equalled $253 billion in 1980[6]., Therefore, it is highly likely that both the direct and indirect costs of obsessive-compulsive disorder continue to increase.

Economic indicators non-withstanding, the broader impact of obsessive-compulsive disorder on social, educational, and occupational function was addressed in a recent study. The investigators found that the symptoms of obsessive-compulsive disorder affected socialization by various means. Lowered self-esteem was observed in 92% of patients sampled, interference with family relationships reported in 73% of patient's sampled, and difficulty maintaining relationships was noted by 62% of patients[7]. Lowered academic achievement was observed in 58% of patients with obsessive-compulsive disorder, indicating that the disorder profoundly impacts educational achievement[7]. Occupational functioning is also affected in patient's with obsessive-compulsive disorder, through: lowered career aspirations, observed in 66% of patients sampled; work interference in 47% of patient's sampled, and ; lost time due to inability to work, reported in 40% of patients[7].

Suicide is the most serious complication of anxiety disorders. Suicide attempts secondary to obsessive-compulsive disorder symptoms have been reported in 13% of patients[7]. Obviously, if a suicide attempt is completed, progress in all three important areas of life function—relationships, educational and vocational function—halt permanently. Harm to family members through related injury, bereavement, lost spousal support, childhood parentification, and impact on the surrounding community is also significant after a completed suicide. In 2008, a total of 36,035 persons died as a result of suicide and in the United States approximately 666,000 persons visited hospital emergency departments for nonfatal, self-inflicted injuries[8]. Although suicidal thoughts do not always lead to a lethal or life threatening suicide attempt, suicidal thoughts and behavior even in the absence of suicide attempt are important. Public health surveillance is performed on suicide-related issues by gathering data at the state level by a national- and state-level survey—the National Survey on Drug Use and Health (NSDUH). Between January 1, 2008–December 31, 2009, the NSDUH obtained data from 92,264 respondents, a representative sample of the civilian, noninstitutionalized U.S. population aged ≥12 years, of various race/ethnicity. In 2008 and 2009, an estimated 8.3 million (annual average) adults aged ≥18 years in the United States (3.7% of the adult U.S. population) reported having suicidal thoughts in the past year[8]. An estimated 2.2 million (annual average) adults in the United States (1.0% of the adult U.S. population) reported having made suicide plans in the past year[8]. An estimated 1 million (annual average) adults in the United States (0.5% of the U.S. adult population) reported making a suicide attempt in the past year[8]. The prevalence of suicidal thoughts, suicide planning, and suicide attempts was significantly higher among young adults aged 18–29 years than it was among adults aged ≥30 years[8]. The prevalence of suicidal thoughts was significantly higher among females than it was among males, but there was no statistically significant difference for suicide planning or suicide attempts[8]. Although the NSDUH did not attempt to gather data according diagnosis, as indicated above, suicide is a primary comorbidity of mood and anxiety disorders. Therefore, the premise that obsessive compulsive disorder has a large impact through its direct and indirect economic costs, as well as its broader social consequences, is significant is ample.

However, the symptoms of obsessive-compulsive disorder are experienced at the level of the individual patient. It is at the level of the individual patient, that anyone can identify with symptoms of obsessive-compulsive disorder. The experience of intrusive, *obsessive* thoughts—wondering if the stove was left on or the front door was left unlocked while driving away from home—and *compulsive* behavior—being compelled to return home and check the stove or the door—is very common.

Obsessive-compulsive disorder is currently defined by the presence of obsessions and compulsions. Obsessions are recurrent, unwelcome thoughts, that may include: fear of dirt, germs, contamination; fear of acting on violent or aggressive impulses; feeling overly responsible for the safety of others; abhorrent religious and sexual thoughts, and/or; inordinate concern with order, arrangement and symmetry. Compulsions are repetitive behaviors that are performed in response to obsessions, in order to lessen the distress caused by obsessions. The short-term gain of reduced anxiety comes at a long-term cost of frequent repetition of these behaviors. Compulsions may affect social and occupational function to a profound degree as described above.

The professional community defines the diagnosis of obsessive-compulsive disorder, using criteria outlined in the DSM-IV[9]. The diagnosis of obsessive-compulsive disorder using modern criteria requires: the presence of obsessions and/or compulsions; recognized as excessive or unreasonable; causing marked distress, time-consumption (>1 hour/day), or interference with functioning. The obsessions and compulsions cannot be due to another Axis I psychiatric disorder, due to substance abuse, substance dependence, substance withdrawal, or due to a medical condition. For example, an individual with obsessive-compulsive disorder may be besot by unwanted and inappropriate sexual thoughts about neighbors, coworkers or family members, and will attempt to "undo" the obsessions by compulsive checking. Similarly, and individual with recurrent obsessions about the fact that they may have harmed individuals, which the patient tries to "undo", by returning over and over to the place where the thought occurred. Alternatively, a patient with obsessive-compulsive disorder may have constant thoughts that they are sinful, which the patient attempts to undo with repetitive prayer. Those who suffer from obsessive-compulsive disorder may be unable to carry out their responsibilities: at work, leading to unemployment; at home, resulting in marital conflict as well as disturbed family relationships, and; in society, leading to social isolation. The disruption of normal social and emotional development in obsessive-compulsive disorder not unlike that experienced in other neurodevelopmental disorders, such as schizophrenia. Like schizophrenia, there is likely both a genetic and environmental contributors to obsessive-compulsive disorder[10]. The altered life trajectory of these illnesses is quite sobering.

The two current effective treatments for patients with obsessive-compulsive include cognitive behavioral therapy (CBT) and pharmacotherapy. CBT consists of a technique called exposure and response prevention, in which patients deliberately and voluntarily expose themselves to fears/ideas, but are discouraged from carrying out compulsive responses. Studies do show successful results for extended periods of time. CBT can fail for various reasons, including, poorly executed treatments; patient or family noncompliance, psychiatric comorbidity such as severe depression or a personality disorder, poor insight (~5% of patients) or severe illness. CBT requires patients that are highly motivated, cooperative, and diligent, and is more likely to be successful when combined with pharmacotherapy. Traditional psychotherapy generally not helpful as a stand-alone therapy for OCD symptoms, although it is appropriate for the ongoing difficulties with adjustment experienced by patients with obsessive-compulsive disorder.

With respect to pharmacotherapy, specific medications have shown some effectiveness in controlling the symptoms of obsessive-compulsive disorder, including: SSRIs (selective serotonin reuptake inhibitors) such as Fluvoxamine, Fluoxetine, Sertraline, Paroxetine, Citalopram, ES Citalopram; SNRIs (serotonin-norepinephrine reuptake inhibitors) such as venlafaxine, and; TCAs (tricyclic antidepressants) such as Clomipramine. Treatment resistance or treatment-refractory obsessive-compulsive disorder is said to occur when patients with obsessive-compulsive disorder fail to benefit from treatment. By conservative estimate, 5% of patients with obsessive-compulsive disorder are treatment resistant. If 5% of Americans have treatment-refractory obsessive-compulsive disorder, then according to the aforementioned monthly or yearly prevalence rates, then 170,000 Americans each month, or 325,000 Americans in their lifetime are afflicted with treatment resistant obsessive-compulsive disorder. Treatment options for these patients are very limited.

2. Orbitofrontal-striatal function

The importance of brain circuits connecting frontal lobe to the basal ganglia was first observed in primates by Alexander and colleagues[11], who reported evidence for an anatomically distinct lateral orbitofrontal circuit loop, comprised of projections from: orbitofrontal cortex to the head of the caudate nucleus and the ventral striatum; to the internal pallidus; to the mediodorsal thalamus; returning from the thalamus to the orbitofrontal cortex. Alexander and colleagues hypothesized: the existence of several relatively specialized fronto-striatal loops; proposed that they were organized in parallel, linking the basal ganglia to the frontal cortex, and; that each circuit played a functional role based on its connections to particular regions of the frontal cortex. Other investigators[12, 13] have suggested that the so-called "limbic" structures (i.e.—hippocampus, anterior cingulate and, basolateral amygdala) ought to include in the lateral orbitofrontal circuit loop circuit, because of their extensive connections to the orbitofrontal cortex. Based on these interconnections, it can be hypothesized that this "greater" lateral orbitofrontal circuit could play a role in emotion, as the function of these so-called "limbic" brain regions play a role in affective states and emotional perception.

The orbitofrontal cortex is a key brain region, not only in emotional behavior, but also for motivation [14-18]. This was first shown by Harlow[19] who provided a naturalistic description of profound changes in behavior of a 19th century railway worker— Phineas Gage—after a charge he was setting, using a tamping rod exploded. He sustained a severe left frontal lobe injury, after the tamping rod was when a was launched through his forehead and out his skull. Reported changes in Gage's behavior following the accidental orbitofrontal cortex damage included not only inappropriate emotional responses, but also, impulsive and poorly thought out decisions, characteristic of behavioral changes in patients with orbitofrontal cortex lesions[20, 21 22].

Since learning-based motivation requires the integration of complex brain systems that include orbitofrontal cortex, researchers have hypothesized that difficulties "unlearning" reinforced behaviors may be associated trouble with sensing change between behavior-reward relationships. Impairment in the unlearning of established reward-motivated behaviors are also observed in animals and humans with orbitofrontal cortex lesions [23 24 25]. Furthermore, patients with focal lesions either in the striatum or the ventral palladium, (an area it projects to) demonstrate behaviors very consistent with those observed in obsessive-compulsive disorder[26, 27].

The results of functional imaging research have provided complementary evidence to the lesion studies demonstrating that the orbitofrontal cortex is a key brain region involved in learning and motivation. The human brain's awareness of expecting a reward and the likelihood that a reward will occur is requires an intact orbitofrontal cortex[28-30]. If the orbitofrontal cortex is not intact, a person's behavior may seem impulsive or they may appear to have poor judgment.

The orbitofrontal cortex may have anatomically and functionally segregated orbitofrontal-thalamic striatal circuits. This idea of Alexander and colleagues is supported by research indicating that the lateral orbitofrontal cortex may have a distinct and separate function from medial orbitofrontal cortex, in that the lateral orbitofrontal cortex was activated when suppressing a response already associated with a reward [31]. This would imply that dysfunction of the lateral orbitofrontal cortex prevents inhibition of behavior reinforced previously by a reward.

3. Evidence for orbitofrontal-striatal dysfunction in obsessive-compulsive disorder

The current most popular model proposed by researchers to explain the neurobiological foundation of obsessive-compulsive disorder focuses on abnormalities in cortical-striatal-thalamic circuitry—the orbitofrontal-striato-thalamic circuits in particular[32-34].

3.1 Evidence from neuroimaging studies

Using techniques that measure brain glucose metabolism, fluorodeoxyglucose positron-emission tomography (FDG PET), investigators demonstrated increased cerebral glucose metabolism present bilaterally in the cerebral hemispheres and orbitofrontal gyrii, as well as both caudate heads, in patient with OCD patients[35 36]. The findings were replicated[37-42] in FDG-PET studies examining patients both at rest, and while provoking symptoms, although not all studies produced positive findings[43-45]. A meta-analysis[46] confirmed abnormalities were present in the orbital gyrus and the head of the caudate in patients with obsessive-compulsive disorder. The results of PET studies are an important piece of supportive evidence of the orbitofrontal-striato-thalamic model.

3.2 Evidence from deep brain stimulation research

Another strong piece of evidence supporting this model is the symptomatic improvement of patients with obsessive-compulsive disorder undergoing capsulotomy. Focal lesioning during a surgical procedures for neuropsychiatric disorders has been known as "psychosurgery". Historically, these procedures have been thought not to be discriminate in terms of neuroanatomical location or groups of patients treated[47-49]. Furthermore, informed consent is thought not to be properly obtained, a process which requires careful assessment of an individual's capacity to weigh the risks and benefits of an experimental medical or surgical procedure[50]. Consequently, psychosurgery is not viewed in a positive light in the popular media [51].

Neurosurgery for psychiatric disorders is a highly invasive treatment. However, it is important to view these interventions in the proper historical context. Prior to 1950, psychiatric illness was essentially untreatable, as no specific medications existed for the treatment of severe psychiatric disorders. Since these illnesses were disabling and lethal, the treatments pursued were aggressive and invasive. These interventions included malarial pyrotherapy described by Epstein in 1936[52], hypoglycemic coma described by Sakel in 1937[53], electroconvulsive therapy, described by Bini in 1938[54], as well as neurosurgery. Historically (and currently) the use of neurosurgery has only been used only for intractable psychiatric illnesses[55].

Burckhardt first published a report of the first (unsuccessful surgical attempts to treat severe psychosis in 1891[56]. The first neuroanatomical models describing both function and structural of mood and behavioral regulation were published by Papez in 1937[57]. At this time, a hypothesis was proposed by researchers that abnormal mood and behavioral regulation was caused by dysfunctional thalamo-cortical communication[58], leading to the use of the prefrontal leucotomy (popularly known as the prefrontal lobotomy), a procedure that disrupted white matter tracts connecting these regions. Because the ability of surgeons to localize and severing specific frontal lobe white matter tracks, lesions were indiscriminantly large. After 1950, pharmacologic interventions were identified that drastically reduced the symptoms of psychiatric disorders. The pharmacology revolution of the mid-twentieth-century resulted in the discovery of medications effective: for mania described by Cade in 1949 and Schou and colleagues in 1954[59, 60]; for psychosis described by Bower in 1954[61], and Winkelman in 1954[62], and; for depression described by Bailey and colleagues in 1959[63], Kiloh and colleagues in 1960[64], and Kuhn in 1958[65].

In the early 1960s, investigators reported that stimulation of different brain area induced hypomania, dysphoria, and anhedonia. These early findings suggested the possible efficacy of DBS in treatment refractory psychiatric disorders. One of the earliest anatomically specific psychosurgery consists of ablation of the anterior limb of the internal capsule—the anterior capsulotomy—was found to be efficacious in severely refractory obsessive-compulsive disorder. The first anterior capsulotomies were performed in Europe in the late 1940's. During the procedure, symmetric bilateral lesions are made in the anterior limb of the internal capsule, which is quite near to the ventral striatum. This lesion, whether made by heat (thermocoagulation during neurosurgery or a thermocapsulotomy) or by minimally invasive gamma irradiation (a gamma-capsulotomy), interrupts the passage of white matter fibers between the prefrontal cortex and the subcortical nuclei, the striatum, and the dorsomedial thalamus. A recent prospective study of 35 patients with obsessive-compulsive disorder who underwent thermocapsulotomy showed that that 70% had "satisfactory outcomes" after 3 years[66].

The recent development of deep brain electrode placement at the ventral capsule/ventral striatum (VC/VS) target is also a very strong piece of evidence supporting this model. Deep brain stimulation is a reversible, neurosurgical procedure. Deep brain stimulation is an invasive neurosurgical intervention being used to treat psychiatric disorders in an investigative fashion. The disorders currently being examined include treatment-resistant major depressive diosrder, treatment-resistant obsessive-compulsive disorder, Tourette's Syndrome, Alzheimer's dementia, and addictions. The actual treatment consists of implanting one or more electrode leads into a particular brain regions through burr holes in the skull using a proprietary stereotactic neurosurgical techniques. Neuroimaging-guided implantation calculates the route to the target using a three-dimensional coordinate system based on external landmark. Current commercially available leads have four electrodes, 1-2 mm in length, separated by 4-5 mm, the complete electrode 10–20 mm in length. The leads connect to subcutaneous extension wires that are tunnelled surgically to pulse generators implanted in the chest. The pulse generators contain a battery and hardware/software that drives the neurostimulation. A programmer can set the programs in the neurostimulator using a handheld computer with a wireless connection.

In the 1960s electrical stimulation of the ventrolateral thalamus was noted to stop tremor. Prolonged electrical stimulation at different targets was found to be effective for treatment-refractory movement disorders, epilepsy, chronic pain and tremor. Investigators then delivered high frequency cathodic (positive) electrical stimulation directly at the surgical target, in order to mimic the effect of a surgical lesion [67, 68], leading to the development of technology first used clinically in Parkinson's disease, essential tremor, and extrapyramidal dyskinesias. Currently, there are many numerous published reports demonstrating the safety and efficacy of DBS for intractable movement disorders[69, 70].

In fact, the efficacy and safety data from studies in patients with movement disorders led the FDA to approve the use of obsessive-compulsive disorder for essential tremor and Parkinson's disease. The FDA eventually approved the use of DBS for dystonia under a Humanitarian Device Exemption (HDE). The results of a recent open label clinical trial of DBS using the VC/VS target suggested that DBS for intractable obsessive-compulsive disorder had encouraging therapeutic effects, with probable benefit even 3 years after surgery [71]. The specificity of this lesion is the strongest piece of evidence supporting the dysfunction of orbitofrontal-striato-thalamic circuits as a likely etiology of obsessive-compulsive disorder.

3.3 Summary and conclusions

Obsessive-compulsive disorder is a serious neuropsychiatric illness. Treatment-resistant obsessive-compulsive disorder is less common, but highly debilitating. The evidence for the role of orbitofrontal-striato-thalamic circuits in mediating emotion, learning, and reward-focused behavior is strong. The evidence that these important brain systems are dysfunctional in patients with obsessive-compulsive disorder is also strong. Expanding knowledge about these brain circuits will provide a rich area for further research and is necessary to develop effective treatments for obsessive-compulsive disorder.

4. References

[1] Weissman MM, Bland RC, Canino GJ, et al. The cross national epidemiology of obsessive compulsive disorder. The Cross National Collaborative Group. J Clin Psychiatry 1994; 55 Suppl:5-10.
[2] Robins LN, Helzer JE, Weissman MM, et al. Lifetime prevalence of specific psychiatric disorders in three sites. Arch Gen Psychiatry 1984; 41:949-58.
[3] Karno M, Golding JM, Sorenson SB, Burnam MA. The epidemiology of obsessive-compulsive disorder in five US communities. Arch Gen Psychiatry 1988; 45: 1094-9.
[4] Regier DA, Myers JK, Kramer M, et al. The NIMH Epidemiologic Catchment Area program. Historical context, major objectives, and study population characteristics. Arch Gen Psychiatry 1984; 41:934-41.
[5] DuPont RL RD, Shiraki S, Rowland CR. . Economic costs of obsessive-compulsive disorder. . Med Interface. 1995 8:102-9.
[6] National Health Statistics Group. National Health Care Expenditures Data. In: Centers for Medicare and Medicaid Services OotA, ed, 2010.
[7] Hollander E, Kwon JH, Stein DJ, Broatch J, Rowland CT, CA. H. Obsessive-compulsive and spectrum disorders overview and quality of life issues. J Clin Psychiatry. 1996; 57 3-6.
[8] Alex E. Crosby M, Beth Han, MD, PhD, LaVonne A. G. Ortega, MD, Sharyn E. Parks, PhD, Joseph Gfroerer, BA. Surveillance Summaries Suicidal Thoughts and Behaviors Among Adults Aged ≥18 Years —United States, 2008–2009 Morbidity and Mortality Weekly Report 2011; 60.
[9] Association AP. DSM-IV, 1994.
[10] Ting JT, Feng G. Neurobiology of obsessive-compulsive disorder: insights into neural circuitry dysfunction through mouse genetics. Curr Opin Neurobiol 2011; 21: 842-8.
[11] Alexander GE, DeLong MR, Strick PL. Parallel organization of functionally segregated circuits linking basal ganglia and cortex. Annual Review of Neuroscience 1986; 9:357–381.
[12] Lawrence AD, Sahakian BJ, Robbins TW. Cognitive functions and corticostriatal circuits: insights from Huntington's disease. Trends in Cognitive Sciences 1998 2,: 379–388.
[13] Phillips ML, Drevets WC, Rauch SL, Lane R. Neurobiology of emotion perception I: the neural basis of normal emotion perception. Biological Psychiatry 2003; 54:504–514.
[14] Rolls ET. The functions of the orbitofrontal cortex. Brain and Cognition 2004; 55:11–29.
[15] Rolls ET, Hornak J, Wade D, McGrath J. Emotion-related learning in patients with social and emotional changes associated with frontal lobe damage. Journal of Neurology, Neurosurgery and Psychiatry 1994; 57:1518–1524.
[16] Elliott R, Deakin B. Role of the orbitofrontal cortex in reinforcement processing and inhibitory control: evidence from functional magnetic resonance imaging studies in healthy human subjects. International Review of Neurobiology 2005; 65 89–116.

[17] Elliott R, Dolan RJ, Frith CD. Dissociable functions in the medial and lateral orbitofrontal cortex: evidence from human neuroimaging studies. Cerebral Cortex 2000; 10:308–317.

[18] Kringelbach ML. The human orbitofrontal cortex: linking reward to hedonic experience. . Nature Reviews Neuroscience 2005. ; 6:691–702.

[19] Harlow JM. Recovery from the passage of an iron bar through the head. Publications of the Massachusetts Medical Society 1868. ; 2:327–347.

[20] Eslinger PJ, Damasio AR. Neurology. Severe disturbance of higher cognition after bilateral frontal lobe ablation: patient EVR. 1985. ; 35:1731–1741.

[21] Bechara A, Damasio AR, Damasio H, Anderson SW. Insensitivity to future consequences following damage to human prefrontal cortex. Cognition 1994.; 50:7–15.

[22] Damasio H, Grabowski T, Frank R, Galaburda AM, Damasio AR. The return of Phineas Gage: clues about the brain from the skull of a famous patient. Science 1994.; 264,:1102–1105.

[23] McEnaney KW, Butter CM. Perseveration of responding and nonresponding in monkeys with orbital frontal ablations. . Journal of Comparative Physiological Psychology 1969. ; 68:558–561.

[24] Jones B, Mishkin M. Limbic lesions and the problem of stimulus—reinforcement associations. . Experimental Neurology 1972.; 36,:362–377.

[25] Rolls ET, Hornak J, Wade D, McGrath J. Emotion-related learning in patients with social and emotional changes associated with frontal lobe damage. . Journal of Neurology, Neurosurgery and Psychiatry 1994. ; 57:1518–1524.

[26] Rapoport JL, Wise SP. Obsessive-compulsive disorder: evidence for basal ganglia dysfunction. Psychopharmacology Bulletin 1988.; 24:380–384.

[27] Laplane D, Levasseur M, Pillon B, et al. Obsessivecompulsive and other behavioural changes with bilateral basal ganglia lesions. A neuropsychological, magnetic resonance imaging and positron tomography study. . Brain 1989.; 112 699-725.

[28] Tremblay L, Schultz W. Relative reward preference in primate orbitofrontal cortex. Nature 1999. ; 398:704-708.

[29] Tremblay L, Schultz W, . Reward-related neuronal activity during go–no-go task performance in primate orbitofrontal cortex. . Journal of Neurophysiology 2000.; 83:1864–1876.

[30] Hikosaka K, Watanabe M. Long- and short-range reward expectancy in the primate orbitofrontal cortex. . European Journal of Neuroscience 2004. ; 19:1046–1054.

[31] Elliott R, Dolan RJ, Frith CD. Dissociable functions in the medial and lateral orbitofrontal cortex: evidence from human neuroimaging studies. . Cerebral Cortex 2000; 10:308–317.

[32] Saxena S, Brody AL, Schwartz JM, Baxter LR. Neuroimaging and frontal–subcortical circuitry in obsessive-compulsive disorder. British Journal of Psychiatry 1998; Supplement:26–37.

[33] Saxena S, Bota RG, Brody AL. Brain–behavior relationships in obsessive-compulsive disorder. Seminars in Clinical Neuropsychiatry 2001a; 6:82–101.

[34] Graybiel AM, Rauch SL. Toward a neurobiology of obsessive-compulsive disorder. Neuron 2000; 28:343–347.
[35] Baxter Jr. LR, Phelps ME, Mazziotta JC, Guze BH, Schwartz, J.M., , Selin CE. Local cerebral glucose metabolic rates in obsessive-compulsive disorder. A comparison with rates in unipolar depression and in normal controls. . Archives of General Psychiatry 1987.; 44,:211–218.
[36] Baxter Jr. LR, Schwartz JM, Mazziotta JC, et al. Cerebral glucose metabolic rates in nondepressed patients with obsessive-compulsive disorder. . American Journal of Psychiatry 1988. ; 145, :1560–1563.
[37] Nordahl TE, Benkelfat C, Semple WE, Gross M, King AC, Cohen RM. Cerebral glucose metabolic rates in obsessive compulsive disorder. . Neuropsychopharmacology 1989. ; 2:23–28.
[38] Swedo SE, Schapiro MB, Grady CL, et al. Cerebral glucose metabolism in childhood-onset obsessive-compulsive disorder. . Archives of General Psychiatry 1989.; 46, :518–523.
[39] Sawle GV, Hymas NF, Lees AJ, Frackowiak RS. Obsessional slowness. Functional studies with positron emission tomography. Brain. 1991. ; 114 2191-2202.
[40] McGuire PK, Bench CJ, Frith CD, Marks IM, Frackowiak RS, Dolan RJ. Functional anatomy of obsessive-compulsive phenomena. British Journal of Psychiatry 1994.; 164:459–468.
[41] Rauch SL, Jenike MA, Alpert NM, et al. Regional cerebral blood flow measured during symptom provocation in obsessive-compulsive disorder using oxygen 15-labeled carbon dioxide and positron emission tomography. Archives of General Psychiatry 1994.; 51:62–70.
[42] Cottraux J, Gerard D, Cinotti L, et al. A controlled positron emission tomography study of obsessive and neutral auditory stimulation in obsessive-compulsive disorder with checking rituals. . Psychiatry Research 1996. ; 60,:101–112.
[43] Martinot JL, Allilaire JF, Mazoyer BM, et al. Obsessive-compulsive disorder: a clinical, neuropsychological and positron emission tomography study. Acta Psychiatrica Scandinavica 1990.; 82:233–242.
[44] Perani D, Colombo C, Bressi S, et al. [18F]FDG PET study in obsessive-compulsive disorder. A clinical/metabolic correlation study after treatment. British Journal of Psychiatry 1995.; 166:244–250.
[45] Busatto GF, Zamignani DR, Buchpiguel CA, et al. A voxel-based investigation of regional cerebral blood flow abnormalities in obsessivecompulsive disorder using single photon emission computed tomography (SPECT). Psychiatry Research 2000.; 99:15–27.
[46] Whiteside SP, Port JD, Abramowitz JS. A meta-analysis of functional neuroimaging in obsessive-compulsive disorder. Psychiatry Research 2004. ; 132:69–79.
[47] Cohrs S, Tergau F, Riech S, et al. High-frequency repetitive transcranial magnetic stimulation delays rapid eye movement sleep. Neuroreport 1998; 9:3439-43.
[48] Mosimann UP, Rihs TA, Engeler J, Fisch H, Schlaepfer TE. Mood effects of repetitive transcranial magnetic stimulation of left prefrontal cortex in healthy volunteers. Psychiatry Res 2000; 94:251-6.

[49] Nobel Prize. Vol. 10-16-2011.
[50] Appelbaum PS, Roth LH, Lidz C. The therapeutic misconception: informed consent in psychiatric research. Int J Law Psychiatry 1982; 5:319-29.
[51] Kesey K. One flew over the cuckoo's nest : a novel. Harmondsworth, Eng. ; New York: Penguin Books, 1976:311 p.
[52] Epstein NN. Artificial fever as a therapeutic procedure. Cal West Med 1936; 44:357–58.
[53] Sakel M. The origin and nature of the hypoglycemic therapy of the psychoses. Bull. N. Y. Acad. Med. 1937; 13:97–109.
[54] Bini L. Experimental researches on epileptic attacks induced by the electric current. Am. J. Psychiatry 1938; 94:172–74.
[55] Hariz MI BP, Zrinzo L. Deep brain stimulation between 1947 and 1987: the untold story. Neurosurgery Focus 2010; 29.
[56] Burckhardt G. On cortical resection as a contribution to the operative treatment of psychosis. Psychiatrie psychischgerichtliche Medizin 1891; 47:463–548.
[57] Papez JW. A proposed mechanism of emotion. Arch. Neurol. Psychiatry 1937; 38:725–43.
[58] Moniz E. Prefrontal leucotomy in the treatment of mental disorders. Am. J. Psychiatry 1937; 93:1379–85.
[59] Cade JFJ. Lithium salts in the treatment of psychotic excitement. Med. J. Austr 1949; 2:349–52.
[60] Schou M J-NN, Stromgren E, Voldby H. The treatment of manic psychoses by the administration of lithium salts. J. Neurol. Neurosurg. Psychiatry 1954; 17:250–60.
[61] Bower WH. Chlorpromazine in psychiatric illness. N. Engl. J. Med. 1954; 251:689–92.
[62] Winkelman NW. Chlorpromazine in the treatment of neuropsychiatric disorders. J. Am. Med. Assoc. 1954; 155:18–21.
[63] Bailey S.D. BL, Gosline E., Kline N.S., Park I.H., . Comparison of iproniazid with other amine oxidase inhibitors, including W-1544, JB-516, RO 4-1018, and RO 5-0700. Ann. N. Y. Acad. Sci. 1959; 80:652–68.
[64] Kiloh LG CJ, Latner G.. A controlled trial of iproniazid in the treatment of endogenous depression. J. Mental Sci. 1960.; 106:1139–44.
[65] Kuhn R. The treatment of depressive states with G22355 (imipramine hydrochloride).. Am. J. Psychiatry 1958; 115:459–64.
[66] Liu K, Zhang H, Liu C. Stereotactic treatment of refractory obsessive compulsive disorder by bilateral capsulotomy with 3 years follow-up. J Clin Neurosci. 2008; 15:622-629.
[67] Benabid AL, Chabardes S, Torres N, et al. Functional neurosurgery for movement disorders: a historical perspective. Prog Brain Res 2009; 175:379-91.
[68] Benabid AL, Pollak P, Gervason C, et al. Long-term suppression of tremor by chronic stimulation of the ventral intermediate thalamic nucleus. Lancet 1991; 337:403-6.
[69] Deuschl G, Schade-Brittinger C, Krack P, et al. A randomized trial of deep-brain stimulation for Parkinson's disease. N Engl J Med 2006; 355:896-908.

[70] Mueller J, Skogseid IM, Benecke R, et al. Pallidal deep brain stimulation improves quality of life in segmental and generalized dystonia: results from a prospective, randomized sham-controlled trial. Mov Disord 2008; 23:131-4.
[71] Greenberg B, Malone D, al. e. "Three-Year outcomes in deep brain stimulation for highly resistant Obsessive-Compulsive Disorder."Neuropsychopharmacology (2006). 1-10.

Part 2

Neurodegenerative Diseases: In Search of Therapies

6

In Search of Therapeutic Solutions for Alzheimer's Disease

Ricardo B. Maccioni[1,2*], Gonzalo Farías[1,2],
Leonel E. Rojo[1,3,4] and José M. Jiménez[1,2]
[1]*Laboratory of Cellular and Molecular Neurosciences,*
International Center for Biomedicine,
[2]*University of Chile, Las Encinas, Ñuñoa, Santiago,*
[3]*Arturo Prat University,*
[4]*Rutgers University (SEBS),*
[1,2,3]*Chile*
[4]*USA*

1. Introduction

Alzheimer's Disease (AD) is the most frequent cause of dementia in the elderly. Prevalence is about 10% in populations of 65 years and older and it increases rapidly as life expectancy increases. Estimates indicate that there are around 36 million cases around the world, while associated costs are higher than US$ 600 billion (Wimo & Prince, 2010). The histopathological characteristics of AD are represented by two main lesions: senile plaques and neurofibrillary tangles (NFTs). The formers main component is the amyloid beta peptide (Aβ peptide) adopting β-sheet structures, while the latter have the hyperphosphorylated tau protein and pathological forms of tau as a major component (Maccioni et al., 2001). AD has triggered a plethora of hypotheses to explain its pathogenesis, possibly strengthened by the fact that no cure has yet been found for this devastating disease since its first description by Alois Alzheimer in 1907. Although, significant advances have been made in neuroscience in the last few decades, the data has not provided effective therapeutic solutions for AD.

1.1 Many hypotheses, one disease, no cure

Many hypotheses have been postulated on the physiopathology of AD (Maccioni and Perry, 2009). During the last two decades, the central paradigm was the **amyloid hypothesis**, based on events triggered by the Aβ cascade: as the unique driving force in neurodegeneration. The hypothesis proposes that accumulation of Aβ in the brain primarily influences pathogenesis of the disease and the rest of the processes in AD are results of the imbalance between production and degradation of Aβ (Hardy & Selkoe, 2002). Nevertheless, recent clinical trials based on this hypothesis have been inconclusive. In fact, the amyloid cascade hypothesis has resulted in misleading approaches to find therapeutic alternatives until

*Corresponding Author

recently (Hardy, 2009). These targets include Aβ vaccines, antibodies against β-amyloid, γ-secretase inhibitors and drugs that block direct Aβ aggregation (Extance, 2010; Gandy, 2010; Rinne et al., 2010).

In this context, new paradigms have been proposed that consider all the implications of the disease and valid therapeutic targets are now emerging. AD is a complex illness involving many risk factors. In fact, during its progress, oxidative stress as well as innate immune system activation appears to play a role. Considering AD as a result of multifactorial events, it is plausible that a concatenated series of damage signals affect brain cells, mainly microglial cells, thus triggering an abnormal response in neuro-immunomodulation with consequent effects on neurons. A common molecular feature of these anomalous signals leads to tau self-aggregation into oligomers as a final event (Maccioni et al., 2010).

1.2 The neuroimmunomodulation hypothesis of AD

During the past few years, increasing sets of evidence support the major role of deregulation of interaction patterns between glial cells and neurons in the pathway toward neuronal degeneration. Neurons and glial cells, together with brain vessels, constitute an integrated system for brain function. Inflammation is a process intimately related to the onset of several neurodegenerative disorders, including Alzheimer's disease (AD). Several hypotheses have been postulated to explain the pathogenesis of AD, but none provide insight into the early events that trigger metabolic and cellular alterations in neuronal degeneration (Rojo et al., 2008).

A study of the factors resulting in AD, has led us to postulate the **neuroimmunomodulation hypothesis**, which focus on pathological events in the neuron-glia cross-talks. Data suggests an important role of the immune system in regulating the progression of the brain aging and neurodegenerative diseases, where the crosstalk between these systems determines the progression of pathological event (Lucin & Wyss-Coray, 2009). In this context, the microglia, the resident macrophages of the CNS, are key factors in the regulation of local cellular environment relative to inflammation. The persistence of activated microglia long after acute injury and in chronic disease suggests that these cells have an innate immune memory of tissue injury and degeneration. Microglial phenotype is also modified by systemic infections or inflammation. Systemic inflammation is associated with a decline in function in patients with chronic neurodegenerative disease, both acutely and in the long term (Perry et al., 2010).

The idea that alterations in the brain immunomodulation are critical for AD pathogenesis provides the most integrative view on this cognitive disorder, considering that converging research lines have revealed the involvement of inflammatory processes in AD. Studies on microglia and neuronal cultures, together with experiments in animal models, and the clinical evidence, suggest that a series of endogenous damaged signals that include, among other factors, Aβ oligomers, oxygen free radicals, iron overload, cholesterol levels in neuronal rafts, folate deficiency, head injury, LDL species and homocysteine trigger the activation of microglial cells. Inflammatory cytokines play a dual role: either promoting neurodegeneration or neuroprotection. This equilibrium is shifted toward the neurodegenerative phenotype upon the action of several risk factors that trigger innate damage signals and activate microglia and then release of inflammatory cytokines (Figure 1) (Fernandez et al., 2008; Maccioni et al., 2009).

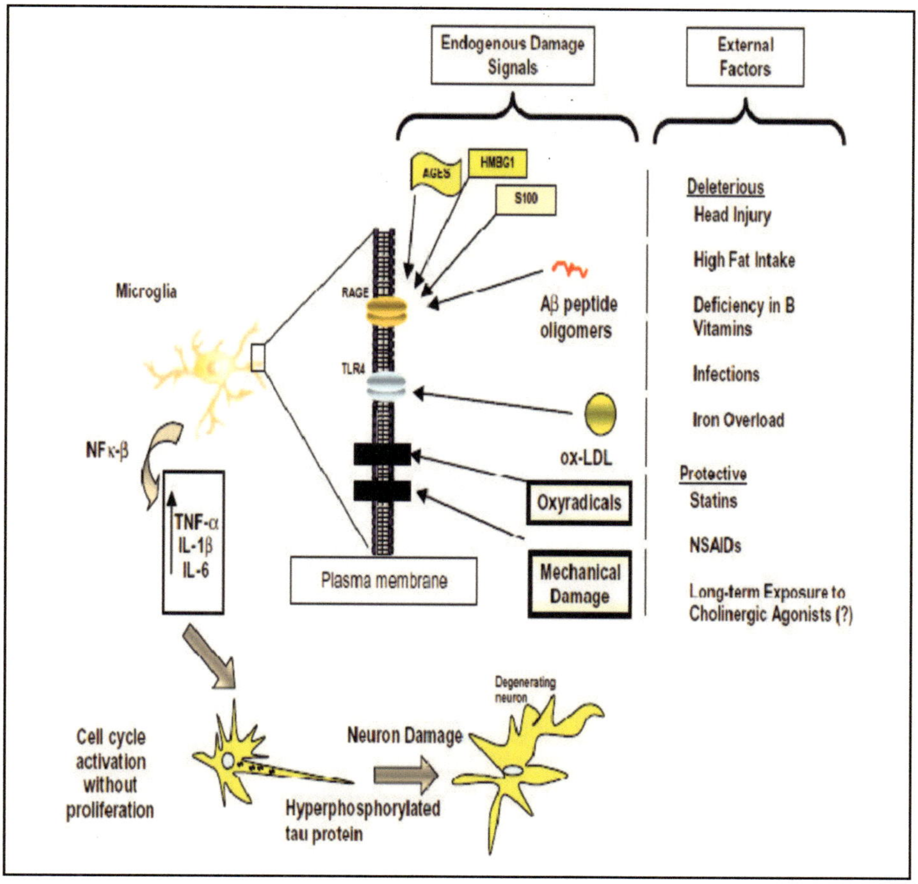

Fig. 1. **Neuroinmmunomodulation Hypothesis in the pathogenesis of AD**. Schematic representation of the hypothetical roles of endogenous danger/alarms signals built into the innate immune system in the early stages of the pathogenesis of AD. Consequentially, (as it may apply to different individuals), danger signals can trigger innate immune system alarm mechanisms resulting in the production of tumor necrosis factor alpha (TNF-*a*), interleukin-1*β* IL-1*β*) and interleukin-6 (IL-6). These signals would then mediate neuronal damage, reflected in alterations such as tau hyperphosphorylation and paired helical filaments formation.

The progression of AD, encompasses increased damage in brain parenchyma preceding the onset of symptoms. Suggesting that tissue distress trigers damage signals and drives neuroinflammation. These signals via toll-like receptors, or receptors for highly glycosylated end products, or other glial receptors activate sensors of the native immune system, inducing the anomalous release of cytokines and promoting the neurodegenerative cascade, a hallmark of brain damage that correlates with cognitive decline. We show that this activation induces NFκ-β expression with the consequent release of cytokine mediators such as TNF-α, IL-6 and IL-1β. An over expression of these mediators may trigger signaling cascades in neurons leading to activation of protein GSK3β, cdk5 kinases, along with inhibition of phosphatases such as PP1, resulting in hyperphosphorylation and self-aggregation of tau protein into neurotoxic oligomeric species. The aggregation of tau protein is the final pathway and key event in the Alzheimer´s pathogenesis (Morales et al., 2010, Farias et al., 2011).

The evidence correlating inflammation and tau phosphorylation has been provided by neuropathology markers and mouse transgenic models with Alzheimer disease. Activated microglia has been found in the postmortem brain tissues of various human tauopathies including Alzheimer's disease (AD) and frontotemporal dementia (FTD) (Gebicke-Haerter, 2001). The administration of LPS, in order to generate systemic inflammation, significantly induced tau hyperphosporylation in the triple transgenic mouse model of AD (3xTg) and rTg4510 mice. In line with this evidence, microglial activation also preceded tangle formation in a murine model of tau pathology (Yoshiyama et al., 2007). Also immunosuppression of young P301S Tg mice with FK506 attenuated tau pathology and increased lifespan, thereby linking neuroinflammation to early progression of tauopathies (Yoshiyama et al., 2007, Lee et al., 2010; Kitazawa et al., 2005). In addition, the neurodegenerative lesions caused by human truncated tau promote inflammatory response manifested by upregulation of immune-molecules (CD11a,b, CD18, CD4, CD45 and CD68) , the morphological activation of microglial cells and leukocyte infiltration in a rat model of tauopathy (Zilka et al., 2009).

On the other hand, it has been demonstrated that proinflammatory cytokines, such as interleukin-1 (IL-1), interleukin-6, and nitric oxide, released from astrocytes can accelerate tau phosphorylation and formation of neurofibrillary tangles (NFTs) in vitro (Li et al., 2003, Quintanilla et al., 2004, Saez et al., 2004). Likewise, the activation of microglia and the microglial-derived proinflammatory cytokine TNFα can induce accumulation of aggregation-prone tau molecules in neurites via reactive oxygen species (Gorlovoy et al., 2009). Recently, it has been identified the fractalkine receptor (CX3CR1) as a key microglial pathway in protecting against AD-related cognitive deficits that are associated with aberrant microglial activation and elevated inflammatory cytokines. In vitro experiments demonstrated that microglial activation elevates the level of active p38 MAPK and enhances tau hyperphosphorylation within neurons which can be blocked by administration of an interleukin-1 receptor antagonist and a specific p38 MAPK inhibitor. This finding suggests that CX3CR1 and IL-1/p38 MAPK pathway may serve as novel therapeutic target for human tauopathies (Bhaskar et al., 2010, Cho et al., 2011).

Other sources of evidence to support neuroimmunomodulation theory are epidemiological data that show individuals consuming nonsteroidal anti-inflammatory drugs (NSAIDs) have a lower risk of AD (McGeer et al, 2006). In fact, patients receiving systemic NSAIDs developed significantly less AD manifestations, suggesting that ameliorating inflammation in the brain helps to prevent or slow down the onset of AD (McGeer et al., 1996). However, controlled randomized clinical trials with common NSAIDs have not shown a positive effect in the decline of AD (Rojo et al, 2008).

Genetic and epidemiological evidence has implicated increased TNFα production as a risk factor for AD. In fact, excess TNF is present in the CSF of individuals with Alzheimer's disease (AD). Recently, Tobinick and colleagues have demonstrated that perispinal administration of etanercept, a potent anti-TNF fusion protein, produced sustained clinical improvement in a 6-month, open-label pilot study in patients with AD ranging from mild to severe. Subsequent case studies have documented rapid clinical improvement following perispinal etanercept in both AD and primary progressive aphasia, providing evidence of rapidly reversible, TNF-dependent, pathophysiological mechanisms in AD and related disorders. Although, some researchers undermine results by their methodologies, perispinal etanercept for AD needs further studies to be validated and gives us new perspectives to support the critical role of immune system in AD (Tobinick & Gross, 2008a, 2008b; Tobinick, 2009).

1.3 Integrated efforts toward prevention, diagnosis and treatment of AD

In the field of prevention of AD, studies have indicated that dietary factors, antioxidants, exercise along with healthy styles of life contribute to diminish risk factors for AD. In addition, the search for dietary supplements, phytocomplexes, and nutraceuticals from natural sources have suggested novel preventive alternatives against AD, based on preclinical and clinical trials outcomes. These include molecular complexes with either anti-inflammatory, antioxidant or antiamyloidogenic properties. The extraordinary properties of polyphenolic extracts may help as coadjuvant in the AD therapy. Investigations are directed to design powerful nutraceuticals, which derive from distinct sources and could be consumed by the high-risk population. This provides data on the action of *Shilajit* and other nutraceuticals as a new tools for prevention.

Innovative approaches are critical in order to improve early detection of AD, which in turn is critical to find therapeutic solutions, and for monitoring new drugs developments against the disease. Our laboratory has developed an integrated strategy to establish reliable diagnosis tools with a high efficacy. We found that different benzimidazoles that tag aggregates of tau protein, serve as specific markers for PET neuroimaging, in order to monitor advances of the disease. Clinical studies are underway to validate PET images that differentiate stages of AD and controls. As a complementary approach, we have studied tau in cerebrospinal fluid (CSF) and also in peripheral blood platelets, providing promising biological non-invasive markers for AD (Maccioni et al, 2006; Neumann et al, 2011).

Until now there are no drugs available as an efficient therapy of AD. Current therapeutic targets focus on avoiding formation of tau aggregates in neurons, modulation of the innate immune system, chelating heavy metals and diminishing the burden of the amyloidogenic molecular variants. On one hand, clinical trials include: tau aggregation inhibitors (like methylene blue with promising results in phase II trials), tau kinase inhibitors, microtubule stabilizers and unfolded protein respond modulators. As mentioned earlier, it has been demonstrated that the relationship between anti-inflammatory molecules (NSAIDs) and the prevalence of AD in longitudinal studies. Unfortunately, no specific drug has a positive effect in the treatment of the disease in controlled double-blind trials. It could be hypothesized that we still do not understand enough about the specific molecules and its receptors involved in the immune system's cross-talk between neuron and glial cells. We believe that the next generation of drugs should focus on these specific targets.

2. New tools for AD diagnosis

More than a century has passed since Dr. Alois Alzheimer described the case of Auguste D, a 51 years old patient with a history of progressive cognitive impairment. Histopathology of brain tissue demonstrated the presence of senile plaques, neurofibrillary tangles (NFTs) and arteriosclerotic changes (Alzheimer, 1907). Today, despite the importance of AD (the world´s primary cause of neurodegenerative dementia) the advances in knowledge of clinical and pathophysiological aspects, and the definitive diagnosis of AD still depends mainly on histopathological analysis.

In contrast, the reliability of standard clinical evaluation is limited, and in most cases allow us only to diagnose the disease as "possible" or "probable" AD (McKhann et al., 1984). This shortcoming of standard clinical methods is specially relevant in early and unusual presentations of AD and has driven the interest to develop both biochemical and imaging tests to support the diagnosis (Dubois et al., 2007; Wiltfang et al., 2007). In this regard the 2011 diagnostic criteria (McKhann et al., 2011) considered the contribution of markers for the pathophysiological process of AD. These criteria divided biomarkers of AD in two classes: a) Biomarkers of brain Aβ deposition –i.e. CSF Aβ 1-42 levels and brain Positron Emission Tomography (PET) amyloid imaging - and b) biomarkers of downstream neuronal degeneration or injury –i.e. CSF total and phosphorylated forms of tau; 18 fluorodeoxyglucose (FDG) PET imaging of temporo - parietal cortex; and atrophy on Nuclear Magnetic Resonance (NMR) imaging in medial, basal, and lateral temporal lobe, and medial parietal cortex. The contribution of AD biomarkers has made possible to raise the concept of **preclinical AD** as a diagnostic category based on AD biomarkers modifications without definite cognitive decline (Sperling et al., 2011).

2.1 Role of biomarkers in AD

A biomarker corresponds to an indicator of the presence or extent of disease, which is directly associated with the clinical features and prognosis of the disease. Biomarkers for cognitive impairment and dementia have been proposed by several research groups in recent years (Maccioni et al., 2004). The consensus report of "The Ronald and Nancy Reagan Research Institute of the Alzheimer's Association" and the "National Institute on Aging working group" on "Molecular and Biochemical Markers of Alzheimer's Disease" (The Ronald and Nancy Reagan Research Institute of the Alzheimer's and Association National Institute on Aging Working Group, 1998) listed the specific criteria and features for an ideal biomarker of AD. It "should detect a fundamental feature of neuropathology and be validated in neuropathologically-confirmed cases: it should have a sensitivity >80% for detecting AD and a specificity of >80% for distinguishing other dementias: it should be reliable, reproducible non-invasive, simple to perform, and inexpensive". Based on these criteria, scientists have been able to discard many substances that are unsuitable as biomarkers and do not contribute or improve AD diagnosis (Mulder et al., 2000).

Biochemical markers for AD have been extensively sought in bodily fluids, and key proteins of AD neuropathology -Aβ, tau and hyperphosphorylated tau isoforms- have been evaluated as potential markers for AD diagnosis and for follow up in clinical trials.

2.1.1 Biomarkers in CSF

As CSF is in close contact with nerve tissue, it is not surprising that CSF has been considered a reliable indicator of brain tissue environment. Most published literature described two main types of CSF-based biomarkers :

- **Amyloid beta (Aβ) levels:** in AD CSF the concentration of Aβ, fraction 1-42 (Aβ 1-42) is regularly reduced to less than 50% of its normal value and has been considered a reliable marker of AD, with a sensitivity of 78% and 81-83% specificity (Wiltfang et al., 2007). Low levels of Aβ 1-42 can also predict the onset of cognitive decline in older women without dementia (Gustafson et al., 2007). Meanwhile, since the fraction Aβ 1-40 is the major constituent of total CSF Aβ, Aβ 1-42 / Aβ 1-40 ratio has also been evaluated and has been proposed as a better marker than isolated Aβ 1-42 levels (Wiltfang et al., 2007).
- **Tau and phosphorylated tau**: Tau protein is aggregated in paired helicoidal filaments (PHFs) and neurofibrillary tangles (NFTs) of AD brains and has been proposed as a pathogenic protein in the disease (Maccioni, 2011). Tau levels are also increased in CSF of AD patients, so they have been studied as suitable biomarkers. In patients with mild cognitive impairment -that in many cases will progress to dementia- CSF tau levels can differentiate those that correspond to depressive syndromes from those that will effectively progress to AD (Schönknecht et al., 2007).

Under pathogenic conditions tau undergoes several modifications that include phosphorylation, truncation, glycation, etc. (Farías et al., 2011), so these forms of modified tau have also been evaluated as biological markers. Tau phosphorylated at threonine 181 (p-tau 181) demonstrates to be useful for differentiating control and AD subjects from subjects with dementia with Lewy Bodies, being a better marker of AD that Aβ 1-42 and total tau (Vanderstichele et al., 2006). Hyperphosphorylated tau increases in AD subjects, as well as in those with mild cognitive impairment that will progress to AD (Andersson et al., 2007; Maccioni et al., 2006).

Aβ 1-42, total tau and p-tau may serve as useful markers to predict progression from mild cognitive impairment to AD (Diniz et al., 2008). CSF p-tau levels also may have a role monitoring response to treatment (Degerman et al., 2007) However, the real value of CSF markers to predict progression of cognitive decline is disputed and may be less robust than cognitive assessment to predict conversion from mild cognitive impairment to AD (Gomar et al., 2011). Apolipoprotein E (Apo E) ε4 genotype may be related to levels of biomarkers in CSF, since increased levels of total tau and p-tau and decreased Aβ 1-42 have been described in CSF of patients with severe involvement of episodic memory and Apo E-ε4 (+) (Andersson et al., 2007).

Although Aβ and tau levels in CSF are the most studied and validated biological markers of AD with enough stability, (Slats et al., 2011); the mayor pitfall of CSF biological markers is the necessity of invasive techniques such as lumbar puncture to obtain samples. Adverse effects are present at 11.7% of subjects, being with clinically significant at 3.97%, including post lumbar puncture headache 0.98% - 5% (Maccioni et al., 2006; Peskind et al., 2005). As a way to face problems of CSF analyses, new and non-invasive biomarkers available in blood, saliva and urine are currently under investigation.

2.1.2 Peripheral biomarkers

Levels of Aβ have been studied in plasma of AD patients. However, Aβ 1-40 levels are not specific for AD and in fact are affected by age (Luchsinger et al., 2007). On the other hand, plasma Aβ 1-42 may be altered early in the disease, but since this marker is not reliable enough, correlation with neuroimages or other biological markers is needed (Blasko et al. 2008). Other metabolic and nutritional markers have also been studied, including levels of folic acid and vitamin B12 but results are conflicting so far (Irizarry et al. 2005; Isobe et al. 2005; Köseoglu & Karaman 2007; Serot et al. 2005; Seshadri et al. 2002).

2.1.3 Apo E polymorphisms

Apo E is a plasma protein involved in cholesterol transport. In the CNS, Apo E is also involved in growth and repair of the nervous system during development and after injury. The Apo E gene has 3 alleles: ε2, ε3, ε4. The ε4 allele is associated with an increased risk of AD (Rojo et al., 2006), ε4 allele is present among 40 to 50% AD subjects (Farrer et al.,1997). Actually Apo E- ε4 is considered as a risk factor of AD (Mayeux et al., 1998).

2.1.4 Inflammatory markers

Proinflammatory molecules have been studied as potential peripheral markers of AD. As stated in preceding paragraphs, in the context of neuroimmunomodulation hypothesis, there is consistent evidence that inflammatory mechanisms play an important role in AD pathophysiology. However, results are inconsistent. In patients with AD elevated levels of plasma soluble CD-40 and a decrease in TGF-β1 have been described, while assessments of IL-1, IL-2, IL-6 and TNF-α have yielded conflicting results (Rojo et al, 2008).

2.1.5 Altered p53

Alterations in the tertiary folding of p53 protein can be recognized in fibroblasts from patients with AD (Uberti et al., 2006). This altered protein is also present in blood mononuclear cells of AD patients. Measurements of these p53 variants by cytofluorometry and immunoprecipitation techniques may serve as AD biomarker with high sensitivity and specificity (90% and 77% respectively) (Lanni et al, 2008; Uberti et al., 2008).

2.1.6 Platelets Amyloid Precursor Peptide (APP)

APP is a transmembrane protein that, by proteolytic cleavage, generates Aβ, the major component of senile plaques. Therefore, APP could be a useful biomarker in AD. Platelets carry more than 95% of circulating APP, containing all the necessary machinery for APP metabolism, so there has been postulated that changes in platelets metabolism that include, but are not limited to APP processing, may correlate to brain pathophysiological processes of AD (Hochstrasser et al., 2011; Neumann et al., 2011; Zainaghi et al., 2007).

Several APP fractions can be resolved by electrophoresis and immunoblot techniques of platelets extracts. Analyses have found a reduction in 130 kDa APP isoforms in relation to 110 kDa APP in AD patients. These alterations in platelets APP ratio are related to severity and progression of the disease (Borroni et al., 2006) High sensitivity and specificity -around 80 to 95%- have been described for this technique (Borroni et al., 2006; Padovani et al., 2002). Platelets APP ratio may be altered early in AD and can be used to detect the conversion of mild cognitive impairment to AD (Borroni et al., 2003; Borroni et al., 2006) and also to monitor treatment responses (Borroni et al., 2001; Liu et al.,2005). However this method is not quantitative and there are important differences in reported data between different studies; this is likely due to differences in multiple steps of sample management and processing.

2.1.7 Platelets tau

Our group has recently demonstrated that platelets also contain tau protein. High molecular weight forms of tau that probably correspond to oligomeric protein can be resolved by electrophoresis and immunoblot with tau specific antibodies. The ratio of high molecular weight tau to normal weight tau in platelets is increased in AD patients so this kind of analysis may represent a novel biomarker for AD (Neumann et al., 2011).

2.2 Disease specific radiotracers: New avenues to pathology-specific imaging technologies

The development of new NMR and PET imaging technologies has become a topic of major interest for both clinical and fundamental neuroscientists over the past few years, as it presents the unparalleled possibility of visualizing pathological processes in the brain parenchyma in a non-invasive and real time manner. Major progresses have been achieved in this field in the past decade mainly due to the use of functional NMR technologies and innovative PET tracers. However, although these current neuroimaging technologies provide precise information on structural and functional aspects of the brain, they have failed to provide information on the specific pathological processes and structural alterations occurred in different neurodegenerative diseases, including Alzheimer. Therefore, the development of new pathology-specific imaging technologies is still an urgent need. This would allow us to make a more accurate diagnose of brain disorders and also to efficiently monitor a number of experimental therapies currently under investigation.

Regarding AD-specific PET tomography, researchers have focused their attention mainly on obtaining maps of the proposed hallmark lesions of this disease, i.e., the senile plaques (SP) and the neurofibrillary tangles (NFT´s) formed by hyperphosphorylated tau. After the publication of Klunk et al. (Klunk et al., 2004) reporting the potential application of Pittsburgh Compound-B as a specific radiotracer for the amyloid deposits in the human brain, a new era in the development of *in vivo* AD neuroimaging seems to have started. Almost at the same time Verhoeff et al. (Verhoeff et al., 2004) published a similar study with another PET radiotracer. None of these studies probed to be applicable for diagnosis of early stages of AD. However they helped us to understand the clinical significance of visualizing cerebral amyloid burden in AD diagnosis. Not long after reports on amyloid-specific PET tracers were published, several groups including ours pointed out the relevance of addressing this challenge from a seemingly more relevant a -and perhaps more efficient-perspective, which is visualizing aggregated forms of tau protein. (Rojo et al., 2007a, 2007b). Recently we reported the potential of benzimidazoles derivatives as pathology-specific PET tracers (Rojo et al., 2010). This work led us to discover that FDA-approved drugs, such as **Lansoprazole and Astemizole** (Rojo et al., 2010), were promising candidates for AD-specific radiotracers (Figures 2 and 3).

The existence of a pathology-specific neuroimaging technology of AD would also allow a rational evaluation of the biological effects of a number of experimental pharmacological therapies available presently, as well as other promising tools to treat AD patients, and methods for clinical trial of anti-tau therapeutic approaches (Rojo et al., 2011).

Molecular structure of these ligands varies from large proteins and peptides such as the A peptide and radio-active monoclonal antibodies to small molecules derived from Congo red, Chrisamina-G, tioflavine-T, and acridine orange (Figure 2). Recent studies have demonstrated that is possible to obtain images of plaques and NFTs *in vivo* whether separately or simultaneously. So far, the most successful molecules have been those with a relatively low molecular weight (Figure 2) (Mathis et al., 2005). It has been shown that some benzimidazole and quinoline derivatives tag aggregated forms of tau *in vitro* and in the context of human brain (Mathis et al., 2005; Okamura et al., 2004; Okamura et al., 2005; Rojo et al., 2007a). This could serve as the milestone for developing neuroimaging technologies to visualize NFTs in the brain of AD patients and those affected with mild cognitive impairments (MCI). We believe that in the future, significant progress will be achieved in this area due to the recent discovery of different benzimidazoles and benzothiazoles with high affinity for brain aggregates of tau protein (Rojo et al., 2007b). Another important step in this area is the search of FDA-approved drug with similar structural features to those of Thioflavine T and other benzothiazole compounds. This implies the possibility of skipping expensive, cumbersome and time-consuming safety studies in humans for their approval in AD diagnosis *in vivo* (Rojo et al, 2010, 2011).

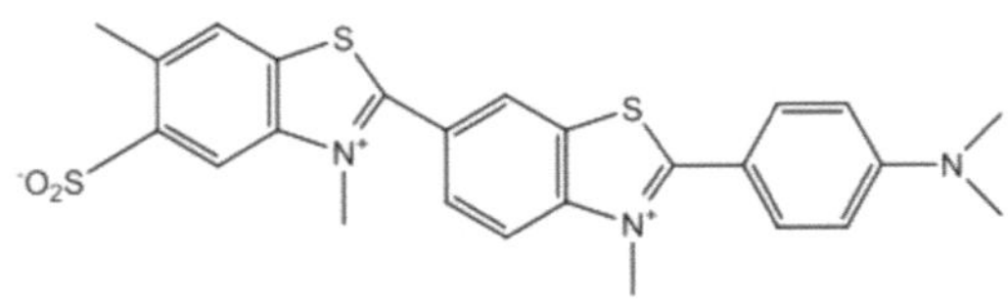

ThS

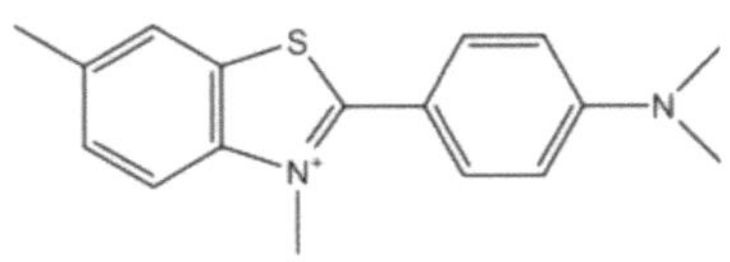

ThT

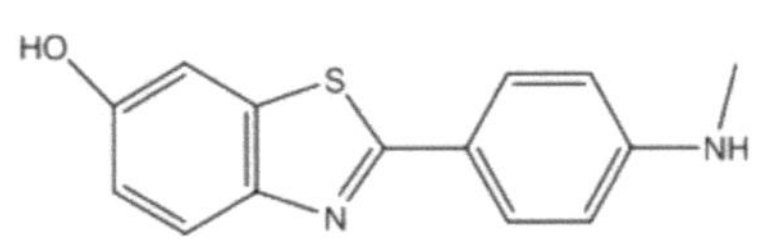

PIB

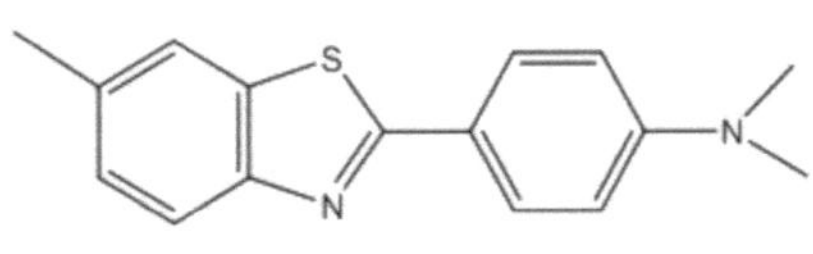

6-ME-BTA-2

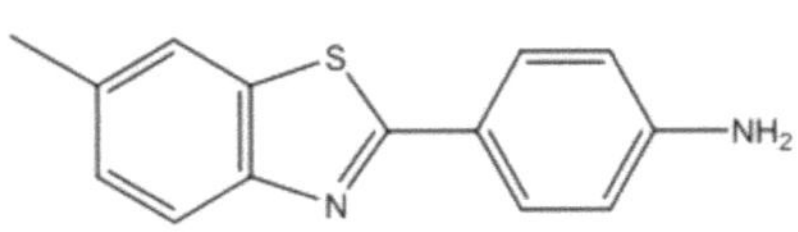

6-ME-BTA-0

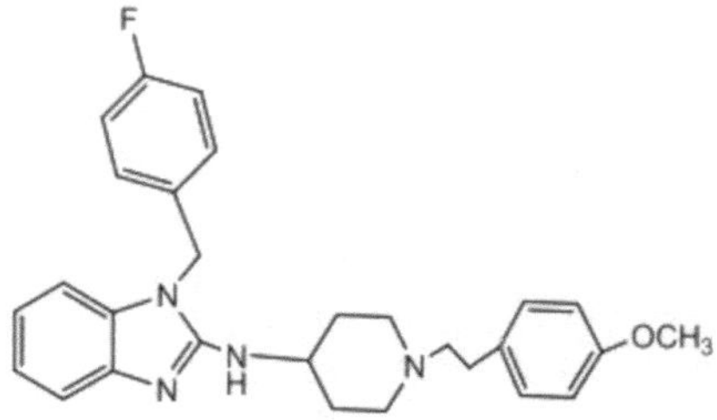

Astemizole

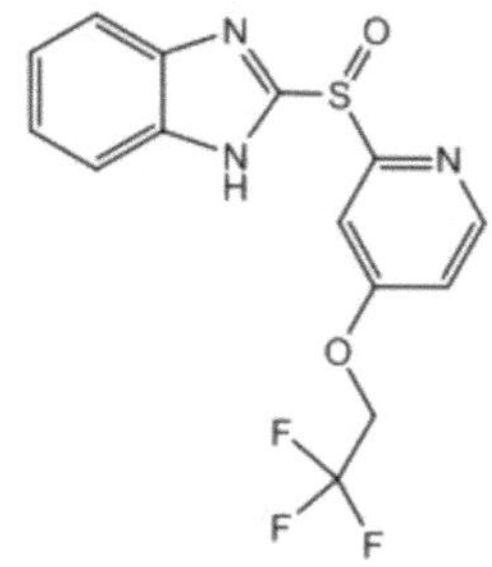

Lansoprazole

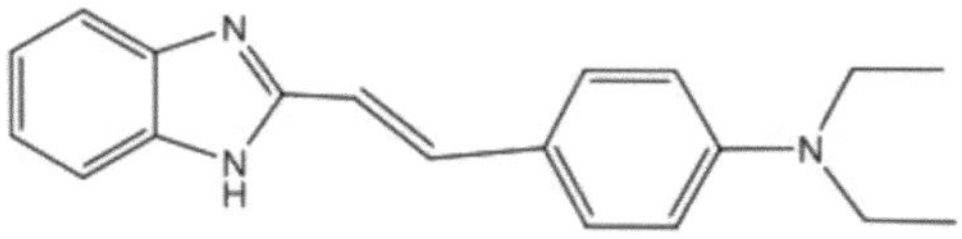

BF 126

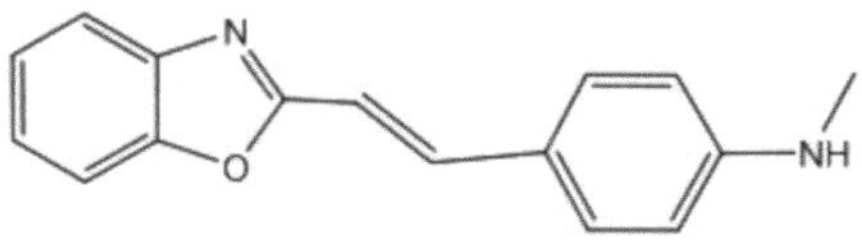

BF 168

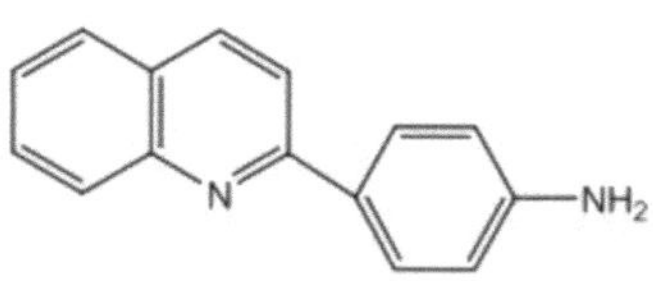

BF-170

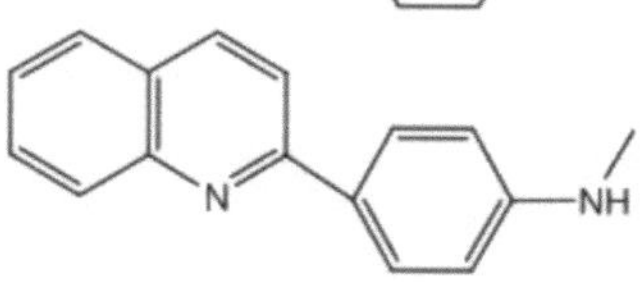

BF-158

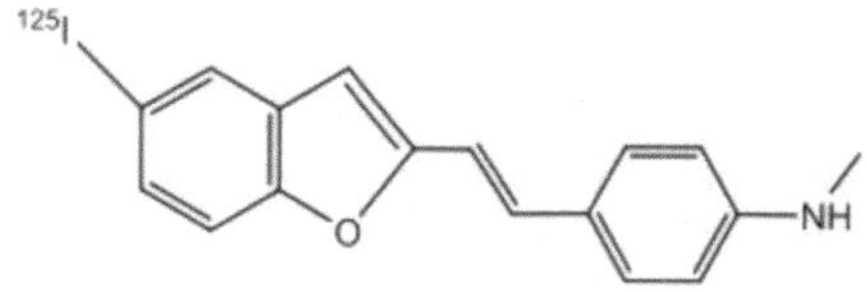

BF 180

Fig. 2. **Benzimidazole and benzothiazole derivatives proposed as potential biomarkers for PET imaging in AD**. Several small molecules have been proposed as PET tracers for both amyloid and tau aggregates. In this figure ThS shows the proposed structures for Thioflavine S (ThS); Thioflavine T (ThT); Pittsburgh compound (PIB), and other amyloid specific radiotracers such as 6-ME-BTA 2 and 6ME-BTA-0. Also here the figure shows the NFTs-specific proposed PET tracers Astemizole, Lansoprazole, BF-126, BF 170, and BF-158.

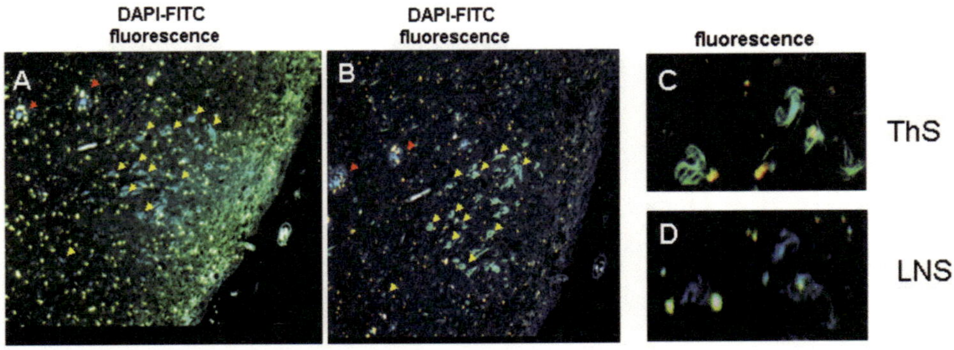

Fig. 3. **Neuropathological staining of brain sections from the entorhinal cortex of AD patients.** Senile plaques (red arrowheads) and NTFs (yellow arrowheads) can be clearly tagged by Thioflavine-S (B, C). Lansoprazole (A,D) tagged NFTs and neurite-like structures in the core of senile plaques.

3. Novel approaches toward prevention and treatment of Alzheimer´s Disease (AD)

AD is the most common type of dementia characterized by the formation of two main protein aggregates in the brain: senile plaques (SP) consisting of the amyloid-β peptide and neurofibrillary tangles (NFT´s), consisting of the microtubule-associated protein tau. Tau accumulates in a hyperphosphorylated state forming intracellular deposits named to as paired helical filaments which generate the NFT´s (Maccioni et al., 2010). Formally approved during the past two decades, pharmacological treatments for AD are mainly based on restoring the levels of acetylcholine transmission in the brain being essentially symptomatic therapies. The anticholinesterase (anti-ChE) agents currently used such as rivastigmine, donepezil and galantamine failed in providing a substantial improvement in the mental health condition of AD patients (Aizenstein, 2008). Cholinesterase inhibitors appear to increase phosphorylated tau in AD (Chalmers et al., 2009). Anti-ChE drugs are being used for symptomatic treatment of mild to moderate AD. Tacrine was the first anti-ChE which showed positive clinical results, however, it is not in use any more due to severe hepatotoxicity. Through the progress of AD, brain cholinergic neurotransmission becomes significantly diminished, thus limiting clinical efficacy of the above mentioned anti-ChE agents. A new drug is being used Cerebrolysine™, based on a combination of peptides and administered intravenously, has shown discrete results. On the other hand the drug memantine, which modulates NMDA-related pathways in brain, has shown moderate results in cases of mild to advanced stages of AD. Moreover, nonsteroidal anti-inflammatory drugs (NSAIDs) appears as promising treatments according to epidemiological studies are able to reduce the risk of developing AD. A 2006 pilot study showed small but significant improvements in various cognitive rating scales in patients with AD after treatment with etanercept (Tobinick et al. 2008a,b; Navarrete et al., 2011). A further study, administering to a single AD patient via perispinal infusion, showed rapid and significant improvement of Alzheimer's symptoms. Nowadays over 300 compounds are being tested for AD at different stages of development, 175 of them are being evaluated at the level of clinical trials. However, the finalized studies have shown only negative results, creating great concern among the medical community who is still expecting an efficacious therapy for AD.

Another compound is huperzine A, an acetylcholinesterase inhibitor that occurs naturally in a species of moss that has been used in China for centuries for the treatment of blood disorders (Wang et al., 2011). The herb has been used to treat AD in China since the late 1990's and is sold in the US as a dietary supplement to help maintain memory (Rafii et al., 2011). The first synthetic approach to produce huperzine A, was recently published aiming to replace the only natural source of huperzine A, the plant *Huperzia serrata,* which produces small amounts of huperzine A. Another drug, memantine directed to NMDA receptors, shows only moderate actions in advanced cases of the disease (Raina et al., 2008). On the other hand, recent anti-amyloid strategies, have failed in their efficacy or safety on their last development phases (Holmes et al., 2008). Statins appear to reduce the burden of NFTs, but clinical studies are not conclusive (Rojo et al., 2006; Boimel et al., 2009). In the whole context, tau based therapies represent a potential therapeutic target, specifically those that that diminish its aggregation, or alter its hyperphosphorylation (Alvarez et al., 2001). To those agents, several anti-tau miscellaneous strategies such as normal microtubule-stabilizing agents can be added to the new search for anti-AD drugs (Maccioni, 2011; Navarrete et al., 2011). Thus, a combination of molecules such as anti-tau agents will be determinant for a substantial control of AD in the future.

3.1 Searching for innovative tau aggregation inhibitors

Physiologically tau stabilizes the microtubule structure, but in the neurons of patients with AD the microtubule system is believed to be disrupted, with the concomitant axonal transport deficits and degeneration (Farias et al., 2011). Several lines of evidence have shown that tau aggregation is the main event involved in the neurodegenerative process, due to the conversion of either soluble tau or oligomers into insoluble filaments. This process correlates with the clinical progression of AD and cognitive impairment (Maccioni and Perry, 2009). The identification of mutations in the tau gene in hereditary frontotemporal dementia revealed that tau dysfunction is central to neurodegeneration (Nakashima et al., 2005). Thus, improvement in the cognition of a transgenic model displaying both NFT´s and SP depends on blockage of tau filaments formation (Zhang et al., 2005). Cellular models where tau is overexpressed evidence the cytotoxicity of formed intracellular aggregates. In the search of new molecules for the treatment of AD, many drugs focused on Aβ aggregation have failed in stopping the progression of the disease (Navarrete et al., 2011). The immunization against Aβ was effective in reducing amyloid plaque load, but it had little effect on improving cognitive functions. A recent study shows the failure of a phase III clinical trial with a γ-secretase inhibitor (Carlson et al., 2011; Lleo and Saura, 2011). Therefore it seems timely to consider alternative drug discovery strategies for AD based on approaches directed at reducing misfolded tau and compensating for the loss of normal tau function.

3.2 Tau hypothesis in the context of the AD clinic

Nowadays AD etiopathogenesis is not yet established, despite different and numerous hypotheses (Maccioni and Perry, 2009). Nevertheless, there is agreement about its early onset, as a result of the convergence of a set of genetic and/ or environmental factors, which vary on time among patients and increase along with age (Glatz et al., 2006; Maccioni et al., 2010). At different stages of the process, a cascade of pathological events is triggered, where factors such as Aβ oligomers, iron overload or oxygen free radicals modify microglial cells thus inducing anomalous signaling to neuronal cells (Fernandez et al., 2008; Maccioni et al., 2009; Morales et al., 2010). These events finally result in alterations of cellular signposting and biochemical abnormalities that lead to cellular dysfunction, the lack of neurotransmission, cellular death and clinical expression of dementia.

For years, the dominant hypothesis was that of the amyloid cascade, which sets the amyloid precursor protein (APP) metabolism dysfunction on the central nervous system, as the responsible agent for extraneuronal formation of SP (Hardy & Selkoe, 2002; Hardy, 2009). The major problem of this postulate is that a significant number of cognitively healthy elderly people, also exhibit abundant amyloid plaques, without a cognitive function impairment. Besides, diverse clinical assays carried out with different antiamyloid molecules, despite consistent effects on preclinical stages, have not shown cognitive and/or functional benefits in treated patients at advance stages of AD (Aizenstein et al., 2008; Panza et al., 2011). In this context, as the amyloid cascade hypothesis does not allow to explain the integrity of AD pathogenesis, the interest on tau hypothesis and neurofibrillary tangles, that sets tau protein abnormal phosphorylation as the possible responsible for these tangles formation, and the consequent neuronal death, has increased (Navarrete et al., 2011). On the other hand, recent data suggest an eventual connection between APP and tau protein, even though they have been treated as different contexts to physiopathologically explain AD (Alvarez et al., 2001; Otth et al., 2003; Czapski et al., 2011; Fuentes and Catalan, 2011).

Studies in APP mice crossed to mutant tau mice, injection of Aβ into the brain of these tau mutant mice and studies on neuronal cells, support the notion that Aβ aggregates can drive neurofibrillary pathology (Otth et al., 2003; Hernandez et al., 2009; Kocherhans et al., 2010). Such investigations bear out the notion that although AD may be considered a primary Aβ amyloidoses and a secondary tauopathy, tau pathology is the major factor that contributes to neurodegeneration (Maccioni et al., 2010).

Moreover, FDA approved drugs over last decades are seemingly drugs which only reinforce cholinergic neurotransmission, as donepezil, galantamine and rivastigmine, or that moderates glutamate/NMDA receptor memantine are approved. After several years of clinical experience on drugs use, it can be concluded that AD treatment with cholinesterase inhibitors and memantine, is essentially symptomatic (Raina et al., 2008). Even though it may result in a moderate improvement, lacks of a real clinical output relative to cognition measurements and global evaluation of dementia. Recent *in vitro* and *in vivo* data suggest that cholinergic drugs may even have negative impact on e amiloyd-β-peptide and tau behavior. According to recent studies, patients treated with ChEIs had accumulated significantly more phospho-tau in their cerebral cortex compared to untreated patients. This data suggests the possibility that increased tau phosphorylation may influence long-term clinical responsiveness to ChEIs (Chalmers et al., 2009; Fuentes and Catalan, 2011).

Efforts to develop drugs more focused on AD underlying pathology, have considered different agents, called disease modifiers, and linked to diverse etiopathogenic hypotheses. Antiamyloid strategies, such as active or passive immunization, or secretase inhibitors, have been predominant, nevertheless they have failed on the efficacy or security on their last development phases (Lleo and Saura, 2011). Actually, molecules that could restrict tau aggregates and the consequent formation of neurofibrillary tangles, have already begun to be explored on clinical trials, having the consideration that the last mentioned lesions, are the responsible for most of the AD cognitive impairments. However, the only anti-tau therapies that have reached the human clinical trial stage are lithium, methylene blue and NAP (Nakashima et al., 2005; Medina et al., 2011; Navarrete et al., 2011).

3.3 Anti-tau miscellaneous strategies

A variety of intracellular proteins have been implicated in regulating both tau aggregation and folding, or potentially mediate clearance of the misfolded and aggregated tau protein. In this context, the ubiquitin ligase C-terminus of heat shock cognate70-interacting protein (CHIP) can polyubiquitinate tau and may play a crucial role in preventing accumulation of phospho-tau and NFTs (Staff et al., 2008). Studies suggest that modulation of CHIP and the ubiquitin preoteasome system could alter tau pathology. Finally, heat shock proteins have been suggested as possible modifiers of tau pathology. HSP90 inhibitors that induce a heat shock response reduce tau phosphorylation at certain sites and are currently being tested in humans as anti-cancer agents (Dickey, 2007, reviewed in Fuentes and Catalan, 2011). Thus, on the basis of information that sequestration of tau results in loss of the normal microtubule-stabilizing function, normal microtubule-stabilizing agents have been tested in several tau mouse models. Paclitaxel administered to the tau mouse model, in a micellar formulation increased microtubule stability and rendered these polymers less dynamics. After three-months of paclitaxel treatment, transgenic mice showed rise on fast axonal transport and of the microtubules bundles in neuronal cells (Zhang et al., 2005). Authors also showed motor function improvement in comparison to the not-treated mice. Considering that paclitaxel does not cross hematoencephalic barrier, its action would be mediated through retrograde transport to spinal motoneurons among other possible explanations.

Moreover, there is another compound named to as NAP, a derivative octapeptide of a natural neurotrophic protein, which cross the blood brain barrier, and has shown to promote microtubules assembly (Matsuoka et al., 2008). Nasal administration for several months to elderly mice that had developed tau aggregations and Aβ deposition, resulted in reduction on tau phosphorylation and Aβ levels, with a cognitive function improvement (Matsuoka et al., 2008). A similar approach has also been employed in an animal model of tauopathy, anti-tau pathologically phosphorylated immunotherapy, where a diminished charge of NFT´s was observed, and the presence of serum antibodies, without evidence of clinical deficits or encephalitis.

3.4 AD prevention: The emergence of natural products in the control of tau pathology

The development of small-molecules that inhibit the aggregation of tau appears as a valid therapeutic target for treatment of AD and as a consequence of the failure on drugs directed against the amyloid and the cumulative evidence in favor of tau hypothesis, current therapeutic strategies are aimed at searching for compounds that can either inhibit the formation of pathological tau filaments or disaggregate them. This hypothesis has been favored by current findings on the compound *methylthioninium chloride* (known as methylene blue), a previously described inhibitor of tau aggregation. A recent study with this compound in phase II clinical trial shows an 81% reduction of cognitive decline with the use of the compound as compared to placebo (Wischik et al., 1996; Medina et al., 2011).

Compounds described for their anti-aggregating capacity in the formation of amyloid aggregates are the polyphenols. In this context, synthetic polyphenols have proved effectiveness in the inhibition of heparin-induced tau aggregation. Following this approach, the current therapeutic strategies are aimed to look for natural phytochemicals and polyphenolic extracts that can be able to either inhibit or disaggregate tau filament formation (Bastianetto et al., 2008; Kim et al., 2010; Cornejo et al., 2011). It has been suggested that naturally occurring phytochemicals have the potential to prevent AD based on their anti-amyloidogenic, anti-oxidative and anti-inflammatory properties. Despite this, there are few phytocomplexes emerging in order to prevent tau aggregation. Only a cinnamon extract and a grape seed polyphenolic extract have been described for this purpose (Peterson et al., 2009). Fulvic acid is one of the most interesting phytocomplex molecules (Goshal et al., 1990). This is a mixture of polyphenolic acid compounds resulting from the long-term microbial degradation of lignin, among other sources. It has several nutraceutical properties, and is one of the most interesting naturally-occurring phytochemicals for their extremely high antioxidant properties and apparent neuroprotective effect. For instance, the interaction of prion protein with fulvic acid and its inhibitory effect on the content of β-sheet structure and the formation of protein aggregates has been described in detail. Only a few polyphenolic molecules have emerged to prevent tau aggregation, and natural drugs targeting against tau have not been approved yet (Peterson et al., 2009; Cornejo et al., 2011). Fulvic acid, a humic substance, has several nutraceutical properties with potential activity to protect cognitive impairment. In this work we provide evidence to show that aggregation process of tau protein, forming paired helical filaments (PHFs) *in vitro*, is inhibited by fulvic acid affecting the length of fibrils and their morphology (Cornejo et al., 2011; Carrasco et al., unpublished results). In addition, we investigated whether fulvic acid is capable of disassembling preformed PHFs. We showed by mean of analysis of aggregation, atomic force microscopy (AFM) and electron microscopy that the fulvic acid is an active compound against pre formed fibrils affecting the whole structure by diminishing length of PHFs and probably acting at the hydrophobic level, as we observed by mean of atomic force techniques. Thus, fulvic acid is likely to provide a new insight to develop potential treatments for AD based on natural products. These observations allowed us to conclude that fulvic acid inhibits heparin-induced tau aggregation *in vitro*. On the other hand, fulvic acid promotes the disassembling of tau preformed fibrils. Thus, fulvic acid could provide a new insight for developing treatments based on natural products for AD (Cornejo et al., 2011; Farias et al, unpublished observations).

4. Conclusion

A major hallmark of AD is the presence of NFT´s containing tau protein. The neuroimmunomodulation theory of AD together with the revitalized tau hypothesis on Alzheimer´s pathogenesis provided a fundamental paradigm to understand this disease. This is very important considering that the slow progress in therapeutic approaches has been the result of a lack of a solid paradigm on this devastating disease. In this context, beside the anticholinesterases, most researchers have focused on drugs that affect the production of β-amyloid or disassembly of senile plaques, with very limited results. Therefore, tau became a major target for future therapeutic approaches. As tau clearly presents a potential therapeutic target in AD, there is a high rise on new drugs investigation that would early interfere with the cascade that leads to tangles formation, and that might contribute to control neuronal degeneration and cognitive impairment. A critical step in the design of potential strategies to control AD is to find reliable biomarkers for its early diagnosis. Despite many efforts in this direction no markers to detect AD at the pre-symptomatic level are available. After the acceptance of tau/amyloid biomarker in the CSF, research is directed to establish a non-invasive marker technology. Innovative studies point to an *in vivo* PET technology based on neuroimaging of NFT´s and tau filaments by using lansoprazole as a radiotracer, and blood biomarkers based on altered tau and amyloid variants in platelets.

Considering the scenario in which new synthesized drugs and novel therapeutic approaches have failed in their clinical trials, new hopes come from the search of natural products and phytocomplexes. Polyphenols have been described for their anti-aggregating capacity in the formation of amyloid aggregates. Most recently, synthetic polyphenols have proved effectiveness in the inhibition of heparin-induced tau aggregation. Following this approach, the current therapeutic strategies are aimed at looking for natural phytochemicals and polyphenolic extracts able to either inhibit or disaggregate tau filament formation. In addition, it has been suggested that naturally occurring phytocomplexes have the potential to prevent AD based on their neuroprotective, anti-oxidative and anti-inflammatory properties. Despite this, there are few natural complexes emerging in order to prevent tau aggregation. These include cinnamon and grape extracts, the anti-oxidant resveratrol, and recently fulvic acid. The combination of vitamins essential for brain health such as folic acid, vitamins B6 and 12 with natural compounds such as natural extracts from plants, flavones, flavonoids and the natural product *shilajit* offer an interesting approach toward the therapy of Alzheimer´s disease.

5. Acknowledgement

We acknowledge support from Fondecyt 1110373 and grants from CORFO 10ANT8051, the Alzheimer´s Association and the International Center for Biomedicine (ICC) to Prof. RB Maccioni. RBM and LR are grateful to Kishan Nakrani for his assistance with writing this publication.

6. References

Aizenstein H, Nebes R, Saxton J, Price J, Mathis Ch, Tsopelas N, Ziolko S, James J, Snitz B, Houck P, Bi W, Cohen A, Lopresti B, DeKosky S, Halligan E & Klunk W. (2008). Frequent amyloid deposition without significant cognitive impairment among the elderly. *Arch Neurol.* (65)11:1509-1517

Alvarez A. J.P. Muñoz, and R.B. Maccioni (2001) "A cdk5/p35 stable complex is involved in the beta-amyloid induced deregulation of Cdk5 activity in hippocampal neurons". *Experimental Cell Research* 264: 266-275.

Alzheimer, A. (1907). Uber eine eigenartige Erkrankung der Hirnrinde. *Allgemeine Zeitschrife Psychiatrie.* vol. 64 pp. 146-148.

Andersson, C., Blennow, K., Johansson, S., Almkvist, O., Engfeldt, P., Lindau, M., & Eriksdotter-Jönhagen, M. (2007). Differential CSF biomarker levels in APOE-epsilon4-positive and -negative patients with memory impairment. *Dement Geriatr Cogn Disord*. 23(2):87-95.

Bastianetto, S., Krantic, S., & Quirion, R. (2008) Polyphenols as potential inhibitors of amyloid aggregation and toxicity: Possible significance toAlzheimer's disease. *Mini Rev Med Chem* 8, 429-435.

Bhaskar, K., Konerth, M., Kokiko-Cochran, O.N., Cardona, A., Ransohoff, R.M., & Lamb, B.T. (2010).Regulation of tau pathology by the microglial fractalkine receptor. *Neuron.*; 68(1):19-31.

Blasko, I., Jellinger, K., Kemmler, G., Krampla, W., Jungwirth, S., Wichart, I., Tragl, K.H., Fischer, P. (2008). Conversion from cognitive health to mild cognitive impairment and Alzheimer's disease: prediction by plasma amyloid beta 42, medial temporal lobe atrophy and homocysteine. *Neurobiol Aging*. ;29(1):1-11.

Boimel M, Grigoriadis N, Lourbopoulos A, Touloumi O, Rosenmann D, Abramsky O, Rosenmann H. (2009) Statins reduce the neurofibrillary tangle burden in a mouse model of tauopathy.*J Neuropathol Exp Neurol.*; 68(3): 314-325.

Borroni, B, Colciaghi, F., Pastorino, L., Pettenati, C., Cottini, E., Rozzini, L., Monastero, R., Lenzi, G.L., Cattabeni, F., Di Luca, M., & Padovani, A. (2001). Amyloid precursor protein in platelets of patients with Alzheimer disease: effect of acetylcholinesterase inhibitor treatment. *Arch Neurol*. Mar; *58*(3), 442-446.

Borroni, Barbara, Colciaghi, F., Caltagirone, C., Rozzini, L., Broglio, L., Cattabeni, F., Di Luca, M., & Padovani, A. (2003). Platelet amyloid precursor protein abnormalities in mild cognitive impairment predict conversion to dementia of Alzheimer type: a 2-year follow-up study. *Arch Neurol*. Dec; *60*(12): 1740-1744.

Borroni, B., Di Luca, M., & Padovani, A. (2006). Predicting Alzheimer dementia in mild cognitive impairment patients. Are biomarkers useful? *Eur J Pharmacol.*;*545*(1): 73-80.

Butler, D., Bendiske, J., Michaelis, M. L., Karanian, D. A. & Bahr, B. A. (2007). Microtubule-stabilizing agents prevent protein accumulation-induced loss of synaptic markers. *Eur J Pharmacology*. 562 (1-2):20-27.

Carlson, C., Estergard, W., Oh, J., Suhy, J., Jack, C.R. Jr, Siemers, E., & Barakos, J. Prevalence of asymptomatic vasogenic edema in pretreatment Alzheimer's disease study cohorts from phase 3 trials of semagacestat and solanezumab. (2011). *Alzheimers Dement*. ;7(4):396-401.

Cornejo A, Jiménez JM, Caballero L, Melo F, Maccioni RB. Fulvic Acid Inhibits Aggregation and Promotes Disassembly of Tau Fibrils Associated with Alzheimer's Disease. *J Alzheimers Dis*. 2011 Jul 22. [Epub ahead of print

Chalmers KA, Wilcock G, Vinters HV, Perry EK, Perry R, Ballard C & Love S. (2009). Cholinesterase inhibitors may increase phosphorylated tau in Alzheime´s disease. *J Neurol*. 256: 717-720

Cho SH, Sun B, Zhou Y, Kauppinen TM, Halabisky B, Wes P, Ransohoff RM, Gan L. (2011). CX3CR1 modulates microglial activation and protects against plaque-independent cognitive deficits in a mouse model of Alzheimer's disease. *J Biol Chem*. Sep 16; 286(37):32713-32722.

Czapski, G.A., Gąssowska, M., Songin, M., Radecka, U.D., & Strosznajder, J.B. (2011). Alterations of cyclin dependent kinase 5 expression and phosphorylation in amyloid precursor protein (APP)-transfected PC12 cells. *FEBS Lett.*;585(8):1243-8.

Degerman-Gunnarsson, M., Kilander, L., Basun, H., & Lannfelt, L. (2007) Reduction of phosphorylated tau during memantine treatment of Alzheimer's disease. *Dement Geriatr Cogn Disord.* 24(4):247-52.

Dickey CA. (2007). The high-affinity HSP90-CHIP complex recognizes and selectively degrades phosphorylated tau client proteins. *J Clin Invest.* 117:648-658.

Diniz, B.S., Pinto Júnior, J.A., & Forlenza, O. (2008). Do CSF total tau, phosphorylated tau, and beta-amyloid 42 help to predict progression of mild cognitive impairment to Alzheimer's disease? A systematic review and meta-analysis of the literature. *World J Biol Psychiatry.* 9(3), 172-182.

Dubois, B., Feldman, H.H., Jacova, C., Dekosky, S.T., Barberger-Gateau, P., Cummings, J., Delacourte, A., Galasko, D., Gauthier, S., Jicha, G., Meguro, K., O'brien, J., Pasquier, F., Robert, P., Rossor, M., Salloway, S., Stern, Y., Visser, P.J., & Scheltens, P. (2007). Research criteria for the diagnosis of Alzheimer's disease: revising the NINCDS-ADRDA criteria. *Lancet Neurol.* Aug;6(8):734-46.

Extance, A. (2010) Alzheimer's failure raises questions about disease-modifying strategies. *Nat Rev Drug Discov.*; 9(10):749-51.

Farías, G, Cornejo, A., Jiménez, J., Guzmán, L., & Maccioni, R B. (2011). Mechanisms of Tau Selfaggregation and Neurotoxicity *Current Alzheimer research* 8: 608-614

Farrer, L.A., Cupples, L.A., Haines, J.L., Hyman, B., Kukull, W.A., Mayeux, R., Myers, R.H., Pericak-Vance, M.A., Risch, N., & van Duijn, C.M. (1997). Effects of age, sex, and ethnicity on the association between apolipoprotein E genotype and Alzheimer disease. A meta-analysis. APOE and Alzheimer Disease Meta Analysis Consortium. *JAMA.* Oct 22-29;278(16):1349-56.

Fernández, J.A., Rojo, L., Kuljis, R.O., & Maccioni, R.B. (2008) The damage signals hypothesis of Alzheimer's disease pathogenesis. *J Alzheimers Dis.* ; 14(3): 329-33.

Fuentes P and Catatalan J (2011) A Clinical Perspective: Anti Tau's Treatment in Alzheimer´s Disease *Current Alzheimer Research,* 2011, 8, 686-688

Gandy S. (2010) Testing the amyloid hypothesis of Alzheimer's disease in vivo. *Lancet Neurol.* ;9(4):333-335.

Gebicke-Haerter, PJ. (2001) Microglia in neurodegeneration: molecular aspects. *Microsc Res Tech.* Jul 1;54(1):47-58.

Ghosal, S., Lal, J., Jaiswal, A.K., & Bhattacharya, S.K. (1993) Effects of *shilajit* and its active constituents on learning and memory in rats. *Phytother Res.* 7, 29-34.

Glatz D., Rujescu D., Tang Y., Berendt FJ., Hartmann AM., Faltraco F., Rosenberg C., Hulette C., Jellinger K., Hampel H., Riederer P., Möll HJ., Andreadis A., Henkel K. & Stamm S. (2006). The alternative splicing of tau exon 10 and its regulatory proteins CLK2 and TRA2-BETA1 changes in sporadic Alzheimer's disease. *J. Neurochemistry.* 96, Issue 3, pages 635–644.

Gomar, J.J., Bobes-Bascaran, M.T., Conejero-Goldberg, C., Davies, P., Goldberg, T.E., & for the Alzheimer's Disease Neuroimaging Initiative. (2011). Utility of Combinations of Biomarkers, Cognitive Markers, and Risk Factors to Predict Conversion From Mild

Cognitive Impairment to Alzheimer Disease in Patients in the Alzheimer's Disease Neuroimaging Initiative. *Arch Gen Psychiatry.*; 68(9):961-969.

Gorlovoy, P., Larionov, S., Pham, T.T., & Neumann, H. (2009) Accumulation of tau induced in neurites by microglial proinflammatory mediators. *FASEB J.*; 23(8):2502-13.

Ghosal, I (1990) "Chemistry of *shilajit*, an immunomodulatory Ayurvedic rasayan", *Pure and Applied Chemistry*, vol. 62, no. 7, pp. 1285-1288.

Gustafson, D.R., Skoog, I., Rosengren, L., Zetterberg, H., & Blennow, K. (2007). Cerebrospinal fluid beta-amyloid 1-42 concentration may predict cognitive decline in older women. *J Neurol Neurosurg Psychiatry.*; 78(5):461-4.

Hampel H., Hampel H, Ewers M, Bürger K, Annas P, Mörtberg A, Bogstedt A, Frölich L, Schröder J, Schönknecht P, Riepe MW, Kraft I, Gasser T, Leyhe T, Möller HJ, Kurz A, Basun H. (2009). Lithium trial in Alzheimer´s disease: a randomized, single-blind, placebo-controlled multicenter 10-week study. *J Clin Psychiatry.* 70:922-931

Hardy J, & Selkoe DJ. (2002). The amyloid hypothesis of Alzheimer's disease: progress and problems on the road to therapeutics. *Science*; 297(5580):353-6.

Hardy J. (2009). The amyloid hypothesis for Alzheimer's disease: a critical reappraisal. *J Neurochem.*; 110(4):1129-34.

Hernandez, P., Lee, G., Sjoberg, M., & Maccioni, R.B. (2009). Tau phosphorylation by cdk5 and Fyn in response to amyloid peptide Abeta (25-35): involvement of lipid rafts. *J Alzheimers Dis.* 16(1):149-56.

Hochstrasser, T., Ehrlich, D., Marksteiner, J., Sperner-Unterweger, B., & Humpel, C. (2011). Matrix Metalloproteinase-2 and Epidermal Growth Factor are Decreased in Platelets of Alzheimer Patients. *Curr Alzheimer Res.* (In press).

Holmes C, Boche D, Wilkinson D, Yadegarfar G, Hopkins V, Bayer A, Jones RW, Bullock R, Love S, Neal J, Zotova E, Nicoll J. (2008) Long-term effects of $A\beta_{42}$ immunisation in Alzheimer´s disease: follow-up of randomized, placebo-controlled phase I trial. *Lancet.* 372:216-23.

ICAD 2008: Alzheimer's Association International Conference on Alzheimer's Disease.

Irizarry, M.C., Gurol, M.E., Raju, S., Diaz-Arrastia, R., Locascio, J.J., Tennis, M., Hyman, B.T., Growdon, J.H., Greenberg, S.M., & Bottiglieri, T. (2005). Association of homocysteine with plasma amyloid beta protein in aging and neurodegenerative disease. *Neurology*; *65*(9):1402-1408.

Isobe, C., Murata, T., Sato, C., & Terayama, Y. (2005). Increase of total homocysteine concentration in cerebrospinal fluid in patients with Alzheimer´s disease and Parkinson´s disease. *Life Sci.*; *77*(15): 1836-1843.

Kim, J., Lee, H.J., & Lee, K.W. (2010). Naturally occurring phytochemicals for the prevention of Alzheimer's disease. *J Neurochem.*; 112(6):1415-30.

Kitazawa, M., Oddo, S., Yamasaki, T.R., Green, K.N., & LaFerla, F.M. (2005). Lipopolysaccharide-induced inflammation exacerbates tau pathology by a cyclin-dependent kinase 5-mediated pathway in a transgenic model of Alzheimer's disease. *J Neurosci.* Sep 28;25(39):8843-53.

Klunk,W.E., Engler, H., Nordberg, A., Wang, Y., Blomqvist, G., Holt, D.P., Bergström, M., Savitcheva, I., Huang, G.F., Estrada, S., Ausén, B., Debnath, M.L., Barletta, J., Price, J.C., Sandell, J., Lopresti, B.J., Wall, A., Koivisto, P., Antoni, G., Mathis, C.A., &

Långström, B. (2004) Imaging brain amyloid in Alzheimer's disease with Pittsburgh Compound-B. *Ann Neurol.*; 55(3):306-19.

Kocherhans, S., Madhusudan, A., Doehner, J., Breu, K.S., Nitsch, R.M., Fritschy, J.M., & Knuesel, I. (2010). Reduced Reelin expression accelerates amyloid-beta plaque formation and tau pathology in transgenic Alzheimer's disease mice. *J Neurosci.* ;30(27):9228-40.

Köseoglu, E., & Karaman, Y. (2007). Relations between homocysteine, folate and vitamin B12 in vascular dementia and in Alzheimer disease. *Clin. Biochem.* Aug; *40*(12), 859-863.

Lanni, C., Racchi, M., Mazzini, G., Ranzenigo, A., Polotti, R., Sinforiani, E., Olivari, L., Barcikowska, M., Styczynska, M., Kuznicki, J., Szybinska, A., Govoni, S., Memo, M., & Uberti, D. (2008). Conformationally altered p53: a novel Alzheimer's disease marker. *Mol Psychiatry; 13*(6), 641-647.

Lee, D.C., Rizer, J., Selenica, M.L., Reid, P., Kraft, C., Johnson, A., Blair, L., Gordon, M.N., Dickey, C.A., & Morgan, D. (2010) LPS- induced inflammation exacerbates phospho- tau pathology in rTg4510 mice. *J Neuroinflammation.* Sep 16; 7:56.

Li, Y., Liu, L., Barger, S.W., & Griffin, W.S. (2003). Interleukin-1 mediates pathological effects of microglia on tau phosphorylation and on synaptophysin synthesis in cortical neurons through a p38-MAPK pathway. *J Neurosci.*;23(5):1605-11.

Liu, H.C., Chi, C.W., Ko, S.Y., Wang, H.C., Hong, C.J., Lin, K.N., Wang, P.N., & Liu, T.Y. (2005). Cholinesterase inhibitor affects the amyloid precursor protein isoforms in patients with Alzheimer's disease. *Dement Geriatr Cogn Disord. 19*(5-6), 345-348.

Lleó, A. & Saura CA. (2011). γ-secretase substrates and their implications for drug development in Alzheimer's disease. *Curr Top Med Chem.* 11(12):1513-27.

Luchsinger, J, Tang, M., Miller, J., Green, R., Mehta, P.D., & Mayeux, R. (2007). Relation of plasma homocysteine to plasma amyloid beta levels. *Neurochem Res.*; *32*(4-5):775-781.

Lucin, K.M. & Wyss-Coray, T. (2009). Immune activation in brain aging and neurodegeneration: too much or too little? *Neuron.*; 64(1):110-22.

Maccioni, R.B., Muñoz, J.P. & Barbeito, L. (2001). The molecular bases of Alzheimer's disease and other neurodegenerative disorders. *Arch Med Res*; 32(5):367-81.

Maccioni, R.B., Lavados, M., Maccioni, C.B., & Mendoza-Naranjo, A. (2004). Biological markers of Alzheimer's disease and mild cognitive impairment. *Curr Alzheimer Res.*; 1(4):307-14.

Maccioni, R.B., Lavados, M., Guillon, M., Mujica, C., Bosch, R., Farias, G., & Fuentes, P. (2006). Anomalously phosphorylated tau and Abeta fragments in the CSF correlates with cognitive impairment in MCI subjects. *Neurobiol Aging, 27*(2), 237-244.

Maccioni, R. B. & Perry, G. (2009) *Current hypotheses and research milestones in Alzheimer´s Disease*, Ed. Springer, ISBN 978-0-387-87994-9, New York, USA.

Maccioni, R.B., Rojo, L.E., Fernández, J.A., & Kuljis, R.O. (2009) The role of neuroimmunomodulation in Alzheimer's disease. *Ann N Y Acad Sci.*; 1153:240-246.

Maccioni, R.B., Farías, G., Morales, I. & Navarrete, L. (2010) The revitalized tau hypothesis on Alzheimer's disease. *Arch Med Res.*;41(3):226-31.

Maccioni, R.B. (2011) Tau Protein and Alzheimer's Disease. *Curr Alzheimer Res.* 8(6): 607-608.

Mathis, C.A., Klunk, W.E., Price, J.C., & DeKosky, S.T. (2005). Imaging technology for neurodegenerative diseases: progress toward detection of specific pathologies. *Arch Neurol, 62*(2): 196-200.

Matsuoka, Y. et al. (2008). A neuronal microtubule-interacting agent, NAPVSIPQ, reduces tau pathology and enhances cognitive function in a mouse model of Alzheimer´s disease. *J Pharmacol Exp Ther.* 325:146-153.

Mayeux, R., Saunders, A.M., Shea, S., Mirra, S., Evans, D., Roses, A.D., Hyman, B.T., Crain, B., Tang, M.X., & Phelps, C.H. (1998). Utility of the apolipoprotein E genotype in the diagnosis of Alzheimer's disease. Alzheimer's Disease Centers Consortium on Apolipoprotein E and Alzheimer's Disease. *N Engl J Med.* Feb 19;338(8):506-11.

McGeer, P.L., Schulzer, M., & McGeer, E.G. (1996). Artritis and anti-inflammatory agents as possible protective factors for Alzheimer's disease: a review of 17 epidemiologic studies. *Neurology* 47: 425–432.

McGeer, P.L., Rogers, J., & McGeer, E.G. (2006). Inflammation, antiinflammatory agents and Alzheimer disease: the last 12 years. *J. Alzheimer's Dis.* 9 (3 Suppl): 271–276.

McKhann, G., Drachman, D., Folstein, M., Katzman, R., Price, D., & Stadlan, E. M. (1984). Clinical diagnosis of Alzheimer´s disease: report of the NINCDS-ADRDA Work Group under the auspices of Department of Health and Human Services Task Force on Alzheimer´s Disease *Neurology*; 34(7):939-44.

McKhann, G.M., Knopman, D.S., Chertkow, H., Hyman, B.T., Jack, C.R. Jr., Kawas, C.H., Klunk, W.E., Koroshetz, W.J., Manly, J.J., Mayeux, R., Mohs, R.C., Morris, J.C., Rossor, M.N., Scheltens, P., Carrillo, M.C., Thies, B., Weintraub, S., & Phelps, C.H. (2011). The diagnosis of dementia due to Alzheimer's disease: Recommendations from the National Institute on Aging-Alzheimer's Association workgroups on diagnostic guidelines for Alzheimer's disease. *Alzheimer's and Dementia*, 7(3), 263-269.

Medina, D.X., Caccamo, A., & Oddo, S. (2011). Methylene blue reduces Abeta levels and rescue early cognitive deficits by increasing proteasome activity. *Brain Pathol.*; 21(2):140-9.

Morales, I., Farías, G., & Maccioni R.B. (2010). Neuroimmunomodulation in the pathogenesis of Alzheimer's disease. *Neuroimmunomodulation*. 17(3):202-4.

Mulder, C., Scheltens, P., Visser, J.J., van Kamp, G.J., & Schutgens, R.B. (2000). Genetic and biochemical markers for Alzheimer's disease: recent developments. *Ann Clin Biochem*. Sep;37 (Pt 5):593-607.

Nakashima H, Ishihara T, Suguimoto P et al. (2005). Chronic lithium treatment decreases tau lesions by promoting ubiquitination in a mouse model of tauopathies. *ActaNeuropathol*. 110 (6):547-556

Navarrete, L. P., Pérez, P., Morales, I. & Maccioni, R.B. (2011) Novel drugs affecting tau behavior in the treatment of Alzheimer´s disease and tauopathies. Special thematic Issue "Tau protein and Alzheimer´s disease. New paradigms and future challenges" *Current Alzheimer´s Research* 8: 678-685

Neumann, K., Farías, G., Slachevsky, A., Perez, P., & Maccioni, R.B. (2011). Human Platelets Tau: A Potential Peripheral Marker for Alzheimer's Disease. *J Alzheimer´s Dis.* Jan 1; 25:103-109.

Okamura, N., Suemoto, T., Shimadzu, H., Suzuki, M., Shiomitsu, T., Akatsu, H., Yamamoto, T., Staufenbiel, M., Yanai, K., Arai, H., Sasaki, H., Kudo, Y., & Sawada, T. (2004). Styrylbenzoxazole derivatives for in vivo imaging of amyloid plaques in the brain. *J Neurosci, 24*(10): 2535-2541.

Okamura, N., Suemoto, T., Furumoto, S., Suzuki, M., Shimadzu, H., Akatsu, H., Yamamoto, T., Fujiwara, H., Nemoto, M., Maruyama, M., Arai, H., Yanai, K., Sawada, T., & Kudo, Y. (2005). Quinoline and benzimidazole derivatives: candidate probes for in vivo imaging of tau pathology in Alzheimer's disease. *J Neurosci, 25*(47): 10857-10862.

Otth, C., Mendoza-Naranjo, A., Mujica, L., Zambrano, A., Concha, I.I., & Maccioni R.B. (2003). Modulation of the JNK and p38 pathways by cdk5 protein kinase in a transgenic mouse model of Alzheimer's disease. *Neuroreport*;14(18):2403-9.

Padovani, A., Borroni, B., Colciaghi, F., Pettenati, C., Cottini, E., Agosti, C., Lenzi, G.L., Caltagirone, C., Trabucchi, M., Cattabeni, F., & Di Luca, M. (2002). Abnormalities in the pattern of platelet amyloid precursor protein forms in patients with mild cognitive impairment and Alzheimer disease. *Arch Neurol.; 59*(1), 71-75.

Panza, F., Frisardi,V., Imbimbo, B.P., Seripa, D., Paris, F., Santamato, A., D'Onofrio, G., Logroscino, G., Pilotto, A., & Solfrizzi, V. (2011) Anti-β-Amyloid Immunotherapy for Alzheimer's Disease: Focus on Bapineuzumab. *Curr Alzheimer Res.* (In press).

Perry, V.H., Nicoll, J.A. & Holmes, C. (2010). Microglia in neurodegenerative disease. *Nat Rev Neurol.*, 6(4):193-201.

Peskind, E.R., Riekse, R., Quinn, J.F., Kaye, J., Clark, C.M., Farlow, M.R., Decarli, C., Chabal, C., Vavrek, D., Raskind, M.A., & Galasko, D. (2005). Safety and acceptability of the research lumbar puncture. *Alzheimer Dis Assoc Disord.*; 19(4):220-5.

Peterson, D.W., George, R.C., Scaramozzino, F., LaPointe, N.E., Anderson, R.A., Graves, D.J., & Lew, J. (2009). Cinnamon extract inhibits tau aggregation associated with Alzheimer's disease *in vitro*. *J Alzheimers Dis.* 17(3):585-97.

Phiel CJ, Wilson CA, Lee VM & Klein PS. (2003). GSK-3alpha regulates production of Alzheimer´s disease amyloid – beta peptides. *Nature*. 423(6938): 435-439

Quintanilla, R.A., Orellana, D.I., González-Billault, C., & Maccioni, R.B. (2004) Interleukin-6 induces Alzheimer-type phosphorylation of tau protein by deregulating the cdk5/p35 pathway. *Exp Cell Res.* ; 295(1):245-57.

Rafii MS, Walsh S, Little JT, Behan K, Reynolds B, Ward C, Jin S, Thomas R, Aisen PS; Alzheimer's Disease Cooperative Study. A phase II trial of huperzine A in mild to moderate Alzheimer disease. *Neurology*. 2011 Apr 19;76(16):1389-94

Raina P, Santaguida P, Ismaila A, *et al* (2008). «Effectiveness of cholinesterase inhibitors and memantine for treating dementia: evidence review for a clinical practice guideline». *Annals of Internal Medicine* 148 (5): pp. 379–397

Rinne, J. O., Brooks, D. J., Rossor, M. N., Fox, N.C., Bullock, R., Klunk, W. E., Mathis, C. A., Blennow, K., Barakos, J., Okello, A. A., Rodriguez Martinez de Liano, S., Liu, E., Koller, M., Gregg, K. M., Schenk, D., Black, R. & Grundman, M. (2010). 11C-PiB PET assessment of change in fibrillar amyloid-beta load in patients with Alzheimer's disease treated with bapineuzumab: a phase 2, double-blind, placebo-controlled, ascending-dose study. *Lancet Neurol.*; 9(4):363-72.

Rojo, L., Sjöberg, M.K., Hernández, P., Zambrano, C., & Maccioni, R.B. (2006). Roles of cholesterol and lipids in the etiopathogenesis of Alzheimer's disease. *J Biomed Biotechnol.* (3): 73976.

Rojo L.E., Chandia, M., Becerra, R., & Maccioni R.B. (2007a). ^{18}F-Lansoprazole, chemical and biological studies towards the development of a new PET radiopharmaceutical. In: *Conference on Clincal PET and Molecular Nuclear Medicine*, vol. 1. Bankok, Thailand.

Rojo, L.E., Avila, M., Chandia, M., Becerra, R., Maccioni RB (2007b). A PET tracer for Amyloid Senile plaques in Alzheimer`s Disease. In: *Annual Congress of the Chilean Society of Pharmacology*, , vol. 1. Iquique, Chile.

Rojo, L.E., Fernández, J.A., Maccioni, A., Jiménez, J.M., & Maccioni, R.B. (2008). Neuroinflammation: implications for the pathogenesis and molecular diagnosis of Alzheimer's disease. *Arch Med Res.;* 39 (1): 1-16.

Rojo, L. E., Alzate, J., Saavedra, I., Davies, P., & Maccioni, R. B. (2010). Selective interaction of lansoprazole and astemizole with tau polimers: potential new clinical use in diagnosis of Alzheimer's disease. *J Alzheimer's Dis.* 19(2): 573-89.

Rojo, L. E., Gaspar, P. A., & Maccioni, R. B. (2011). Molecular targets in the rational design of AD specific PET tracers: tau or amyloid aggregates? *Curr Alzheimer Res.* 8: 652-658

Saez, T.E., Pehar, M., Vargas, M., Barbeito, L., & Maccioni, R.B. (2004). Astrocytic nitric oxide triggers tau hyperphosphorylation in hippocampal neurons. *In Vivo,* 18(3):275-280.

Schönknecht, P., Pantel, J., Kaiser, E., Thomann, P., & Schröder, J. (2007). Increased tau protein differentiates mild cognitive impairment from geriatric depression and predicts conversion to dementia. *Neurosci Lett.;* 416(1):39-42.

Serot, J., Barbé, F., Arning, E., Bottiglieri, T., Franck, P., Montagne, P., & Nicolas, J. (2005). Homocysteine and methylmalonic acid concentrations in cerebrospinal fluid: relation with age and Alzheimer's disease. *J Neurol Neurosurg Psychiatry.; 76*(11), 1585-1587.

Seshadri, S., Beiser, A., Selhub, J., Jacques, P.F., Rosenberg, I.H., D'Agostino, R.B., Wilson, P.W., & Wolf, P.A. (2002). Plasma homocysteine as a risk factor for dementia and Alzheimer's disease. *N Engl J Med.; 346*(7), 476-483.

Slats, D., Claassen, J.A., Spies, P.E., Borm, G., Besse, K.T., Aalst, W.V., Tseng, J., Sjögren, M.J., Olde Rikkert, M.G., & Verbeek, M.M. (2011).Hourly variability of cerebrospinal fluid biomarkers in Alzheimer's disease subjects and healthy older volunteers. *Neurobiol Aging.* (In press).

Sperling, R.A., Aisen, P.S., Beckett, L.A., Bennett, D.A., Craft, S., Fagan, A.M., Iwatsubo, T., Jack, C.R., Jr, Kaye, J., Montine, T.J., Park, D.C., Reiman, E.M., Rowe, C.C., Siemers, E., Stern, Y., Yaffe, K., Carrillo, M.C., Thies, B., Morrison-Bogorad, M., Wagster, M.V., & Phelps, C.H. (2011). Toward defining the preclinical stages of Alzheimer's disease: Recommendations from the National Institute on Aging-Alzheimer´s Association workgroups on diagnostic guidelines for Alzheimer´s disease. *Alzheimer's and Dementia;* 7(3):280-92.

Staff RT et al. (2008). Tau aggregation inhibitor (TAI) therapy with Rember arrest the trajectory of Rcbf decline in brain regions affected by tau pathology in mild to moderate Alzheimer´s disease. *Alzheimers Dement.* 4:T:775.

The Ronald and Nancy Reagan Research Institute of the Alzheimer's Association, National Institute on Aging Working Group. (1998). Consensus report of the Working Group on: "Molecular and Biochemical Markers of Alzheimer's Disease" In *Neurobiology of aging* (Vol. 19, p. 109-116). Presented at the Neurobiology of Aging.

Tobinick, E.L. & Gross, H. (2008a). Rapid cognitive improvement in Alzheimer's disease following perispinal etanercept administration. *J Neuroinflammation*; 5:2.

Tobinick, E.L. & Gross, H. (2008b). Rapid improvement in verbal fluency and aphasia following perispinal etanercept in Alzheimer's disease. *BMC Neurol*. Jul 21;8:27.

Tobinick, E.L. (2009). Tumour necrosis factor modulation for treatment of Alzheimer's disease: rationale and current evidence. *CNS Drugs*; 23(9):713-25.

Uberti, D., Lanni, C., Carsana, T., Francisconi, S., Missale, C., Racchi, M., Govoni, S., & Memo, M. (2006). Identification of a mutant-like conformation of p53 in fibroblasts from sporadic Alzheimer's disease patients. *Neurobiol Aging*; 27(9):1193-201.

Uberti, D., Lanni, C., Racchi, M., Govoni, S., & Memo, M. (2008). Conformationally altered p53: a putative peripheral marker for Alzheimer's disease. *Neurodegener Dis*. 5(3-4):209-11.

Vanderstichele, H., De Vreese, K., Blennow, K., Andreasen, N., Sindic, C., Ivanoiu, A., Hampel, H., Bürger, K., Parnetti, L., Lanari, A., Padovani, A., DiLuca, M., Bläser, M., Olsson, A.O., Pottel, H., Hulstaert, F., & Vanmechelen, E. (2006). Analytical performance and clinical utility of the innotest phospho-tau181p assay for discrimination between Alzheimer's disease and dementia with Lewy bodies. *Clin Chem Lab Med*. 44(12):1472-1480.

Verhoeff, N.P., Wilson, A.A., Takeshita, S., Trop, L., Hussey, D., Singh, K., Kung, H.F., Kung, M.P., & Houle, S. (2004). In-vivo imaging of Alzheimer disease beta-amyloid with [11C]SB-13 PET. *Am J Geriatr Psychiatry, 12*(6), 584-595.

Verhoeff NP. (2005). Acetylcholinergic neurotransmission and the beta-amyloid cascade: implications for Alzheimer´s disease. *Expert Rev Neurother* . 5: 277-284.

Wang Y, Wei Y, Oguntayo S, Jensen N, Doctor B, Nambiar M. Huperzine A Protects Against Soman Toxicity in Guinea Pigs. *Neurochem Res*. 2011 Aug 7. Epub ahead of print

Wiltfang, J., Esselmann, H., Bibl, M., Hüll, M., Hampel, H., Kessler, H., Frölich, L., Schröder, J., Peters, O., Jessen, F., Luckhaus, C., Perneczky, R., Jahn, H., Fiszer, M., Maler, J.M., Zimmermann, R., Bruckmoser, R., Kornhuber, J., & Lewczuk, P. (2007). Amyloid beta peptide ratio 42/40 but not A beta 42 correlates with phospho-Tau in patients with low- and high-CSF A beta 40 load. *J Neurochem.*; 101(4):1053-1059.

Wimo, A. & Prince, M. (2010). The Global Economic Impact of Dementia. Alzheimer's disease International (ADI) World Alzheimer Report 2010.

Wischik CM, Edwards PC, Lai RY, Roth M and Harrington CR (1996) Selective inhibition of Alzheimer disease-like aggregation by phenotiazines. *Proc. Nat. Acad. Sci. USA* 93: 11213-11218

Yoshiyama, Y., Higuchi, M., Zhang, B., Huang, S.M., Iwata, N., Saido, T.C., Maeda, J., Suhara, T., Trojanowski, J.Q., & Lee, V.M. (2007). Synapse loss and microglial activation precede tangles in a P301S tauopathy mouse model. *Neuron.*; 53(3):337-351.

Zainaghi, I.A., Forlenza, O.V., & Gattaz, W.F. (2007). Abnormal APP processing in platelets of patients with Alzheimer's disease: correlations with membrane fluidity and cognitive decline. *Psychopharmacology.*; *192*(4): 547-553.

Zhang, B. et al. (2005). Microtubule-binding drugs offset tau sequestration by stabilizing microtubules and reversing fast axonal transport deficits in a tauopathy model. *Proc Natl Acad Sci.* 102:227-231.

Zilka, N., Stozicka, Z., Kovac, A., Pilipcinec, E., Bugos, O., & Novak, M. (2009). Human misfolded truncated tau protein promotes activation of microglia and leukocyte infiltration in the transgenic rat model of tauopathy. *J Neuroimmunol.*; 209(1-2): 16-25.

7

Bis(12)-Hupyridone, a Promising Multi-Functional Anti-Alzheimer's Dimer Derived from Chinese Medicine

Yifan Han et al†*
Department of Applied Biology and Chemical Technology, Institute of Modern Medicine, The Hong Kong Polytechnic University, Hong Kong

1. Introduction

Alzheimer's disease (AD), clinically characterized by progressive impairments of memory, cognitive functions and behaviors, is a major form of dementia that mainly affects elderly individuals. Alzheimer's Disease International undertook a Delphi study that showed worldwide in 2001 there were 24.3 million people with dementia, most of whom with AD; and the figure will rise to 42.3 million and 81.1 million in 2020 and 2040, respectively (Ferri et al., 2005). It is estimated that dementia causes people over the age of 60 to spend 11.2% of their last years living with disability (Morris and Mucke, 2006).The rapid increase in the number of dementia patients, most of whom AD patients, imminently calls for effective therapeutic prevention and treatment, in particularly, for AD patients (van Marum, 2008).

The plaque of β-amyloid (Aβ) and the neurofibrillary tangle composed mainly by hyper-phosphorylated tau protein are two major pathological hallmarks of AD (Fu et al., 2009). Therefore, in developing anti-AD drugs, preventing the generations of abnormal Aβ and hyper-phosphorylated tau proteins is the major target. For example, bapineuzumab, the antibody of abnormal Aβ, semagacestat and tarenflubil, the modulators of γ-secretase, and tramiprosate, the blocker of Aβ aggregation, have been proposed to treat AD by targeting Aβ cascade (Aisen et al., 2007; Ballard et al., 2011; Green et al., 2009; Thakker et al., 2009). However, all these drugs have failed in randomized controlled trials (Ballard et al., 2011). Although several reasons might be provided to explain why these trials in AD failed, some

* Wei Cui[1], Tony Chung-Lit Choi[1], Shinghung Mak[1], Hua Yu[3], Shengquan Hu[1], Wenming Li[1,4], Zhong Zuo[2]
[1]*Department of Applied Biology and Chemical Technology, Institute of Modern Medicine, The Hong Kong Polytechnic University,*
[2]*School of Pharmacy, Faculty of Medicine, The Chinese University of Hong Kong,*
[3]*School of Chinese Medicine, Hong Kong Baptist University,*
[4]*Departments of Pharmacology and Neurology, Emory University School of Medicine, Atlanta, GA 30322*
[1,2,3]*Hong Kong*
[4]*USA*
† Corresponding Author

scientists suggested the Aβ and tau hypotheses might be invalid (Smith, 2010). They proposed that the complexity of AD would require not one single drug, but multiple drugs or a multifunctional drug to modify the disease progress (Mangialasche et al., 2010; Smith, 2010).

The neuropathology of AD is characterized by a decreased cholinergic transmission caused by the loss of cholinergic neurons. Acetylcholinesterase (AChE) inhibitors, which enhance the function of cholinergic neurons by prolonging the duration in which acetylcholine stays in the synaptic clefts, have shown promising potential in the treatment of AD (Li et al., 2007b). The inhibitors of AChE could stabilize cognitive and behavior functions of AD patents at a steady level for at least 1 year in 50% and up to 2 years in about 24% of treated patients (Wang et al., 2006). Moreover, those AD patients who do not response to one AChE inhibitor could take another (Wang et al., 2006). So far, four AChE inhibitors, namely tacrine (Cognex), donepezil (Atricept), rivastigmine (Exelon) and galantamine (Reminyl), have been approved by the U.S. Food and Drug Administration (FDA) for the treatment of AD (Ellis, 2005; Francis et al., 2005).

Huperzine A, a *Lycopodium* alkaloid discovered from the traditional Chinese medicine *Huperzia serrata* (Qian Ceng Ta) (FIG. 1A), is also widely used in the treatment of AD in China (FIG. 1B). It is a selective AChE inhibitor with much higher potency and longer duration of AChE inhibition than those of tacrine, donepezil, rivastigmine and galantamine (Wang and Tang, 1998; Zhao and Tang, 2002). Double-blind, randomized clinical trials in China have demonstrated that huperzine A induces significant improvement in memory in elderly people and AD patients without any significant side effects (Wang et al., 2009). However, the lack of natural supply of *Huperzia serrata* and difficulty in its chemical synthesis have limited the clinical usage of huperzine A (Zhang and Tang, 2006).

Memantine (Namenda), which is an uncompetitive antagonist of N-methyl-D-aspartate (NMDA) receptors with a fast on-/off-rate, could reduce excessive glutamate-induced excitotoxicity. The success of memantine in clinical trials led to its being approved by FDA to treat moderate to severe AD in 2003 (Lipton, 2006). This encouraging news pointed to a new direction for the development of anti-AD drugs: by boosting the activities of healthy neurons and reducing abnormal brain functions (Gravitz, 2011). However, further studies have indicated that either AChE inhibitors or NMDA receptor antagonists have limited success in reversing AD progress as they are unable to stop neurodegeneration (Roberson and Mucke, 2006).

The effectiveness of multiple drug strategy has been proven. One of the examples is the HIV drug cocktail (Zhang, 2005). Combinations of drugs with different targets, are also widely used in cancer therapy (Hanahan and Weinberg, 2011). The involvement of multi-factorial etiopathogenesis in AD suggests that the treatment of AD may also require multiple drug therapy to target its different pathological aspects (Youdim and Buccafusco, 2005). However, there are some challenges in the use of drug cocktail strategy that works at different therapeutic targets. Different drugs have differences in bioavailability, pharmacokinetics and metabolisms. They may also cross-react with one another which in turn cause serious side-effects. Therefore, the one-compound-multiple-targets strategy, a novel drug development approach pioneered by Prof. Moussa Youdim, has emerged as a practical alterative to overcome these challenges (Youdim and Buccafusco, 2005). Designing a single molecule synergistically targeting two or more therapeutic pathways is reasonably more proficient than the combination of one-compound-one-target drugs because of simple bioavailability and pharmacokinetics. Therefore, many neuroscience research institutes and pharmaceutical companies devote the majority of their resources to search for the effective one-compound-multi-functional agents for the treatment of AD.

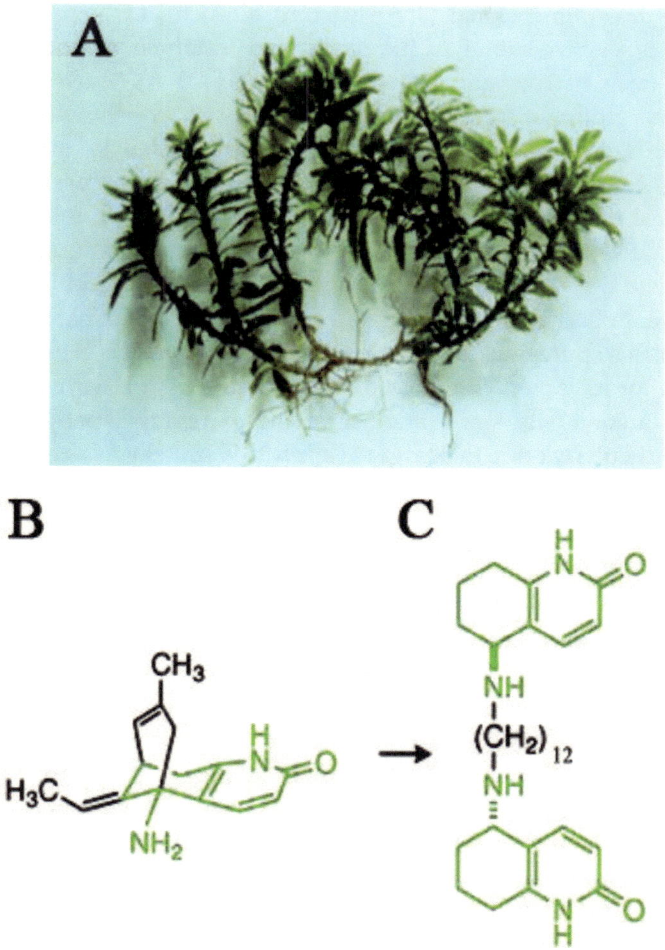

Fig. 1. The chemical structures of huperzine A and bis(12)-hupyridone.
(A) Chinese medicinal herb *Huperzia serrata* (Qian Ceng Ta); (B) The structure of huperzine A; (C) The structure of bis(12)-hupyridone.

Our group has devoted enormous efforts in developing drugs that are better efficacy than current AChE inhibitors the currently available AChE inhibitors over the past years. With the help of scientists from Israel, the USA and China, we have developed a series of novel bis(n)-hupyridones by the homo-dimerization of hupyridone, the ineffective fragments of huperzine A (Carlier et al., 2000; Wong et al., 2003). These dimeric compounds are easy to synthesis and have been shown to be more potent than huperzine A in the inhibition of AChE. In this article, we will review that bis(12)-hupyridone (FIG. 1C), one of our novel dimeric promising anti-AD candidates, possesses multiple functions that include the enhancement of cognitive functions, the protection against neurotoxins, and the promoting of neuronal differentiation for the treatment of AD.

2. Design and synthesis of novel anti-AChE dimers derived from huperzine A

By studying the three-dimensional (3D) structure of AChE of *Torpedo californica* electric organ (*Tc*AChE), one active site of AChE, named the "catalytic anionic site", was found at the bottom of a deep narrow gorge (active-site gorge, 20 Å) (Axelsen et al., 1994), The quaternary amino group of acetylcholine interacts with the indole side chain of the conserved residue Trp84 in a cation-ð interaction at this "catalytic anionic site" (Ma and Dougherty, 1997). Moreover, another active site of AChE, named "peripheral anionic site", was also found near the top of the "active-site gorge", about 14 Å from the "catalytic anionic site" (Harel et al., 1993). The major element of the "peripheral anionic site" is the residue of Trp279. The bivalent ligand strategy is widely used in the synthesis of novel drugs in which identical or different pharmacophores are connected by a suitable linker (Haviv et al., 2005). The advantage of this strategy is the chelate effect, which creates a bifunctional ligand with enhanced affinity for its target. The molecular structure of AChE with one active site in the gorge ("catalytic anionic site") and another at the extremity ("peripheral anionic site") makes AChE a particularly attractive target to apply this strategy.

The crystal structure study of the complex of AChE with (-)-huperzine A has shown important hydrophobic interactions between (-)-huperzine A and Trp84 at the "catalytic anionic site" of AChE (Raves et al., 1997). We initially synthesized hupyridone (5-amino-2(1H)-quinolinones), a fragment which lacks the C6-C8 bridge of (-)-huperzine A. Although this fragment does not show any significant inhibition of AChE, it appears to possess much of the intrinsic functionality of (-)-huperzine A (FIG. 1). It retains the pyridine oxygen atom and NH group of (-)-huperzine A, which form hydrogen bonds to Tyr-130 and Gly-117, respectively (Raves et al., 1997). Most importantly, hupyridone also retains the 5-amino group of (-)-huperzine A, which is essential for the inhibitory activity of (-)-huperzine A by interacting with Try84. We speculated that it is possible to find high-affinity inhibitors of AChE with structure like that of hupyridone. Particularly, the loss of hydrophobic contact in the "catalytic anionic site" could be compensated by additional chelating interactions at the "peripheral anionic site" (Carlier et al., 2000; Wong et al., 2003).

A series of hupyridone dimer or bis(n)-hupyridones, with different alkylene chain lengths, has been synthesized from 7,8-dihydroquinoline-2,5(1H,6H)-dione by the condensation, the reduction and the dimerization (FIG. 2). Computational calculations showed that 12 methylene units were the most approximate chain length. The 3D study of *Tc*AChE-ligand complexes has shown that bis(12)-hupyridone binds more tightly to *Tc*AChE than (-)-huperzine A (Wong et al., 2003). Overlaying the structures of *Tc*AChE with those of bis(12)-hupyridone and (-)-huperzine A also reveals that the dimer makes cation-ð and hydrogen bonding interactions at the "peripheral anionic site" (Trp279), interactions that can contribute to bis(12)-hupyridone's higher affinity compared with (-)-huperzine A (FIG.3) (Wong et al., 2003). It is suggested that the tether of the hupyridone unit of bis(12)-hupyridone provides minimal entropy and substantially compensates for the weaker and/or missing interactions of (-)-huperzine A with AChE (Raves et al., 1997; Wong et al., 2003).

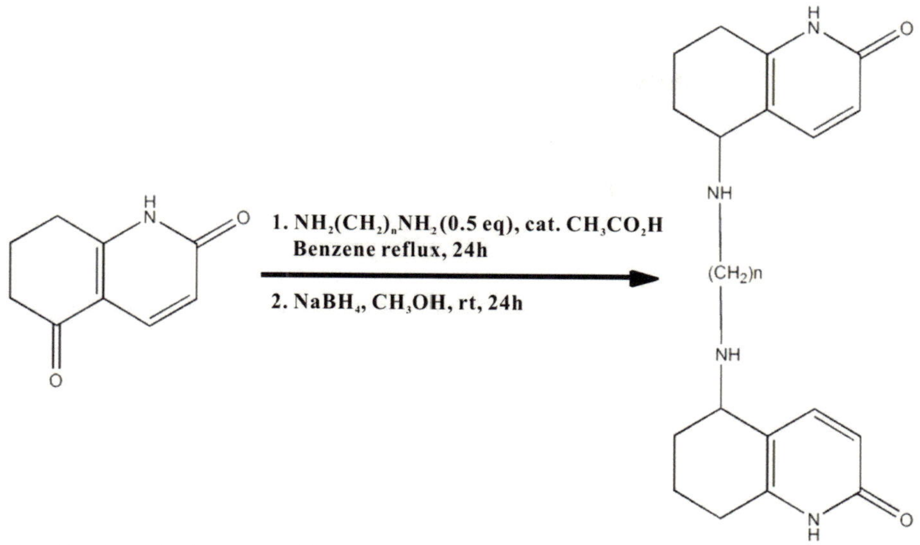

Fig. 2. Synthesis of bis(n)-hupyridones.

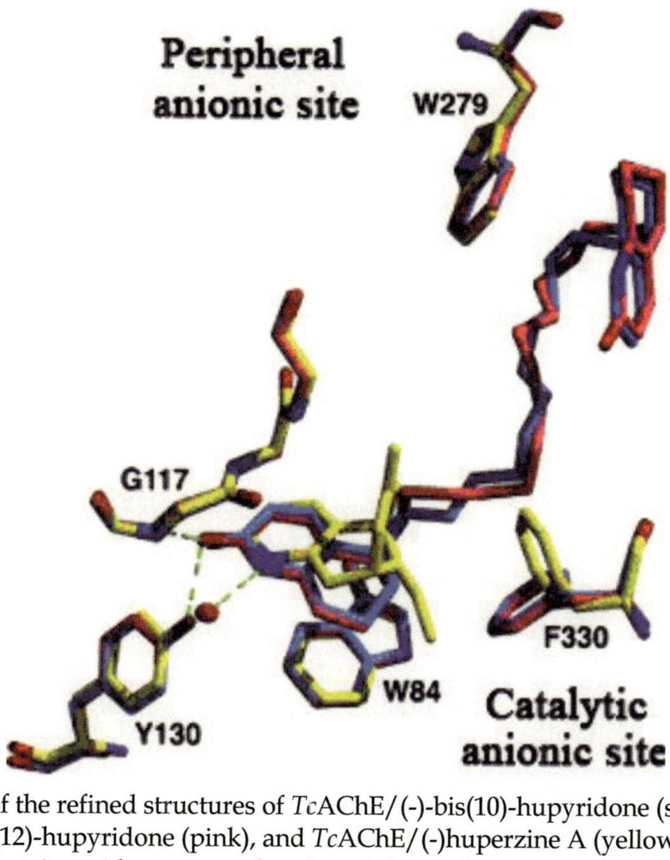

Fig. 3. Overlay of the refined structures of *Tc*AChE/(-)-bis(10)-hupyridone (sky blue), *Tc*AChE/(-)-bis(12)-hupyridone (pink), and *Tc*AChE/(-)huperzine A (yellow). Inhibitors and protein residues are rendered as sticks, and water molecules are shown as red spheres. The figure is modified from the reference (Wong et al., 2003).

3. Multifunctional potencies of bis(12)-hupyridone

3.1 Inhibition of AChE

The anti-cholinesterase activities of bis(n)-hupyridones were further tested *in vitro* and *in vivo*. It has been shown that bis(n)-hupyridones inhibit AChE in a tether-length-dependent manner. The 50% inhibitory concentration (IC_{50}) on AChE by bis(12)-hupyridone was about 52 nM, which was comparable to those of AChE inhibitors used for treating AD (Table 1) (Li et al., 2007b). Furthermore, kinetic analysis of bis(12)-hupyridone suggested that the inhibition pattern was mixed competitive with an apparent K_i value of 28.9 nM. *In vivo* study has also shown that single *p.o.* administration of bis(12)-hupyridone significantly inhibit AChE activity in various brain regions (cortex, hippocampus, striatum) in rats (Table 2) (Li et al., 2007b).

AChE inhibitors	IC_{50} (μM)		Ratio of IC_{50} (BuChE/AChE)	Inhibitory Pattern	K_i (μM)
	AChE	BuChE			
Bis(10)-hupyridone	0.151	1.82	12.1	N.D.	N.D.
Bis(11)-hupyridone	0.084	1.16	13.8	N.D.	N.D.
Bis(12)-hupyridone	**0.052**	**9.6**	**185.0**	**Mixed**	**28.9**
Bis(13)-hupyridone	0.052	16.7	321.0	N.D.	N.D.
Bis(14)-hupyridone	0.24	59.5	148.0	N.D.	N.D.
Huperzine A	**0.082**	**74.43**	**907.7**	**mixed**	**24.9**
Galantamine	**1.995**	**12.59**	**6.3**	**Competitive**	**210.0**
Donepezil	**0.010**	**5.01**	**501.0**	**Noncompetitive**	**12.5**
Tacrine	**0.093**	**0.074**	**0.8**	**Noncompetitive**	**105**

Table 1. Anti-AChE activities of bis(n)-hupyridones and other AChE inhibitors used in the treatment of AD.

The concentrators of inhibitors yield 50% inhibition of enzyme activity. The cortex homogenate was pre-incubated for 5 min with iso-OMPA 0.1 mM. The rate of color production was measured spetrophotometrically at 440 nM. N.D.: not determined, Data are from references (Cheng et al., 1996; Li et al., 2007b; Wang and Tang, 1998).

AChE inhibitor	Dose (μmol/kg)	AChE inhibition (%)			BuChE inhibition (%)
		cortex	hippocampus	striatum	serum
Bis(12)-hupyridone	**90**	**16 ± 5** **	**40 ± 3** **	**28 ± 6** **	**NA**
	45	**12 ± 3** **	**14 ± 4** **	**12 ± 5** *	**NA**
	22	**11 ± 3** **	**4 ± 3**	**5 ± 4**	**NA**

Table 2. Anti-cholinesterase activities of single *p.o.* administration of bis(12)-hupyridone in rats.

Values expressed as percentage of inhibition (*versus* saline control) were the means ± SD. *p < 0.05 and **p < 0.01 *versus* saline group (ANOVA and Dunnett's test). Basal saline control values of cortex, hippocampus and striatum are 1360 ± 70, 1540 ± 150 and 9390 ± 880 A values/g protein, respectively. Basal saline control value of serum is 23 ± 5 A values/g protein. Data are from the reference (Li et al., 2007b).

3.2 Blockade of NMDA receptors

There are increasing evidences that show that the overstimulation of glutamate receptors of the NMDA subtype may be involved in the neuronal loss of AD (Lipton, 2006; Parsons et al., 2007). With the disruption of the neuron-neuron and neuron-glial connections, glutamate might be not only improperly cleared, but also inappropriately released. Meanwhile, energetically compromised neurons become depolarized because they cannot maintain their ionic homeostasis in the absence of energy. The depolarization relieves the normal Mg^{2+} blockade of NMDA receptor-coupled channel, and then excessive stimulation of glutamate receptors occurs (Li et al., 2005; Lipton, 2004). Thus the NMDA receptor has been considered an attractive therapeutic target for the development of anti-stroke drugs. On the other hand, the NMDA receptor, as the major excitatory neurotransmitter receptor in the central nervous system, mediates many important physiological processes, such as synaptic plasticity, and learning and memory (Petrovic et al., 2005; Villmann and Becker, 2007). The therapeutic potential of many powerful NMDA receptor antagonists, such as MK-801, is limited and they fail in clinical trials because of the psychotropic side effects resulting from their interference with normal brain functions (Parsons et al., 2007). NMDA receptor blockers with moderate to low affinity, such as memantine, may inhibit NMDA receptor-mediated pathological but not NMDA receptor-mediated physiological functions. This kind of NMDA receptor antagonists has been at the center of interest in the search for the next generation of neuroprotective drugs for AD (Lipton, 2004; Lipton, 2007).

Using the receptor-ligand binding assay, bis(12)-hupyridone has been found to compete with [^{3}H]MK-801 with a K_i value of 7.7 μM. In the same testing system, memantine and MK-801 competed with [^{3}H]MK-801with a K_i value of 0.8 and 0.04 μM, respectively (Table 3) (Li et al., 2007a) (our unpublished data). These results suggested that bis(12)-hupyridone is a moderate NMDA receptor antagonist and thus might be useful in AD therapy.

NMDA receptor antagonists	[^{3}H]MK-801 binding K_i (μM)
Bis(12)-hupyridone	**7.7**
Memantine	**0.8**
MK-801	**0.04**

Table 3. Bis(12)-hupyridones moderator inhibits NMDA receptors at MK-801 site.

The membrane proteins from rat cerebellar cortex were incubated with 4 nM [^{3}H]MK-801.and treated with the serial concentrations of bis(12)-hupyridone/memantine/MK-801. The K_i values were calculated from the corresponding IC_{50} values, which were measured from the obtained data using at least eight concentrations of each chemical (in duplicates) based on the Cheng-Prusoff equation: $K_i = IC_{50}/(1+[\text{ligand}]/K_d)$. Data are either from our unpublished paper or from the reference (Li et al., 2007a).

3.3 Protection against excitotoxicity

It is well known that the overstimulation of NMDA receptors is essential to the neuronal apoptotic cell death induced by glutamate; and that the blockade of NMDA receptors may prevent neuronal cell death induced by excitotoxicity (Danysz and Parsons, 2003). We thus investigated the neuroprotective effects of bis(12)-hupyridone against excitotoxicity in the primary cerebellar granule neurons (CGNs). We have demonstrated that bis(12)-hupyridone inhibits glutamate-induced apoptosis in a concentration-dependent manner, and its preventive effect is significant even at the low dosage of 1 nM. Further study using fluorescein diacetate/propidium iodide double staining, Hoechst 33324 staining and DNA fragmentation gel assays have shown that bis(12)-hupyridone significantly reverses the glutamate-evoked nuclear condensation, apoptotic bodies and DNA fragmentation, indicating that this dimer is a powerful neuroprotectant against excitotoxicity *in vitro* (FIG. 4 and our unpublished data).

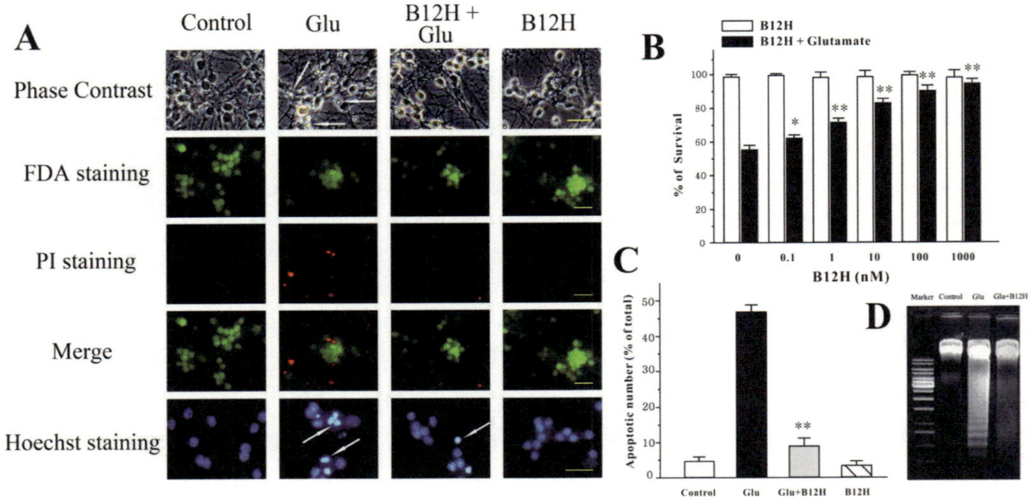

Fig. 4. Bis(12)-hupyridone prevents neuronal death induced by glutamate in primary CGNs. (A) CGNs were pre-incubated with or without 1 μM bis(12)-hupyridone and exposed to 75 μM glutamate 2 h later. At 24 h after glutamate challenge, CGNs were assayed with a phase contrast microscope, fluorescein diacetate/propidium iodide double staining and Hoechst 33324 staining. Apoptotic nuclei were indicated by white arrows. (B) CGNs were pre-incubated with bis(12)-hupyridone at different concentrations as indicated and exposed to 75 μM glutamate 2 h later. Cell viability was measured by MTT assay at 24 h after glutamate challenge. (C) The counts of apoptotic bodies by Hoechst staining. (D) Under the same treatment conditions as (B), DNA fragmentation was extracted from CGNs after 24 h of challenge, and then agarose gel electrophoresis and ethidium bromide staining were used to visualize the DNA extracted from the above samples. B12H: bis(12)-hupyridone; Glu: glutamate. All data, expressed as percentage of control, were the means ± SEM of three separate experiments; $^{*}p < 0.05$, $^{**}p < 0.01$ *versus* glutamate group (ANOVA and Dunnett's test).

3.4 Prevention of ROS-induced neuronal toxicity via regulating the VEGFR-2/Akt pathway

Oxidative stress plays an important role in the pathogenesis of AD as it is the main factor in the neuronal loss of this disease (Shibata and Kobayashi, 2008; Zhu et al., 2007). Although the detailed mechanisms underlying oxidative stress-induced neuronal death remain unknown, drugs with antioxidant properties have therapeutic significance in preventing AD (Pratico, 2008). H_2O_2 is widely used as a toxicant to establish *in vitro* models of oxidative stress-induced neuronal apoptosis as it is an uncharged and freely diffusible molecule (Lee et al., 2007).

Using primary CGNs as a cell model, we have demonstrated that bis(12)-hupyridone at a low concentration (3 nM) prevents H_2O_2-induced apoptosis(Cui et al., 2011c). We have also shown that this protection of bis(12)-hupyridone is a novel activity that is apart from its AChE inhibitory property. The decreased activation of glycogen synthase kinase (GSK) 3β was observed after H_2O_2 exposure, and bis(12)-hupyridone could reverse the altered activation of GSK3β, indicating that bis(12)-hupyridone may exert its neuroprotective effects via signaling molecule(s) upstream of GSK3β. Our further study using the antibody of phosphorylated vascular endothelial growth factor receptor-2 (VEGFR-2) and the inhibitor of VEGFR-2 has demonstrated that bis(12)-hupyridone prevents H_2O_2-induced neuronal apoptosis through regulating the VEGFR-2/Akt signaling pathway (FIG. 5) (Cui et al., 2011c; Liu et al., 2009). We speculated that bis(12)-hupyridone might either directly interact with VEGFR-2 as a potential agonist or indirectly facilitate the activation of VEGFR-2 such as by stabilizing the dimerization or increasing the endogenous VEGF from elevating its translation, transcription or post-transcription (Cui et al., 2011b). Further investigations on the exact role that bis(12)-hupyridone plays in the activation of VEGFR-2 are being undertaken in our laboratory.

3.5 Promoting neuronal differentiation via activating α7nAChR

Currently prescribed drugs that treat AD have shown only modest and symptomatic effects by reducing the degree of impairment without preventing or curing the disease. It is partially because that these drugs cannot induce neurogenesis to compensate for the neurons that have lost their functions (Maggini et al., 2006). Transplantation of stem cells is considered a potential strategy as it may provide neurons to replace those that have been lost in the brains of AD patients, and reverse the progress of neurodegeneration (Zhongling et al., 2009). However, there is one key problem with this strategy as grafted stem cells are not able to differentiate into fully mature neurons in the micro-environments of the brain of AD patients (Waldau and Shetty, 2008). The application of agents capable of promoting neuronal differentiation at the impaired site may be a valid alternative or adjunct to solve that problem.

With the help of rat hippocampus neural stem cells, we have evaluated the effects of bis(12)-hupyridone in promoting neural stem cell differentiation. The percentage of βIII-tubulin positively stained neurons gave evidence that the efficacy of 10 μM bis(12)-hupyridone was similar to that of 0.5 μM retinoic acid, a potent inducer of neuronal differentiation. Moreover, under the same condition, huperzine A was not able to induce differentiation (FIG. 6) (Cui et al., 2011a). Bis(12)-hupyridone therefore might be a promising anti-AD drug candidate to promote differentiation of neural stem cells.

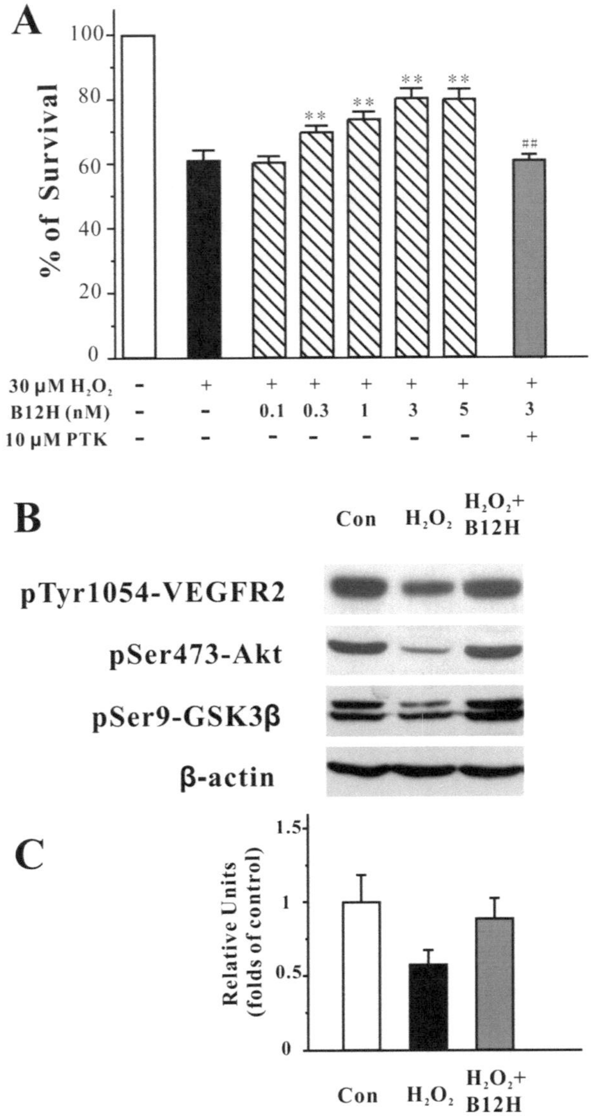

Fig. 5. Bis(12)-hupyridone inhibits H_2O_2-induced neuronal death from reversing VEGFR-2/Akt pathway. (A) The preventive functions of bis(12)-hupyridone against H_2O_2-induced cell death could be abolished by specific VEGFR-2 inhibitors in CGNs. CGNs with or without 30 min PTK787 (PTK, a specific VEGFR-2 inhibitor) pre-treatment were treated with bis(12)-hupyridone at the indicated concentrations for 2 h and then exposed to 30 μM H_2O_2. Cell viability was measured by the MTT assay at 6 h after H_2O_2 challenge. $^{**}p < 0.01$ *versus* H_2O_2 group, and $^{\#\#}p < 0.01$*versus* bis(12)-hupyridone plus H_2O_2 group (Tukey's test). (B) Bis(12)-hupyridone reversed H_2O_2-induced decreasing of pTyr1054-VEGFR-2, pSer473-Akt and pSer9-GSK3β. CGNs were pre-treated with 3 nM B12H for 2 h and then exposed to 30 μM H_2O_2 for 1 h, the total proteins were detected with using the specific antibodies. (C) The ratio to optical density (OD) values of pTry1054-VEGFR-2 over β-actin. The figure is modified from the reference (Cui et al., 2011c).

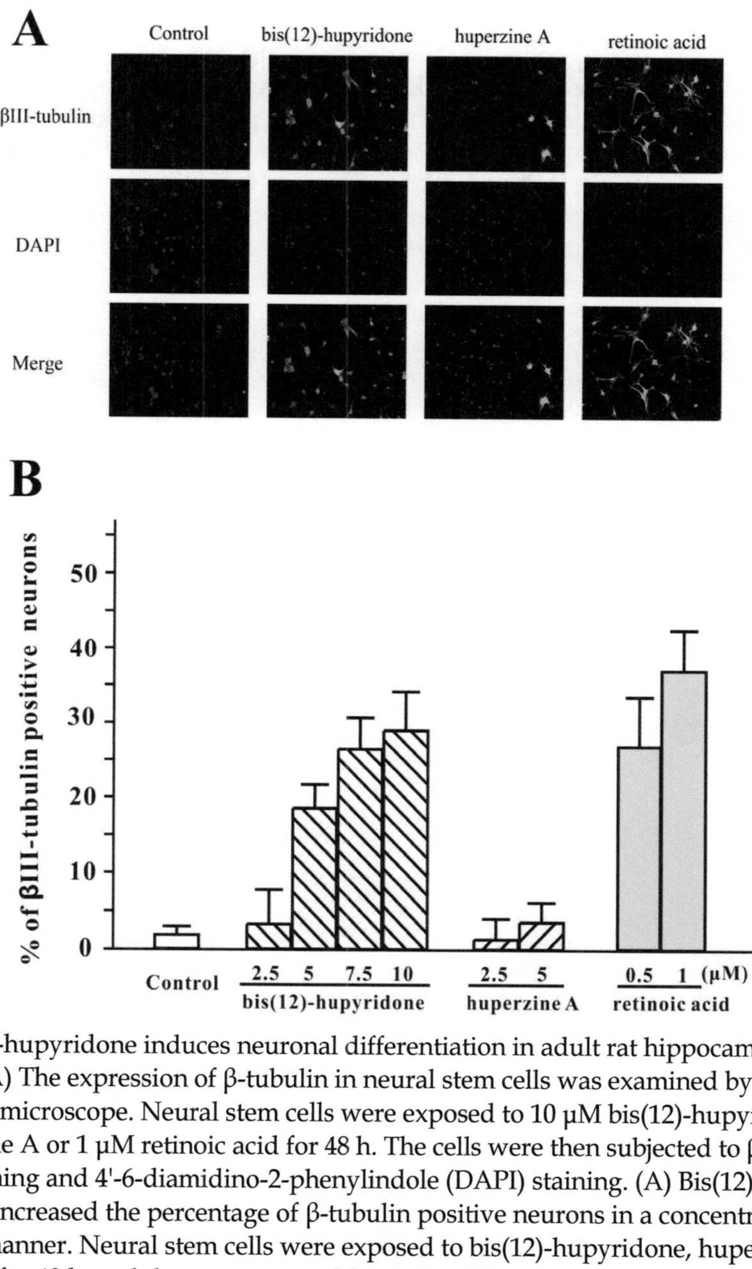

Fig. 6. Bis(12-hupyridone induces neuronal differentiation in adult rat hippocampus neural stem cells. (A) The expression of β-tubulin in neural stem cells was examined by fluorescence microscope. Neural stem cells were exposed to 10 μM bis(12)-hupyridone, 5 μM huperzine A or 1 μM retinoic acid for 48 h. The cells were then subjected to β-tubulin immunostaining and 4'-6-diamidino-2-phenylindole (DAPI) staining. (A) Bis(12)-hupyridone increased the percentage of β-tubulin positive neurons in a concentration-dependent manner. Neural stem cells were exposed to bis(12)-hupyridone, huperzine A or retinoic acid for 48 h, and the percentage of β-tubulin positive neurons was calculated. The figure is modified from the reference (Cui et al., 2011a).

We have also examined the neuronal differentiation promotion effects of bis(12)-hupyridone and its underlying mechanisms in the rat PC12 pheochromocytoma cell line, a well studied cell model of neuronal differentiation (Vaudry et al., 2002). Bis(12)-hupyridone (3 - 30 μM) has been demonstrated to induce neurite outgrowth in a concentration- and time-dependent manner with an efficacy that is three times higher than that of huperzine A in PC12 cells

(Cui et al., 2011a). Furthermore, mitogen-activated protein kinase kinase (MEK) inhibitor and alpha7-nicotinic acetylcholine receptor (α7nAChR) antagonist blocked the neurite outgrowth and the activation of extracellular signal-regulated kinase (ERK) induced by bis(12)-hupyridone, suggesting that bis(12)-hupyridone potently induces pro-neuronal cells into differentiated neurons by activating the ERK pathway via regulating α7nAChR (FIG. 7). As α7nAChR is essential for neuronal differentiation in the rat brain, and loss of α7nAChR impairs the maturation of dendritic neurons in adult hippocampus (Campbell et al., 2010; Le Magueresse et al., 2006), our results provide a novel insight into the possible therapeutic potential of bis(12)-hupyridone in treating AD. To date, as other clinically used anti-AD drugs such as huperzine A, donepezil and tacrine have also shown some activities in inducing neurite outgrowth in different neuronal cell lines *in vitro* (Table 4), it would be quite interesting to compare their effects on promoting neuronal differentiation in certain types of neurons, for example, neural stem cells with bis(12)-hupyridone (Cui et al., 2011a; De Ferrari et al., 1998; Oda et al., 2007; Sortino et al., 2004; Tang et al., 2005).

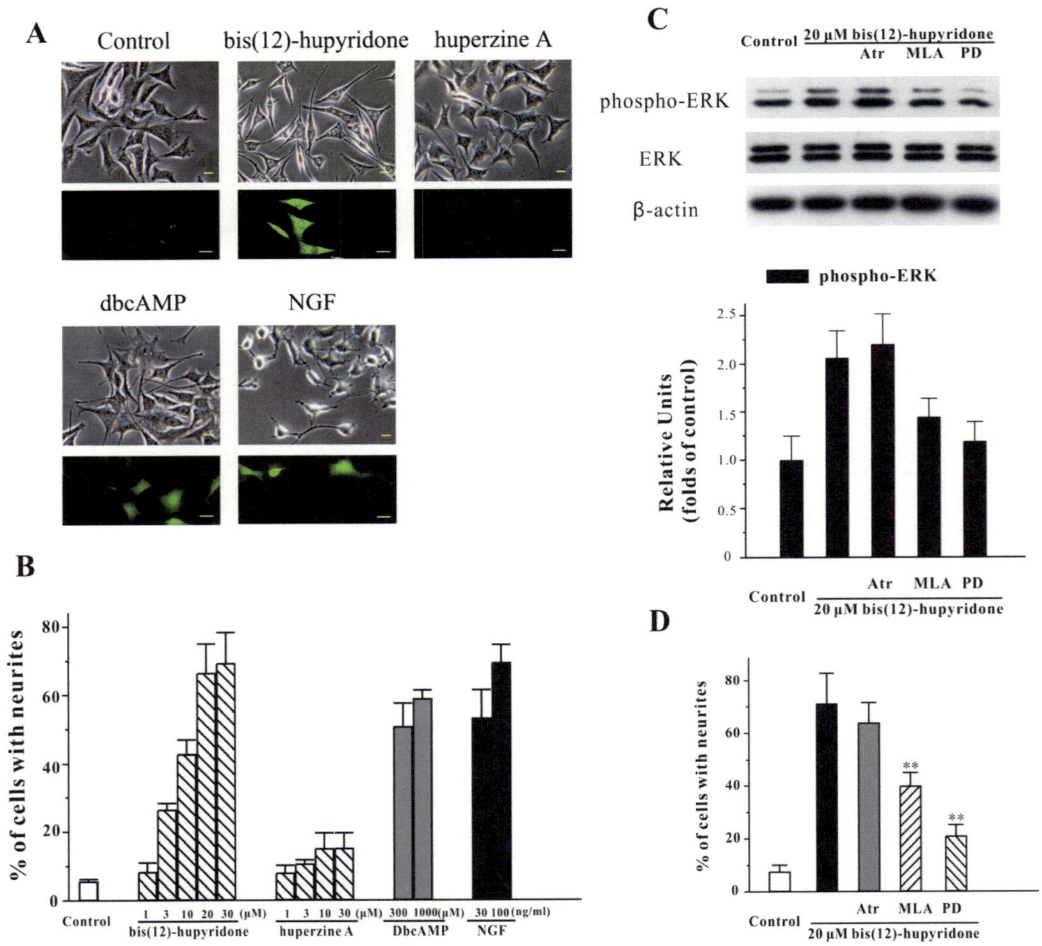

Fig. 7. Bis(12)-hupyridone induces neurite outgrowth from activating α7nAChR in PC12 cells. (A) The effects of bis(12)-hupyridone in promoting neurite outgrowth were evidenced by the

morphological changes and expression of GAP-43. PC12 cells were exposed to 20 μM bis(12)-hupyridone, 30 μM huperzine A, 3 mM dibutyryl cAMP (dbcAMP) or 100 ng/ml nerve growth factor (NGF) for 7 days. The morphological changes of neurites were examined by light microscope, and the expressions of GAP-43 were examined by fluorescence microscope. Scale bar 5 μm. (B) Induction of neurite outgrowth by bis(12)-hupyridone is in a concentration-dependent manner. PC12 cells were exposed to bis(12)-hupyridone, huperzine A, dbcAMP or NGF for 7 days, and the percentage of cells with neurites was measured. (C) The α7nAChR antagonist attenuates the activation of ERK induced by bis(12)-hupyridone. PC12 cells were treated with 0.3 μM methyllycaconitine (MLA, a specific α7nAChR antagonist), 10 μM atropine (Atr, a specific muscarinic acetylcholine receptor antagonist) or 30 μM PD98059 (PD, a specific MEK inhibitor) for 30 min before the administration of 20 μM bis(12)-hupyridone. The total proteins were extracted 30 min after the addition of bis(12)-hupyridone for Western blot analysis with specific antibodies. (D) The α7nAChR antagonist attenuates the neurite outgrowth induced by bis(12)-hupyridone. PC12 cells were incubated with 0.3 μM methyllycaconitine, 10 μM atropine or 30 μM PD98059 for 2 h and treated with 20 μM bis(12)-hupyridone. The percentage of cells with neurites was measured 7 days after treatment with bis(12)-hupyridone. The data, expressed as percentage of control, are the mean ± SEM of three separate experiments, with $^{**}p < 0.01$ *versus* the bis(12)-hupyridone group in employing ANOVA and Dunnett's test. The figure is modified from the reference (Cui et al., 2011a).

Drugs	Concentration (μM)	Neuronal Differentiation Models	Main Effects
Bis(12)-hupyridone	2.5 - 10	rat hippocampus neural stem cells	increased the percentage of βIII-tubulin positively stained neurons, induced neurite outgrowth
	3 - 30	PC12 cells	induced neurite outgrowth, and increased the expression of GAP-43
Huperzine A	10	PC12 cells	increased the number of neurite-bearing cells
Donepezil	1 -10	PC12 cells	potentiated the neurite outgrowth evoked by NGF
	0.1 - 10	SH-SY5Y cells	inhibited cell proliferation, and increased the expression of the neuronal marker MAP-2
Tacrine	10 - 50	Neuro 2A cells	induced neurite outgrowth

Table 4. Anti-AD drugs promote neuronal differentiation *in vitro*.
Data are from references (Cui et al., 2011a; De Ferrari et al., 1998; Oda et al., 2007; Sortino et al., 2004; Tang et al., 2005).

3.6 Enhancement of learning and memory

It is widely accepted that enhancement of learning and memory is beneficial for AD patients; and it has been proven that AChE inhibitors are the most effective agents in promoting cognitive functions in AD therapies. Our novel dimer bis(12)-hupyridone has demonstrated superior AChE inhibition *in vivo*. It is reasonable to expect that this dimer could remedy the impairments of learning and memory in AD patients. To prove this hypothesis, the model of scopolamine-induced performance deficits was used. We have demonstrated that *i.p.* injection of bis(12)-hupyridone (0.088 - 0.352 μmol/kg) significantly shortens the escape latency in Morris water maze after scopolamine administration in rats (Li et al., 2007b). Under the same condition, the relative potency of bis(12)-hupyridone (0.176 μmol/kg) to reverse the increased escape latency was higher than that of huperzine A (0.206 μmol/kg) (FIG. 8) (Li et al., 2007b).

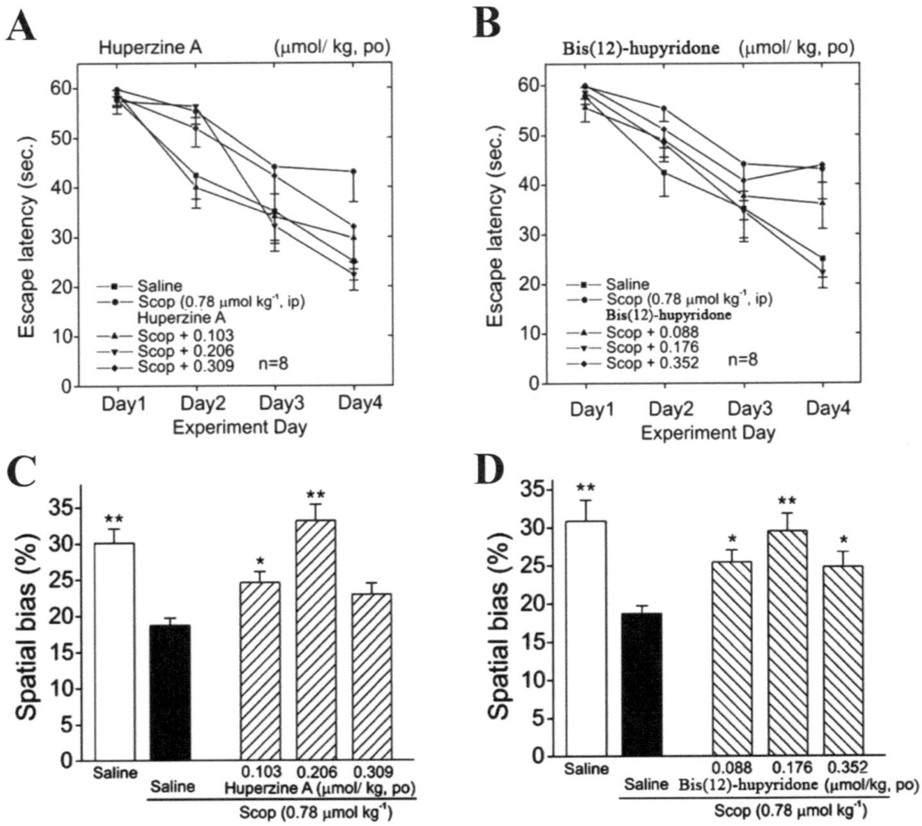

Fig. 8. Memory-enhancing effects of huperzine A and bis(12)-hupyridone on scopolamine-induced performance deficiency. Huperzine A (A) and bis(12)-hupyridone (B) reverse scopolamine-induced performance deficiency in rats. Huperzine A (C) and bis(12)-hupyridone (D) reverse scopolamine-induced decrease in spatial bias (% of total distances swum in the training quadrant during spatial probe trial) in rats. All data were expressed as means ± SD, $^{*}p < 0.05$ and $^{**}p < 0.01$ *versus* scopolamine group in (C) and (D) (ANOVA and Dunnett's test). The figure is modified from the reference (Li et al., 2007b).

3.7 Recovery of ischemic insult

Ischemia-induced insults result from the complex interplay of multiple pathways including excitotoxicity, oxidative stress and impairment of neurogenesis (Van der Schyf et al., 2006a). And some of these pathways are also underlying the impairments of AD progress. Therefore, agents targeting at multiple site for the treatment of stroke may also possess therapeutic effects for AD (Weinreb et al., 2009).

We have demonstrated in the 2-hour middle cerebral artery occlusion (MCAO) rat model, that bis(12)-hupyridone (0.70 - 1.41 μmol/kg, *i.p.*) could improve neurological behavior impairment, and reduce infarct volume as well as brain edema after ischemia. In addition, TUNEL staining assay has shown that bis(12)-hupyridone at quite a low concentration (0.70 μmol/kg, *i.p.*) could prevent cerebral ischemia-induced apoptosis in the penumbra region (FIG. 9, our unpublished paper). Compared with the currently used anti-AD drugs,

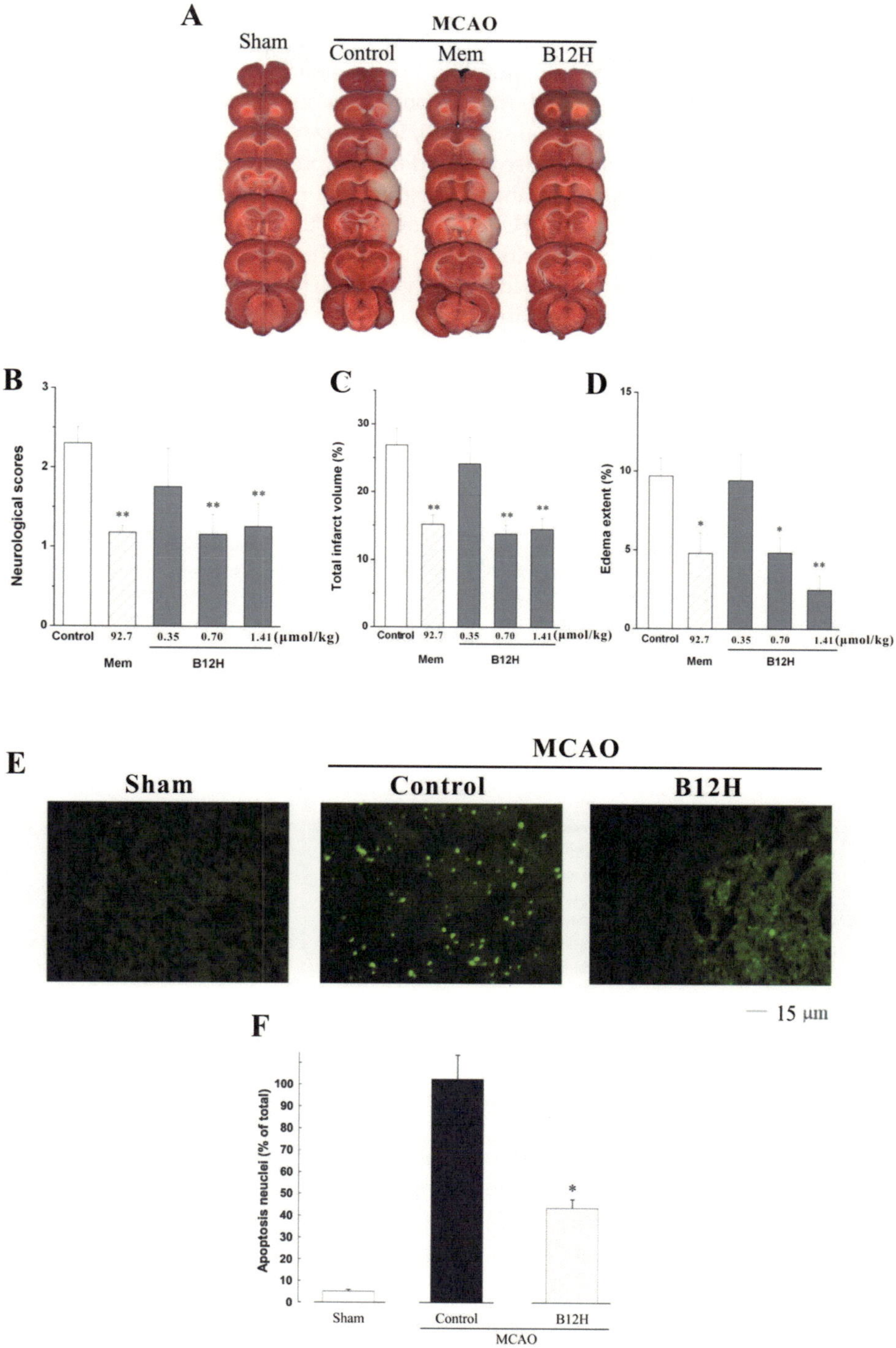

Fig. 9. Bis(12)-hupyridone rescues acute neurological impairments in rats after 2 h of MCAO followed by 24 h of reperfusion. Bis(12)-hupyridone was injected *i.p.* 30 min pre-ischemia and 15 min post-ischemia. (A) Representative photos of 2, 3, 5-triphenyltetrazolium

chloride (TTC)-stained brain slices showed that the enlarged infarct tissue area (pale unstained region) in the ischemic hemisphere of a control rat was reversed in animals treated with bis(12)-hupyridone (0.70 μmol/kg) or memantine (92.7 μmol/kg, 15 min post-ischemia *i.p.*). Bis(12)-hupyridone at both the concentrations of 0.70 and 1.41 μmol/kg reversed the decreases in neurological score (B), total infarction (C) and brain edema (D). (E) Bis(12)-hupyridone (0.70 μmol/kg) also rescued the apoptotic neurons in the penumbral region. Upper insets show representative photographs of terminal deoxynucleotidyl transferase dUTP nick end labeling (TUNEL) staining of the cerebral cortex penumbral zones of sham-treated rats, control animals, and bis(12)-hupyridone-treated rats. The lower panel shows the quantities of the TUNEL-positive cells. The number of TUNEL-positive neurons was randomly and double-blindly counted in three representative photomicrographs of each slice. B12H: bis(12)-hupyridone; Mem: memantine 92.7 μmol/kg. All data were expressed as means ± SEM, $^{*}p < 0.05$, $^{**}p < 0.01$ *versus* control group (ANOVA and Dunnett's test). The figure is adapted from our unpublished paper.

Drugs	Drug Dose (μmol/kg)	Transient Ischemia Model	Drug Treatment	Main Effects
Bis(12)-hupyridone	0.70 - 1.41	2 h middle cerebral artery occlusion followed by 24 h of reperfusion in rats	30 min pre- and 15 min post-ischemia, *i.p.*	attenuated ischemia-induced apoptosis in the penumbra region, improved neurological behavior impairment, and decreased cerebral infarct volume, cerebral edema
Memantine	46.3 - 92.7	3 h middle cerebral artery occlusion followed by 3 h of reperfusion in rat	15 min post-ischemia, *i.p.*	reversed ischemia-induced neurological deficit, reduced infract volumes, attenuated brain edema formation and blood-brain barrier permeability at the periphery
Huperzine A	0.41	45 min middle cerebral artery occlusion followed by 24 h of reperfusion in rats	at the onset and 6 h post-ischemia, *i.p.*	reversed ischemia-induced neurological deficit, reduced infract volumes, and decreased ROS production
Galantamine	7.0	20 min common carotid arteries occlusion followed by 24 h of reperfusion in rats	20 min post-ischemia, *i.p.*	reversed ischemia-induced learning impairment,
Tacrine	2.5 - 5.0	20 min common carotid arteries occlusion followed by 24 h of reperfusion in mice	1 h pre-ischemia, *p.o.*	prevented the reduction of step-down latency in the passive avoidance task, and shortened the escape latency in the Morris water maze task

Table 5. Anti-AD drugs protect transient ischemia induced impairments.
N.A.: not applicable. Data are either from our unpublished paper or from references (Gorgulu et al., 2000; Iliev et al., 2000; Wang et al., 2008; Xu et al., 2000; Zheng et al., 2008).

bis(12)-hupyridone has been shown to have high potency in preventing transient ischemia-induced neuronal impairments (Table 5) (Gorgulu et al., 2000; Iliev et al., 2000; Wang et al., 2008; Xu et al., 2000; Zheng et al., 2008). This high potency makes this dimer a promising drug candidate for the treatment of stroke and AD.

4. The physicochemical and pharmacokinetic properties of bis(12)-hupyridone

To predict the *in vivo* behaviors of bis(12)-hupyridone after dosing, its physicochemical properties have been studied and reported in our previous publication (Yu et al., 2008). As a dihydrochloride salt, bis(12)-hupyridone presents a poor solubility (S_w: 11.16 mg/ml) with two ionization constants (pK_{a1}: 7.5 and pK_{a2}: 10.0) in water. Its solubility can be largely affected by the ionic strength (mainly the concentration of chloride ion) existed in the solution (S: 2.07 mg/ml in saline) and the pH value of the solution (S: 0.75 mg/ml in physiological phosphate buffer saline, pH 7.4). In addition, bis(12)-hupyridone has been determined to be highly lipophilic due to the symmetric chemical structure. The large difference of the oil-water partition coefficients between its neutral form (log P_N : 5.4) and ionized form (log $D_{pH\ 7.4}$: 1.1) suggests that bis(12)-hupyridone might be able to easily cross the biological barriers and reach to the site of effect (i.e. the central nerve system). Further investigated by an *in vivo* study, its maximum inhibition on AChE at mice brain could be reached in 15 min after an intraperitoneal injection (*i.p.*, 5.28 µmol/kg) and the effect could be lasted for more than 4 h (Yu et al., 2008).

Previously, bis(12)-hupyridone has been identified to be quite safe both *in vitro* and *in vivo*. No cytotoxicity was observed in the MTT assay after 3 h incubation of Caco-2 cells with bis(12)-hupyridone (264 µM) (Yu et al., 2011)) In addition, no side-effects were observed for bis(12)-hupyridone after an intravenous (*i.v.*) administration to rats even at a dosage as high as 8.8 µmol/kg which suggests the good compliance of bis(12)-hupyridone to rats (Yu et al., 2009).

The pharmacokinetic properties of bis(12)-hupyridone have been studied and reported (Yu et al., 2009). After the *i.v.* bolus injection (8.8 µmol/kg), bis(12)-hupyridone presents a two-compartmental elimination in rats with a first-order kinetic process. Comparing to huperzine A, bis(12)-hupyridone exhibits a relative faster distribution and elimination ($t_{1/2\alpha}$: 1.7 ± 0.4 min and $t_{1/2\beta}$: 92.9±7.9 min) *in vivo* than those of huperzine A ($t_{1/2\alpha}$: 6.6 ± 1.1min and $t_{1/2\beta}$: 149±96 min) (Wang et al., 2006). The greater distribution volume and mean blood clearance determined (V_d: 7.54 ± 0.88 L/(min kg) and CL: 0.067 ± 0.006 L/(min kg)) suggest the extensive tissue distribution and moderate blood elimination of bis(12)-hupyridone in rats. Furthermore, the previous pharmacokinetic study has revealed that bis(12)-hupyridone could be rapidly absorbed after *i.p.* administration to rats at a dose of 10 or 20mg/kg (t_{max} of 9.33 and 4.75 min, respectively), with an absolute bioavailability of >75% (Yu et al., 2008). It suggests that bis(12)-hupyridone could be well absorbed and most of the administrated drugs could enter into the systematic circulation after extra-vascular injection.

Although all the evidence from *in vitro* and *in vivo* studies suggest a reasonable permeation of bis(12)-hupyridone through the biological barrier after *i.p.* administration, it is somewhat surprising that bis(12)-hupyridone could not detectable in rat blood after oral administration (*p.o.*, 50 mg/kg), which suggesting its poor oral bioavailability. In order to assess its extent of absorption from the gastrointestinal (GI) tract, the mechanisms of bis(12)-hupyridone transport in intestine has been evaluated using Caco-2 cell model(Yu et al., 2011). As reported, bis(12)-hupyridone has been investigated to be a substrate to ATP-binding cassette (ABC) transporters and its directional transport could be regulated by the ABC-transporters mediated efflux. ABC-transporter inhibitors can significantly increase the absorptive transport of bis(12)-hupyridone thus facilitating its oral bioavailability. Since ABC-transporters are widely presented not only in the intestine but also at the blood-brain barrier (Dallas et al., 2006; Murakami and Takano, 2008), combined treatment of bis(12)-hupyridone with ABC-transporter inhibitors might to be developed as an effective approach to improve its transport through the biological barriers and enhance its pharmacological effects at the central nerve system.

5. Conclusion

Based on the unique structure of the AChE enzyme and with the help of the bivalent ligand strategy, we have developed bis(n)-hupyridone, a novel series of dimers derived from the ineffective fragment of huperzine A. These dimers are proven to be potent and selective inhibitors of AChE both *in vitro* and *in vivo*. Bis(12)-hupyridone is a superior representative among these dimers. We have further shown that bis(12)-hupyridone, similar to memantine (an FDA approved anti-AD drug), moderately blocks NMDA receptors at the MK-801 site. Our studies have demonstrated that bis(12)-hupyridone could prevent excitotoxicity-induced neuronal loss and H_2O_2-induced neuronal apoptosis. Moreover, this dimer could promote neuronal differentiation with an efficacy similar to retinoic acid in neural stem cells. *In vivo* studies have shown that bis(12)-hupyridone possesses excellent efficacy in improving learning and memory deficits and protecting against neuronal loss *in vivo*. Our toxicological, physicochemical and pharmacokinetic studies have proved that bis(12)-hupyridone is promising for *in vivo* applications. Based on these novel findings, we conjecture that bis(12)-hupyridone could benefit AD patients by acting on multiple pathological targets concurrently (FIG. 10). As the synergism between anti-AChE, anti-NMDA receptors, anti-ROS, pro-neuronal differentiation might serve as the most effective

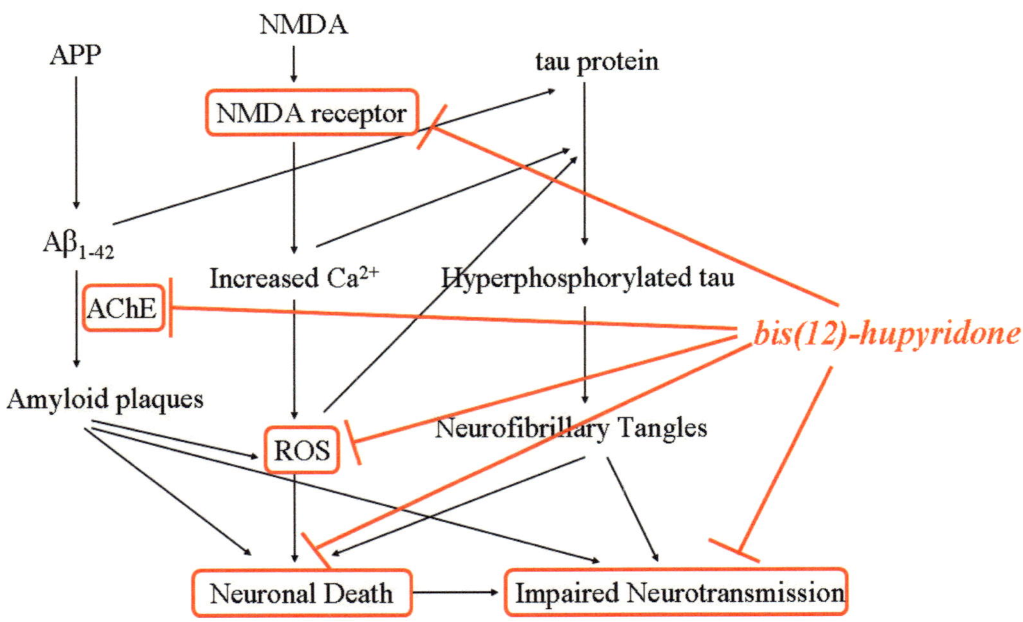

Fig. 10. Bis(12)-hupyridone acts as a multi-functional dimer for the treatment of AD.

therapeutic strategy to prevent and treat neurodegeneration AD, our findings not only provide a new direction for the design of effective compounds with multiple targets for the prevention and the treatment of AD, but also offer novel insights into the molecular basis for the development of potent therapeutic strategies for this disease.

6. Acknowledgement

This work was supported by grants from the Research Grants Council of Hong Kong (PolyU5609/09M, PolyU5610/11, N_PolyU618/07, AoE/B15/01-II), The Hong Kong Polytechnic University (G-U952), the Shenzhen Shuangbai Funding Scheme 2008. We sincerely thank Ms. Josephine Leung for editing our manuscript.

7. References

Aisen, P.S., Gauthier, S., Vellas, B., Briand, R., Saumier, D., Laurin, J., Garceau, D., (2007). Alzhemed: a potential treatment for Alzheimer's disease. *Curr Alzheimer Res*. Vol. 4, No. 4, 473-8.

Axelsen, P.H., Harel, M., Silman, I., Sussman, J.L., (1994). Structure and dynamics of the active site gorge of acetylcholinesterase: synergistic use of molecular dynamics simulation and X-ray crystallography. *Protein Sci*. Vol. 3, No. 2, 188-97.

Ballard, C., Gauthier, S., Corbett, A., Brayne, C., Aarsland, D., Jones, E., (2011). Alzheimer's disease. *Lancet*. Vol. 377, No. 9770, 1019-31.

Campbell, N.R., Fernandes, C.C., Halff, A.W., Berg, D.K., (2010). Endogenous signaling through alpha7-containing nicotinic receptors promotes maturation and integration of adult-born neurons in the hippocampus. *J Neurosci*. Vol. 30, No. 26, 8734-44.

Carlier, P.R., Du, D.M., Han, Y.F., Liu, J., Perola, E., Williams, I.D., Pang, Y.P., (2000). Dimerization of an inactive fragment of huperzine A produces a drug with twice the potency of the natural product *Angew Chem Int Ed Engl*. Vol. 39, No. 10, 1775-1777.

Cheng, D.H., Ren, H., Tang, X.C., (1996). Huperzine A, a novel promising acetylcholinesterase inhibitor. *Neuroreport*. Vol. 8, No. 1, 97-101.

Cui, W., Cui, G.Z., Li, W., Zhang, Z., Hu, S., Mak, S., Zhang, H., Carlier, P.R., Choi, C.L., Wong, Y.T., Lee, S.M., Han, Y., (2011a). Bis(12)-hupyridone, a novel multifunctional dimer, promotes neuronal differentiation more potently than its monomeric natural analog huperzine A possibly through alpha7 nAChR. *Brain Res*. Vol. 1401, No., 10-7.

Cui, W., Li, W., Han, R., Mak, S., Zhang, H., Hu, S., Rong, J., Han, Y., (2011b). PI3-K/Akt and ERK pathways activated by VEGF play opposite roles in MPP(+)-induced neuronal apoptosis. *Neurochem Int*. Vol., No.

Cui, W., Li, W., Zhao, Y., Mak, S., Gao, Y., Luo, J., Zhang, H., Liu, Y., Carlier, P.R., Rong, J., Han, Y., (2011c). Preventing HO-induced apoptosis in cerebellar granule neurons by regulating the VEGFR-2/Akt signaling pathway using a novel dimeric antiacetylcholinesterase bis(12)-hupyridone. *Brain Res*. Vol. 1394, No., 14-23.

Dallas, S., Miller, D.S., Bendayan, R., (2006). Multidrug resistance-associated proteins: expression and function in the central nervous system. *Pharmacol Rev.* Vol. 58, No. 2, 140-61.

Danysz, W., Parsons, C.G., (2003). The NMDA receptor antagonist memantine as a symptomatological and neuroprotective treatment for Alzheimer's disease: preclinical evidence. *Int J Geriatr Psychiatry.* Vol. 18, No. Suppl 1, S23-32.

De Ferrari, G.V., von Bernhardi, R., Calderon, F.H., Luza, S.C., Inestrosa, N.C., (1998). Responses induced by tacrine in neuronal and non-neuronal cell lines. *J Neurosci Res.* Vol. 52, No. 4, 435-44.

Ellis, J.M., (2005). Cholinesterase inhibitors in the treatment of dementia. *J Am Osteopath Assoc.* Vol. 105, No. 3, 145-58.

Ferri, C.P., Prince, M., Brayne, C., Brodaty, H., Fratiglioni, L., Ganguli, M., Hall, K., Hasegawa, K., Hendrie, H., Huang, Y., Jorm, A., Mathers, C., Menezes, P.R., Rimmer, E., Scazufca, M., (2005). Global prevalence of dementia: a Delphi consensus study. *Lancet.* Vol. 366, No. 9503, 2112-7.

Francis, P.T., Nordberg, A., Arnold, S.E., (2005). A preclinical view of cholinesterase inhibitors in neuroprotection: do they provide more than symptomatic benefits in Alzheimer's disease? *Trends Pharmacol Sci.* Vol. 26, No. 2, 104-11.

Fu, H., Dou, J., Li, W., Cui, W., Mak, S., Hu, Q., Luo, J., Lam, C.S., Pang, Y., Youdim, M.B., Han, Y., (2009). Promising multifunctional anti-Alzheimer's dimer bis(7)-Cognitin acting as an activator of protein kinase C regulates activities of alpha-secretase and BACE-1 concurrently. *Eur J Pharmacol.* Vol. 623, No. 1-3, 14-21.

Gorgulu, A., Kins, T., Cobanoglu, S., Unal, F., Izgi, N.I., Yanik, B., Kucuk, M., (2000). Reduction of edema and infarction by Memantine and MK-801 after focal cerebral ischaemia and reperfusion in rat. *Acta Neurochir (Wien).* Vol. 142, No. 11, 1287-92.

Gravitz, L., (2011). Drugs: a tangled web of targets. *Nature.* Vol. 475, No. 7355, S9-11.

Green, R.C., Schneider, L.S., Amato, D.A., Beelen, A.P., Wilcock, G., Swabb, E.A., Zavitz, K.H., (2009). Effect of tarenflurbil on cognitive decline and activities of daily living in patients with mild Alzheimer disease: a randomized controlled trial. *Jama.* Vol. 302, No. 23, 2557-64.

Hanahan, D., Weinberg, R.A., (2011). Hallmarks of cancer: the next generation. *Cell.* Vol. 144, No. 5, 646-74.

Harel, M., Schalk, I., Ehret-Sabatier, L., Bouet, F., Goeldner, M., Hirth, C., Axelsen, P.H., Silman, I., Sussman, J.L., (1993). Quaternary ligand binding to aromatic residues in the active-site gorge of acetylcholinesterase. *Proc Natl Acad Sci U S A.* Vol. 90, No. 19, 9031-5.

Haviv, H., Wong, D.M., Greenblatt, H.M., Carlier, P.R., Pang, Y.P., Silman, I., Sussman, J.L., (2005). Crystal packing mediates enantioselective ligand recognition at the peripheral site of acetylcholinesterase. *J Am Chem Soc.* Vol. 127, No. 31, 11029-36.

Iliev, A.I., Traykov, V.B., Mantchev, G.T., Stoykov, I., Prodanov, D., Yakimova, K.S., Krushkov, I.M., (2000). A post-ischaemic single administration of galanthamine, a cholinesterase inhibitor, improves learning ability in rats. *J Pharm Pharmacol*. Vol. 52, No. 9, 1151-6.

Le Magueresse, C., Safiulina, V., Changeux, J.P., Cherubini, E., (2006). Nicotinic modulation of network and synaptic transmission in the immature hippocampus investigated with genetically modified mice. *J Physiol*. Vol. 576, No. Pt 2, 533-46.

Lee, K.Y., Koh, S.H., Noh, M.Y., Park, K.W., Lee, Y.J., Kim, S.H., (2007). Glycogen synthase kinase-3beta activity plays very important roles in determining the fate of oxidative stress-inflicted neuronal cells. *Brain Res*. Vol. 1129, No. 1, 89-99.

Li, W., Pi, R., Chan, H.H., Fu, H., Lee, N.T., Tsang, H.W., Pu, Y., Chang, D.C., Li, C., Luo, J., Xiong, K., Li, Z., Xue, H., Carlier, P.R., Pang, Y., Tsim, K.W., Li, M., Han, Y., (2005). Novel dimeric acetylcholinesterase inhibitor bis7-tacrine, but not donepezil, prevents glutamate-induced neuronal apoptosis by blocking N-methyl-D-aspartate receptors. *J Biol Chem*. Vol. 280, No. 18, 18179-88.

Li, W., Xue, J., Niu, C., Fu, H., Lam, C.S., Luo, J., Chan, H.H., Xue, H., Kan, K.K., Lee, N.T., Li, C., Pang, Y., Li, M., Tsim, K.W., Jiang, H., Chen, K., Li, X., Han, Y., (2007a). Synergistic neuroprotection by bis(7)-tacrine via concurrent blockade of N-methyl-D-aspartate receptors and neuronal nitric-oxide synthase. *Mol Pharmacol*. Vol. 71, No. 5, 1258-67.

Li, W.M., Kan, K.K., Carlier, P.R., Pang, Y.P., Han, Y.F., (2007b). East meets West in the search for Alzheimer's therapeutics - novel dimeric inhibitors from tacrine and huperzine A. *Curr Alzheimer Res*. Vol. 4, No. 4, 386-96.

Lipton, S.A., (2004). Paradigm shift in NMDA receptor antagonist drug development: molecular mechanism of uncompetitive inhibition by memantine in the treatment of Alzheimer's disease and other neurologic disorders. *J Alzheimers Dis*. Vol. 6, No. 6 Suppl, S61-74.

Lipton, S.A., (2006). Paradigm shift in neuroprotection by NMDA receptor blockade: memantine and beyond. *Nat Rev Drug Discov*. Vol. 5, No. 2, 160-70.

Lipton, S.A., (2007). Pathologically activated therapeutics for neuroprotection. *Nat Rev Neurosci*. Vol. 8, No. 10, 803-8.

Liu, Y., Wen, X.M., Lui, E.L., Friedman, S.L., Cui, W., Ho, N.P., Li, L., Ye, T., Fan, S.T., Zhang, H., (2009). Therapeutic targeting of the PDGF and TGF-beta-signaling pathways in hepatic stellate cells by PTK787/ZK22258. *Lab Invest*. Vol. 89, No. 10, 1152-60.

Ma, J.C., Dougherty, D.A., (1997). The Cationminus signpi Interaction. *Chem Rev*. Vol. 97, No. 5, 1303-1324.

Maggini, M., Vanacore, N., Raschetti, R., (2006). Cholinesterase inhibitors: drugs looking for a disease? *PLoS Med*. Vol. 3, No. 4, e140.

Mangialasche, F., Solomon, A., Winblad, B., Mecocci, P., Kivipelto, M., (2010). Alzheimer's disease: clinical trials and drug development. *Lancet Neurol*. Vol. 9, No. 7, 702-16.

Morris, R., Mucke, L., (2006). Alzheimer's disease: A needle from the haystack. *Nature*. Vol. 440, No. 7082, 284-5.

Murakami, T., Takano, M., (2008). Intestinal efflux transporters and drug absorption. *Expert Opin Drug Metab Toxicol*. Vol. 4, No. 7, 923-39.

Oda, T., Kume, T., Katsuki, H., Niidome, T., Sugimoto, H., Akaike, A., (2007). Donepezil potentiates nerve growth factor-induced neurite outgrowth in PC12 cells. *J Pharmacol Sci*. Vol. 104, No. 4, 349-54.

Parsons, C.G., Stoffler, A., Danysz, W., (2007). Memantine: a NMDA receptor antagonist that improves memory by restoration of homeostasis in the glutamatergic system--too little activation is bad, too much is even worse. *Neuropharmacology*. Vol. 53, No. 6, 699-723.

Petrovic, M., Horak, M., Sedlacek, M., Vyklicky, L., Jr., (2005). Physiology and pathology of NMDA receptors. *Prague Med Rep*. Vol. 106, No. 2, 113-36.

Pratico, D., (2008). Evidence of oxidative stress in Alzheimer's disease brain and antioxidant therapy: lights and shadows. *Ann N Y Acad Sci*. Vol. 1147, No., 70-8.

Raves, M.L., Harel, M., Pang, Y.P., Silman, I., Kozikowski, A.P., Sussman, J.L., (1997). Structure of acetylcholinesterase complexed with the nootropic alkaloid, (-)-huperzine A. *Nat Struct Biol*. Vol. 4, No. 1, 57-63.

Roberson, E.D., Mucke, L., (2006). 100 years and counting: prospects for defeating Alzheimer's disease. *Science*. Vol. 314, No. 5800, 781-4.

Shibata, N., Kobayashi, M., (2008). The role for oxidative stress in neurodegenerative diseases. *Brain Nerve*. Vol. 60, No. 2, 157-70.

Smith, A.D., (2010). Why are drug trials in Alzheimer's disease failing? *Lancet*. Vol. 376, No. 9751, 1466.

Sortino, M.A., Frasca, G., Chisari, M., Platania, P., Chiechio, S., Vancheri, C., Copani, A., Canonico, P.L., (2004). Novel neuronal targets for the acetylcholinesterase inhibitor donepezil. *Neuropharmacology*. Vol. 47, No. 8, 1198-204.

Tang, L.L., Wang, R., Tang, X.C., (2005). Effects of huperzine A on secretion of nerve growth factor in cultured rat cortical astrocytes and neurite outgrowth in rat PC12 cells. *Acta Pharmacol Sin*. Vol. 26, No. 6, 673-8.

Thakker, D.R., Weatherspoon, M.R., Harrison, J., Keene, T.E., Lane, D.S., Kaemmerer, W.F., Stewart, G.R., Shafer, L.L., (2009). Intracerebroventricular amyloid-beta antibodies reduce cerebral amyloid angiopathy and associated micro-hemorrhages in aged Tg2576 mice. *Proc Natl Acad Sci U S A*. Vol. 106, No. 11, 4501-6.

van Marum, R.J., (2008). Current and future therapy in Alzheimer's disease. *Fundam Clin Pharmacol*. Vol. 22, No. 3, 265-74.

Vaudry, D., Stork, P.J., Lazarovici, P., Eiden, L.E., (2002). Signaling pathways for PC12 cell differentiation: making the right connections. *Science*. Vol. 296, No. 5573, 1648-9.

Villmann, C., Becker, C.M., (2007). On the hypes and falls in neuroprotection: targeting the NMDA receptor. *Neuroscientist*. Vol. 13, No. 6, 594-615.

Waldau, B., Shetty, A.K., (2008). Behavior of neural stem cells in the Alzheimer brain. *Cell Mol Life Sci*. Vol. 65, No. 15, 2372-84.

Wang, J., Zhang, H.Y., Tang, X.C., (2009). Cholinergic deficiency involved in vascular dementia: possible mechanism and strategy of treatment. *Acta Pharmacol Sin*. Vol. 30, No. 7, 879-88.

Wang, R., Yan, H., Tang, X.C., (2006). Progress in studies of huperzine A, a natural cholinesterase inhibitor from Chinese herbal medicine. *Acta Pharmacol Sin*. Vol. 27, No. 1, 1-26.

Wang, T., Tang, X.C., (1998). Reversal of scopolamine-induced deficits in radial maze performance by (-)-huperzine A: comparison with E2020 and tacrine. *Eur J Pharmacol*. Vol. 349, No. 2-3, 137-42.

Wang, Z.F., Wang, J., Zhang, H.Y., Tang, X.C., (2008). Huperzine A exhibits anti-inflammatory and neuroprotective effects in a rat model of transient focal cerebral ischemia. *J Neurochem*. Vol. 106, No. 4, 1594-603.

Wong, D.M., Greenblatt, H.M., Dvir, H., Carlier, P.R., Han, Y.F., Pang, Y.P., Silman, I., Sussman, J.L., (2003). Acetylcholinesterase complexed with bivalent ligands related to huperzine a: experimental evidence for species-dependent protein-ligand complementarity. *J Am Chem Soc*. Vol. 125, No. 2, 363-73.

Xu, J., Murakami, Y., Matsumoto, K., Tohda, M., Watanabe, H., Zhang, S., Yu, Q., Shen, J., (2000). Protective effect of Oren-gedoku-to (Huang-Lian-Jie-Du-Tang) against impairment of learning and memory induced by transient cerebral ischemia in mice. *J Ethnopharmacol*. Vol. 73, No. 3, 405-13.

Youdim, M.B., Buccafusco, J.J., (2005). CNS Targets for multi-functional drugs in the treatment of Alzheimer's and Parkinson's diseases. *J Neural Transm*. Vol. 112, No. 4, 519-37.

Yu, H., Li, W.M., Kan, K.K., Ho, J.M., Carlier, P.R., Pang, Y.P., Gu, Z.M., Zhong, Z., Chan, K., Wang, Y.T., Han, Y.F., (2008). The physicochemical properties and the in vivo AChE inhibition of two potential anti-Alzheimer agents, bis(12)-hupyridone and bis(7)-tacrine. *J Pharm Biomed Anal*. Vol. 46, No. 1, 75-81.

Yu, H., Li, W.M., Cheung, M.C., Zuo, Z., Carlier, P.R., Gu, Z.M., Chan, K., Huang, M., Wang, Y.T., Han, Y.F., (2009). Development and validation of an HPLC-DAD method for bis(12)-hupyridone and its application to a pharmacokinetic study. *J Pharm Biomed Anal*. Vol. 49, No. 2, 410-4.

Yu, H., Hu, Y.Q., Ip, F.C., Zuo, Z., Han, Y.F., Ip, N.Y., (2011). Intestinal transport of bis(12)-hupyridone in Caco-2 cells and its improved permeability by the surfactant Brij-35. *Biopharm Drug Dispos*. Vol. 32, No. 3, 140-50.

Zhang, H.Y., (2005). One-compound-multiple-targets strategy to combat Alzheimer's disease. *FEBS Lett*. Vol. 579, No. 24, 5260-4.

Zhang, H.Y., Tang, X.C., (2006). Neuroprotective effects of huperzine A: new therapeutic targets for neurodegenerative disease. *Trends Pharmacol Sci*. Vol. 27, No. 12, 619-25.

Zhao, Q., Tang, X.C., (2002). Effects of huperzine A on acetylcholinesterase isoforms in vitro: comparison with tacrine, donepezil, rivastigmine and physostigmine. *Eur J Pharmacol*. Vol. 455, No. 2-3, 101-7.

Zheng, C.Y., Zhang, H.Y., Tang, X.C., (2008). Huperzine A attenuates mitochondrial dysfunction after middle cerebral artery occlusion in rats. *J Neurosci Res*. Vol. 86, No. 11, 2432-40.

Zhongling, F., Gang, Z., Lei, Y., (2009). Neural stem cells and Alzheimer's disease: challenges and hope. *Am J Alzheimers Dis Other Demen*. Vol. 24, No. 1, 52-7.

Zhu, X., Su, B., Wang, X., Smith, M.A., Perry, G., (2007). Causes of oxidative stress in Alzheimer disease. *Cell Mol Life Sci*. Vol. 64, No. 17, 2202-10.

8

Schisandrin B, a Lignan from *Schisandra chinensis* Prevents Cerebral Oxidative Damage and Memory Decline Through Its Antioxidant Property

Tetsuya Konishi*, Vijayasree V. Giridharan and Rajarajan A. Thandavarayan

Department of Functional and Analytical Food Sciences, Niigata University of Pharmacy and Applied Life Sciences, Niigata, Japan

1. Introduction

Increasing longevity over the coming decades is expected to cause a dramatic increase in the prevalence of dementia. The resources required to care for people with dementia will rise along with the prevalence. Healthcare systems are largely unprepared for the expects rise in prevalence and for the complex care many people with dementia require. It is important to note that prevention may not be "all or none". Current pharmaceutical treatment for dementia can only modestly improve symptoms and cannot cure or prevent dementia. As a result, prevention of dementia through identification and modification of risk facors is critical (Patrick McNamara, 2011)). However, rapidly growing evidence suggest that oxidative stress play a major role in the pathophysiology of neurodegenerative disease (Ali qureshi ali G Syed and Parvez SH, 2004).

Oxidative stress is the result of an imbalance in pro-oxidant/antioxidant homeostasis that leads to the generation of toxic reactive oxygen species (ROS). The brain is considered to be especially vulnerable towards oxidative stress due to several reasons: The brain is the highest utilization of inspired oxygen, the large amount of easily oxidizable polyunsaturated fatty acids, the abundance of redox-active transition metal ions, and the relative dearth of antioxidant defense systems. Free radicals are produced from a number of sources, among which are enzymatic, mitochondrial, and redox metal ion-derived sources. Hence, brain cells are continuously exposed to ROS generated by oxidative metabolism, and in certain pathological conditions, defense mechanisms against oxygen radicals may be weakened and/or overwhelmed (Butterfield DA, Stadtman ER, 1997).

Recent reports have established that the oxidative stress and damages are playing a role in the pathogenesis of a number of neurodegenerative diseases including Alzheimer's disease (AD), Parkinson disease (PD), corticobasal degeneration, Pick's disease and Alexander's disease (Gerst, Siedlak et al. 1999). Since these neurodegenerative diseases are the serious

* Corresponding Author

factor decreasing quality of life in the longevity of society, the prevention of cerebral oxidative stress in an emergent of social task. (Konishi 2009). Many approaches have been reported such as the use of simple antioxidant molecules including antioxidant vitamins or dietary antioxidant to benefit neurodegeneration. (Srinivasan, Pandi-Perumal et al. 2005; Strimpakos and Sharma 2008; Sun, Wang et al. 2010) Among the various neurodegenerative disorders, the dementia is given importance and is focused in the present article.

Dementia is a brain disorder characterized by progressive memory loss and cognitive dysfunction, which occurs in mid to late life (McKhann, Drachman et al. 1984). Dementias and other severe cognitive dysfunction states pose a daunting challenge to existing medical management strategies. An integrative, early intervention approach seems warranted. Accumulating evidence suggests that nutritional and botanical therapies are attractive since they have proven degrees of efficacy and generally favorable benefit-to-risk profiles (Iriti M et al., 2010). Furthermore, several follow-up studies have reported a decreased risk of dementia assiciated with AD with increasing dietary or supplementary intake of antioxidants (Barberger-Gateau, Raffaitin et al. 2007). Thus antioxidant traditional herbal prescriptions was implicated as promised approach to prevent cerebral oxidative injury and prevent decline of brain function (Konishi T, 2009). Since many antioxidant ingredients have been identified in the component herbs of oriential medicine prescriptions, it is interesting to know the isolated antioxidant ingredient is active against cerebral oxidative stress as original herb or related formula. In the present article, we focus our attention onto Schisandrin B as a typical example of such herbal ingredient.

Schisandrin B (Sch B) is the major lignan with dibenzocyclooctadiene structure isolated from the fruit of *Schisandra chinensis* (FS) which is a major component of herb medicine belonging to Magnoliaceae family (Li, Xu et al. 2005) It is also one of three constituent herbs of famous traditional oriental medicine prescription, Shengmai san. Earlier, we have demonstrated the potential of Shengmai san to prevent cerebral oxidative damage and cerebral-ischemia injury in rat model (Xuejiang, Magara et al. 1999; Ichikawa, Wang et al. 2006). We also have reported quite recently the potential of Shengmai san in preventing scopolamine induced cerebral oxidative stress and memory dysfunction (Giridharan, Thandavarayan et al. 2011). Among the three component herbs, FS showed major contribution to the antioxidant activity of Shengmai san (Ichikawa H, 2003) and thus Sch B might be the major antioxidant ingredient characterizing antioxidant property of Shengmai san. In the present article, we discuss the potential of Sch B in preventing brain disorder especially in preventing cerebral oxidative stress and improving memory. It is our attempt to put forth the evidence for involvement of free radicals in pathophysiology of dementia and the potential benefit of treatment with antioxidants and radical scavengers by showing the role of dietary antioxidants Sch B in preventing oxidative stress induced cognitive disorders. We also put forth the behavioral and biochemical evidence for the potential of Sch B as a memory improving agent.

2. Schisandrin B

Sch B is the isolated component from the *Schisandra chinensis.*

Latin name: Fructus Schisandrae
Common name: Chinese magnoliavine fruit
Scientific Name: *Schisandra chinensis, Schisandra sphenanthera*
Chinese Name: Wu wei zi

A **B**

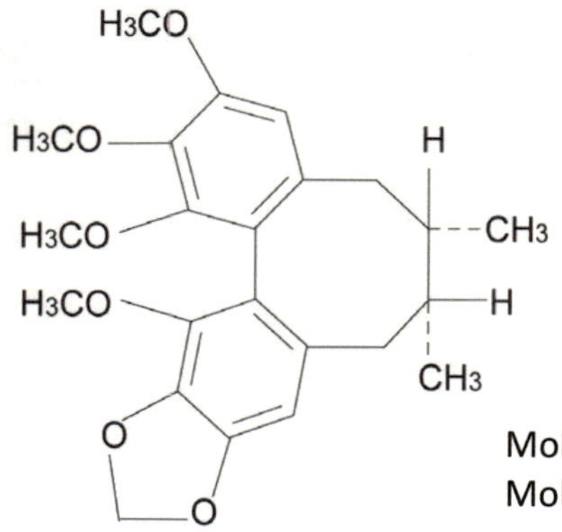

Molecularweight: 400.46
Molecularweight: $C_{23}H_{28}O_6$

Schisandra chinensis **Schisandrin B**

Fig. 1. A) Fruits of *Schisandra chinensis*; B) Structure of Sch B

The Chinese name of FS, Wuweizi is actually comprised of three Chinese words. The first word "Wu" means "five". The second word "Wei" means "taste" and the third word "Zi" refers to "seed". As the name of the herb suggests, FS is a seed with five tastes, which are sour, bitter, sweet, pungent and salty. Schisandra species grow in China, Japan, Eastern Russia, the Himalayas and Korea. FS traditionally used as astringent, to promote fluid production, to relieve pupil dilation, to relieve heat and arrest sweating and vomiting and to relieve diarrhea. The number of lignan isolated from FS includes schisandrin and its derivatives α-, β-, γ-, δ-, ε-schisandrin, psedo-γ-schisandrin, deoxyschisandrin, neoschisandrin, schisandrol and others (Wang, Hu et al. 2008).

3. The antioxidant potential of Sch B in various organs

The isolated lignan from FS reported to possess the protective effects against various organs due to its strong antioxidant potential. The hepato-protective effect of Sch B against carbon tetra chloride toxicity was mediated by both enhancements of mitochondrial glutathione antioxidant status and heat shock proteins. (Zhu, Lin et al. 1999; Tang, Chiu et al. 2003). Further in *in-vitro* model Sch B elicits a glutathione antioxidant response and protects against apoptosis via the redox-sensitive ERK/Nrf2 pathway in AML12 heptocytes. (Leong, Chiu et al. 2011) Sch B treatment increases antioxidant status of the heart and improves cardiac function against the adiramycin, doxorubicin and ischemia/reperfusion induced cardiac dysfunction (Li, Pan et al. 2007; You, Pan et al. 2006; Chiu and Ko 2004). It has been reported that Sch B enhances renal mitochondrial antioxidant status and protects against gentamicin-induced nephro-toxicity in rats. (Chiu, Leung et al. 2008) Furthermore, renal failure induced by the acute oxidant mercuric chloride found to be decreased in Sch B treated rats.(Stacchiotti, Li Volti et al. 2011).

3.1 Neuroprotective potential of Sch B

Chen and coworkers have reported that long-term treatment with Sch B enhances mitochondrial antioxidant status, structural integrity against the cerebral ischemia/reperfusion injury in rat model. The cerebro-protection afforded by Sch B treatment was associated with increases in the levels and activity of mitochondrial antioxidant components (GSH, α-TOC, and Mn-SOD), as well as preservation of mitochondrial structural integrity. Structural integrity as indicated by the decrease in sensitivity to Ca^{2+} stimulated mitochondrial permeability transition *in-vitro,* was further evidenced by decrease in the extents of mitochondrial malondialdehyde (MDA) production, Ca^{2+} loading, and cytochrome c release (Chen, Chiu et al. 2008). Sch B also shown to have protection against L-glutamate induced neurotoxicity and the protection was associated with 1) an inhibition of the increase of intracellular $[Ca^{2+}]$; 2) an improvement in the glutathione defense system, the level of glutathione, and the activity of glutathione peroxidase (GPx); and 3) an inhibition in the formation of cellular peroxide (Kim, Lee et al. 2004). Recently Sch B was reported to have protection against amyloid beta and homocysteine induced neurotoxicity in PC12 *in-vitro* system (Song, Lin et al.; Wang and Wang 2009). In addition we have also reported the potential of Sch B against scopolamine induced cerebral oxidative stress and memory dysfunction. (Giridharan, Thandavarayan et al. 2011). Of all the above reports states the neuroprotective potential of Sch B on the basis of its antioxidant property.

4. Behavioral evidences

4.1 Passive avoidance task (PAT)

We currently showed the neuroprotective effects of Sch B against experimental dementia induced by scopolamine and cisplatin (cDDP) (Giridharan, Thandavarayan et al., 2011). Passive avoidance behavior based on negative reinforcement was used to examine the long term memory (Giridharan, Thandavarayan et al. 2011). In this test, subjects learn to avoid an environment in which an aversive stimulus (such as a foot-shock) was previously delivered.

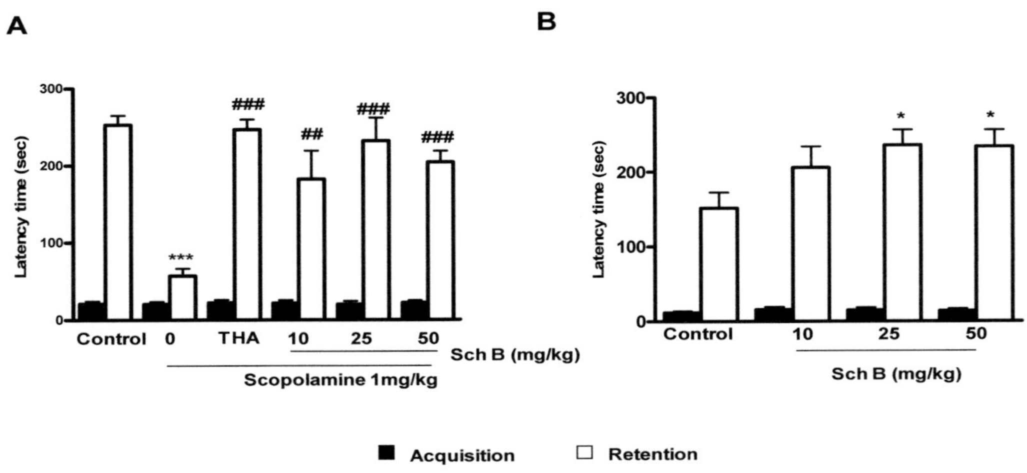

Fig. 2. A. Effects of Sch B on scopolamine-induced memory impairment in the PAT response in mice. For the study on the effect of Sch B on scopolamine-induced memory deficit model, mice were administered Sch B (10,25 and 50 mg/kg) or THA (10 mg/kg, p.o., positive control) 1 h before the acquisition trial. Memory impairment was induced by scopolamine treatment (1 mg/kg, i.p.) and acquisition trials were carried out 30 min after scopolamine treatment. At 24 h after the acquisition trials, retention trials were carried out. Data represents mean ± S.E.M (n=6). *$p<0.05$, *** $p<0.001$, statistically different from control group. ### $p<0.001$, ## $p<0.01$ statistically different from scopolamine-treated group.

The animals can freely explore the light and dark compartments of the chamber and a mild foot shock is delivered in one side of the compartment. Animals eventually learn to associate certain properties of the chamber with the foot shock. The latency to pass the gate in order to avoid the stimulus is used as an indicator of learning and memory. The passive avoidance task is useful for evaluating the effect of novel chemical entities on learning and memory as well as studying the mechanisms involved in cognition.

In the PAT, the anticholinergic agent scopolamine induced increase in step-through latency was finely inhibited by Sch B treatment. Sch B alone treated mice also found to have significant memory improving effect as showing Figure 2B. Sch B at the dose of 25 mg/kg recovered the memory level to 75.4% and thus the activity was comparable to tacrine (THA) (76.9%) the standard drug used in the treatment for AD. (Giridharan, Thandavarayan et al. 2011). We have observed that shengmai san also found to inhibit scopolamine induced memory deficits in PAT model (Giridharan, Thandavarayan et al. 2011).

4.2 Morris water maze test

The acquisition and retention of a spatial navigation task is examined using a Morris Water Maze (Kumar, Seghal et al. 2006). The hippocampal formation plays an important role in memory and learning. The Morris Water Maze (MWM) is a test of spatial learning for rodents that relies on distal cues to navigate from started locations around the perimeter of an open swimming arena to locate a submerged escape platform. Spatial learning is assessed across repeated trials and reference memory is determined by preference for the platform area when the platform is absent (Vorhees and Williams 2006).

The memory enhancing potential of Sch B was also observed in the spatial memory task where Sch B treatment significantly decreased the escape latency and the result was comparable to that of THA. In the probe trail the time spent in the target quardrant was significantly improved the Sch B (Giridharan, Thandavarayan et al. 2011).

A

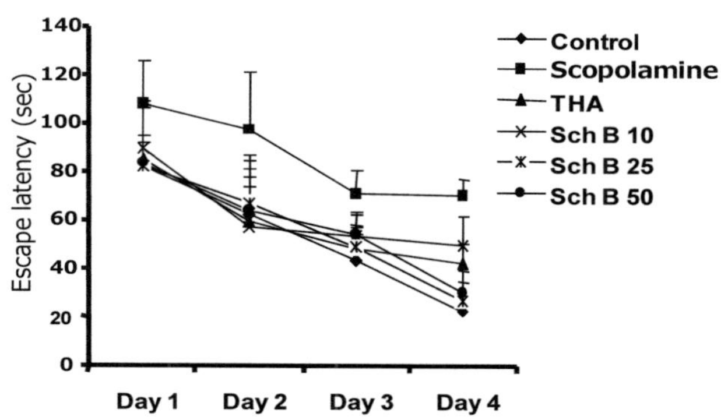

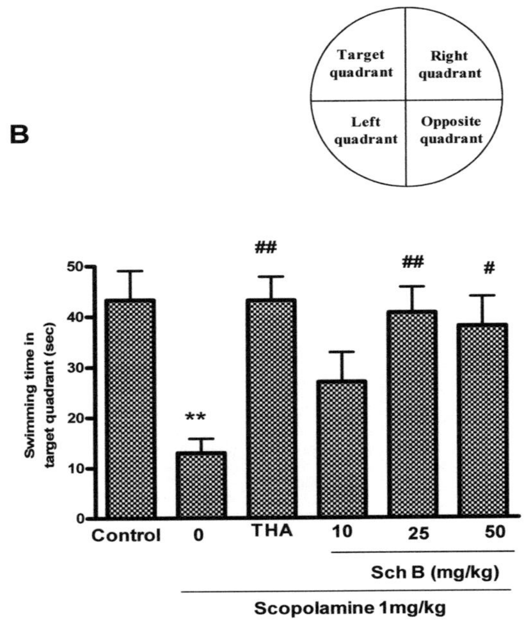

Fig. 3. Effect of Sch B on performance during training trial sessions (A) and probe trial sessions (B) of the MWM in scopolamine-induced memory deficit mice. At 1 h before the training trial session, Sch B (10,25 and 50 mg/kg) or THA (10 mg/kg, p.o., positive control) was administered to mice. Memory impairment was induced by scopolamine treatment (1 mg/kg, i.p.) 30 min after Sch B or THA administration. Data represents mean ± S.E.M (n=6). ** $p<0.01$, statistically different from control group. ## $p<0.01$, # $p<0.05$ statistically different from scopolamine-treated group.

4.3 Elevated plus maze test (EPM)

The elevated plus maze has been described as a simple method for assessing anxiety responses of rodents. There is great diversity in possible applications of the elevated plus maze. The elevated plus maze can be used as a behavioral assay to study the brain sites (e.g., limbic regions, hippocampus, amygdala, dorsal raphe nucleus, etc Furthermore, beyond its utility as a model to detect anxiolytic effects can also be used as a behavioral assay to study the brain sites (e.g., limbic regions, hippocampus, amygdala, dorsal raphe nucleus, etc.) and mechanisms (e.g., GABA, glutamate, serotonin, hypothalamic-pituitary-adrenal axis neuromodulators, etc.) underlying anxiety behavior. (Gonzalez and File 1997; Walf and Frye 2007). Briefly, rodents are placed in the intersection of the four arms of the elevated plus maze and their behavior is typically recorded for 5 min. The behaviors that are typically recorded when rodents are in the elevated plus maze are the time spent and entries made on the open and closed arms.

Behavior in this task (i.e., activity in the open arms) reflects a conflict between the rodent's preference for protected areas (e.g., closed arms) and their innate motivation to explore novel environments. Anti-anxiety behavior (increased open arm time and/or open arm entries) can be determined simultaneously with a measure of spontaneous motor activity (total and/or closed arm entries). As shown in the figure treatment with Sch B at higher dose significantly increased the open arm entry, suggesting its anti-anxiety property (Giridharan , Thandavarayan et al 2011).

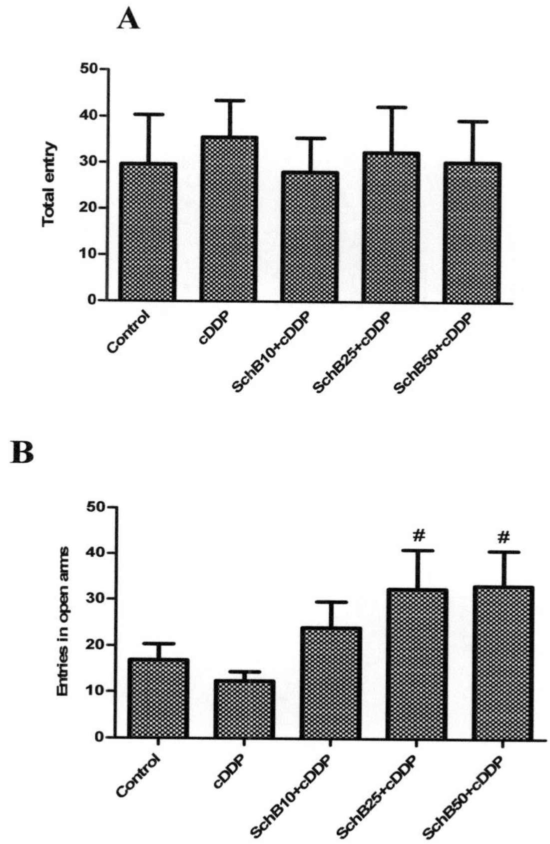

Fig. 4. Effect of Sch B (10, 25, and 50 mg/kg) on the EPM task against cisplatin. (A) Total number of entries (B) Entries in open arms. Data are represented as the mean ± S.E.M (n=8). # $p<0.05$, statistically different from cisplatin-treated group.

5. Biochemical evidences

5.1 Cholinergic relationship with Sch B

For a quarter of a century, the pathogenesis of AD associated dementia has been linked to a deficiency in the brain neurotransmitter acetylcholine (ACh). This was based on the observations of cholinergic system abnormalities leading to intellectual impairment. Subsequently, the 'cholinergic hypothesis' of AD gained considerable acceptance. It stated that a serious loss of cholinergic function in the central nervous system contributed to cognitive symptoms. Over the years, both evidence for and challenges to the relationship between ACh dysfunction and AD have been put forward, and acetylcholinesterase inhibitors (AChEIs) were introduced for the symptomatic treatment of AD. The prevailing view is that the efficacy of AChEIs is attained through their augmentation of ACh-medicated nerve transmission. (Tabet 2006).

Currently, the evidence was provided by us that Sch B act as a cholinesterase inhibitor by decreasing the levels of aceylcholinesterase (AChE) and improving the levels of ACh against scopolamine induced memory deficits animals (Giridharan, Thandavarayan et al. 2011).

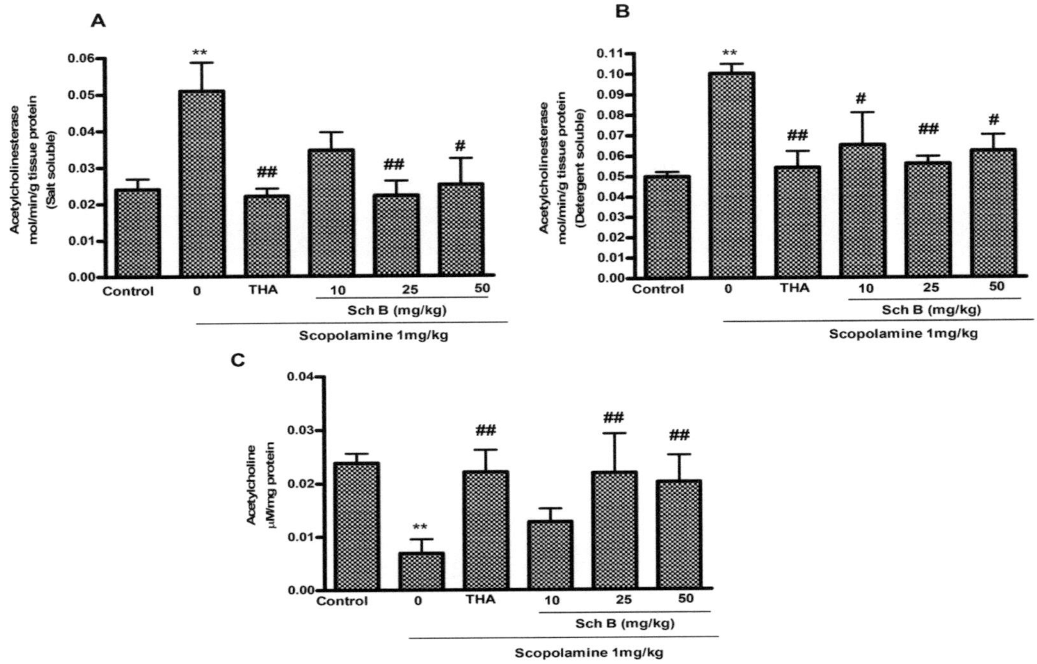

Fig. 5. The effect of Sch B (10,25 and 50 mg/kg) administration for 7 days on AChE activity in SS fraction (A) and DS fraction (B) on ACh levels (C) of brain homogenate in scopolamine-induced memory deficit mice. Data represents mean ± S.E.M (n=6). ** $p<0.01$ statistically different from control group. #p<0.05, ## $p<0.01$, statistically different from scopolamine-treated group.

It is well documented that the AChE occurs in different molecular isoforms having differential localizations in neuronal cells. The two major isoforms are globular monomer (G1) and globular tetramer (G4) of the same monomer subunit. The G1 isoform is reported to present in the cytoplasm of neuronal cells, whereas the G4 isoform is predominantly membrane-bound (Massoulie, Pezzementi et al. 1993). In the present study, both forms were measured according to the method of Das et al (Das, Dikshit et al. 2005). Results showed that the AChE (G1 and G4 isoforms) levels in both salt soluble (SS) and detergent soluble (DS) fractions were significantly increased compared to normal control after scopolamine treatment but Sch B treatment reduced the level in both SS and DS fractions dose-dependently. The percentage reduction of AChE activity in both SS and DS brain homogenates was 56.5 % and 44.3%, respectively, at 25 mg/kg of Sch B and the values are comparable to those of THA (56.8% and 45.97%).

When direct inhibitory action of Sch B was examined on AChE activity *in vitro*, the IC50 values obtained was >500 µM that was far larger than the value of THA (approximately 2 nM). Therefore, the inhibitory effect of Sch B may involve other mechanism than its direct inhibition of the enzyme. Further we analyzed the ACh levels in the brain homogenate of memory deficits mice. We observed that, ACh levels were significantly reduced in scopolamine-treated mice but treatment with Sch B (25 and 50 mg/kg) increased the reduced ACh level as well as THA.

Altogether, our data suggest that the ameliorating effects of Sch B on memory deficit might involve the modulation of ACh level through an inhibition of enzyme. (Giridharan, Thandavarayan et al. 2011).

6. Prevention of oxidative DNA damage by Sch B

Oxidative DNA damage is an inevitable consequence of cellular metabolism, with a propensity for increased levels following toxic insult. Of the molecules subject to oxidative modification, DNA has received the greatest attention as the biomarkers of exposure and effect closest to validation. (Cooke, Evans et al. 2003; Evans, Dizdaroglu et al. 2004). Although ROS can attack a variety of biomolecules, DNA may be the primary target of the free radical damage that contributes to cellular degeneration and aging (Markesbery and Lovell 2006). Indeed, multiple studies show oxidative damage to DNA may be important in cancer and, because of its high oxygen consumption rate, may also be important in neuronal damage associated with aging and neurodegenerative diseases.(Lovell and Markesbery 2007).

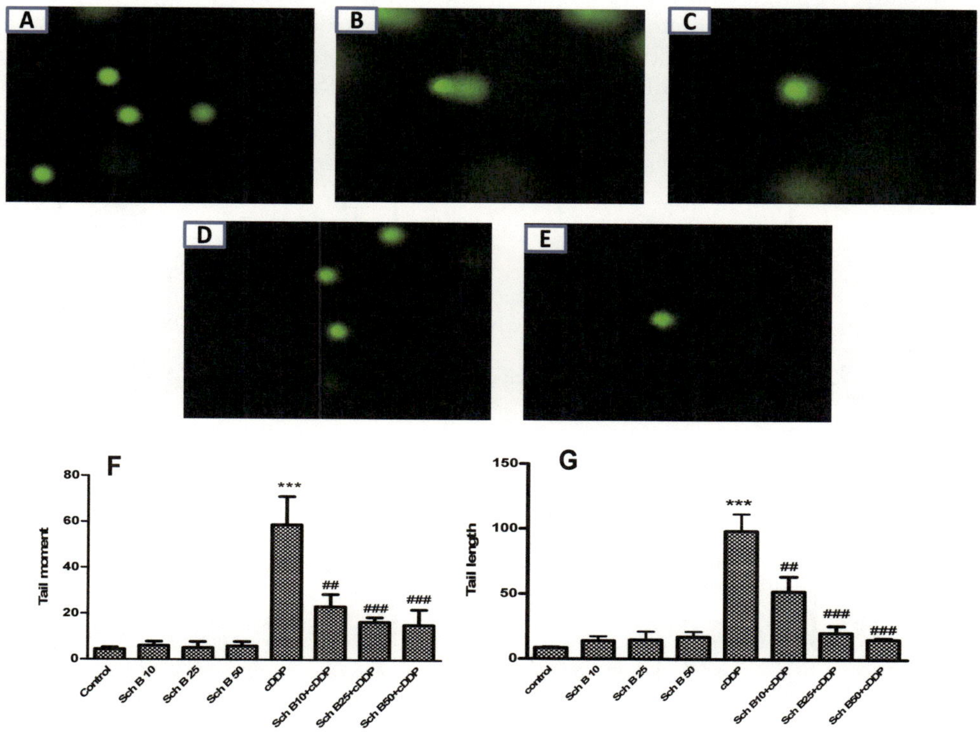

Fig. 6. Photomicrographs showing comets from forebrain stained with SYBR Green-II (A–E) Comet in a (A) normal cell (B) cisplatin-treated cell (C) Sch B 10+ cisplatin -treated cell (D) Sch B 25 + cisplatin -treated cell, (E) Sch B 50+ cisplatin -treated cell (F) Tail moment (G) Tail length. N=4 *** p<0.01 statistically different from control group. ### p<0.001, ## p<0.01 statistically different from cisplatine-treated group.

The protective effect of Sch B was studied by the classical comet assay against chemotherapeutic agent cisplain -induced DNA damage in mouse brain (Fig.6) Treatment with Sch B effectively inhibited the cisplatin induced oxidative DNA damage as measured in terms of tail length and tail moment (Giridharan, Thandavarayan et al, 2011).

7. The antioxidant potential of Sch B

It is stated that along with increased oxidative damage, impaired antioxidant defenses have also been proposed to be prominent features of AD. Usually, the body produces different antioxidants (endogenous antioxidants) to neutralize free radicals and protect the body from different diseases lead by the oxidative injury. Exogenous antioxidants externally supplied to the body through food also plays important role to protect the body. The body has developed several endogenous antioxidant defense systems classified into two groups such as enzymatic and non enzymatic. The enzymatic defense system includes different endogenous enzymes like superoxide dismutase (SOD), catalase (CAT), glutathione peroxidase (GPx), glutathione reductase (GR) and non enzymatic defense system included small antioxidant molecules including vitamin E, vitamin C and reduced glutathione (GSH) (Harris 1992).

The antioxidant system uses GSH, the most abundant non -protein thiol, which buffers free radicals in brain tissue. It eliminates H_2O_2 and organic peroxides by GPx coupled with GSH oxidation to glutathione disulfide (GSSG). GSH is regenerated by redox recycling, in which GSSG is reduced to GSH by GR with consumption of one NADPH. A reduction in level of

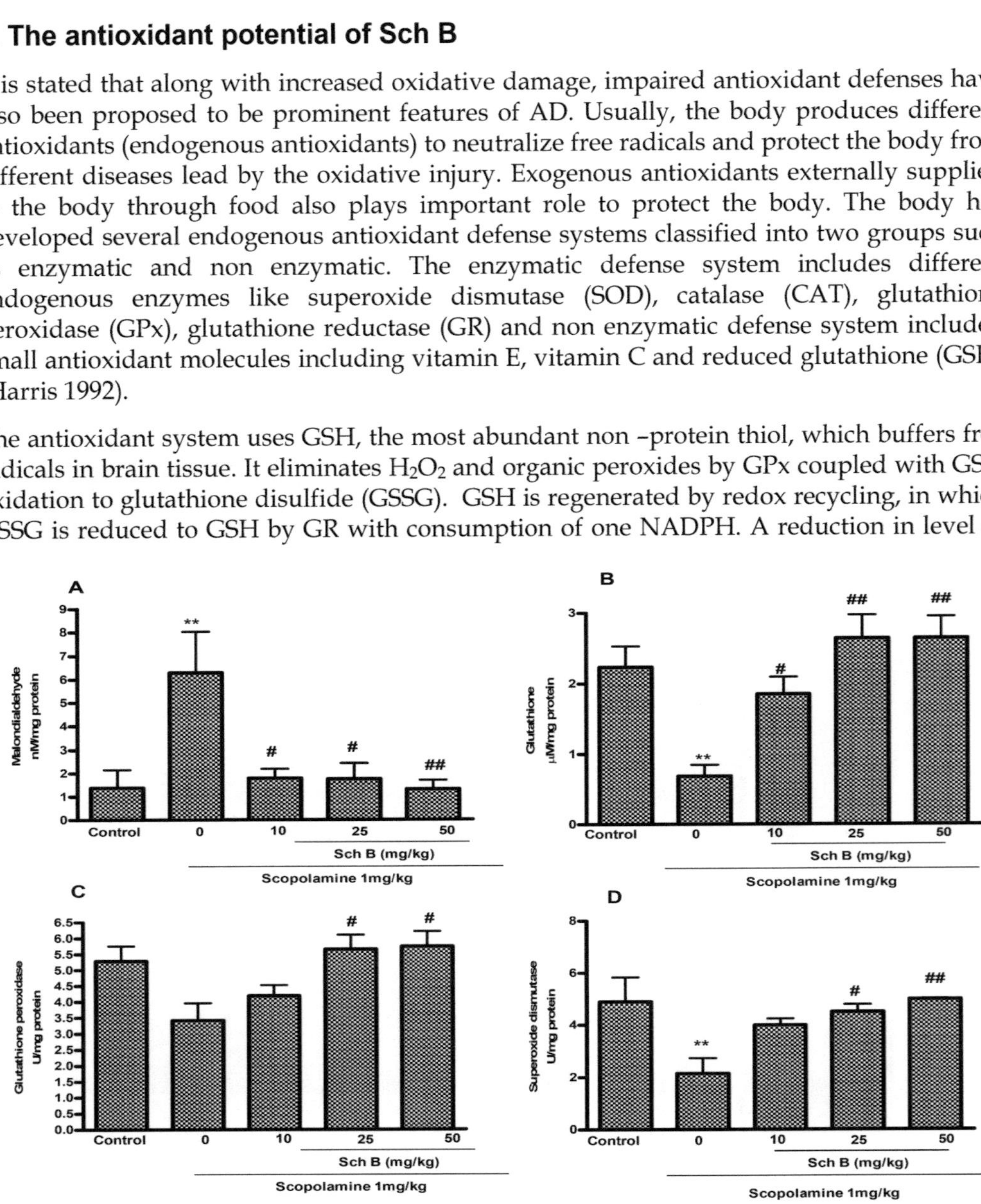

Fig. 7. Effects of acute Sch B (10,25 and 50 mg/kg) treatment on the concentrations of MDA (A) and GSH (B) and activities of GPx (C) and SOD (D) in scopolamine-induced memory deficit mice. Data represents mean ± S.E.M (n=6). $^{**}p<0.01$, statistically different from control group. $^{\#\#}$ $p<0.01$, $^{\#}$ $p<0.05$ statistically different from scopolamine-treated group.

GSH may impair H_2O_2 clearance and promotes OH radical formation, one of the most toxic ROS to the brain leading to oxidative damage. The OH radical induces peroxidation of polyunsaturated fatty acids leading to the formation MDA, an end product of lipid peroxidation (Deshmukh, Sharma et al. 2009). Interestingly, SOD mimics have come to the forefront of antioxidative therapeutics of neurodegenerative disease (Pong 2003).

We have evaluated the antioxidant potential of Sch B against scopolamine and cisplatin induced cerebral oxidative stress. Treatment with Sch B significantly increased the levels of antioxidant enzymes such as GPx and SOD, and cellular GSH levels with parallel decrease in lipid peroxidation levels (Giridharan, Thandavarayan et al, 2011).

8. Conclusion

Oxidative stress is a ubiquitously observed hallmark of neurodegenerative disorders. Neuronal cell dysfunction and cell death due to oxidative stress may causally contribute to the pathogenesis of progressive neurodegenerative disorders, such as AD and PD, as well as acute syndromes of neurodegeneration, such as ischemic and haemorrhagic stroke. Neuroprotective antioxidants are considered a promising approach to slow the progression and limit the extent of neuronal cell loss in these disorders. The clinical evidence demonstrating that antioxidant compounds can act as protective drugs in neurodegenerative disease (Guglielmotto, Giliberto et al.) Recently, cholinesterase inhibitors hybrids such as THA-melatonin developed for the treatment of AD. As AD is considered as multi complex disease with various biochemical targets, multi target-directed ligand strategy is a logical approach for designing a suitable therapy.(Fernandez-Bachiller, Perez et al. 2009; Leon and Marco-Contelles 2011). In the present article, we provided evidence for the multi-factorial role of nutritional antioxidants Sch B which behaves as neuro-protective agent, anti-cholinergic agent, and also as potential antioxidants. Further studies are needed to know more precise molecular mechanism of Sch B function as neuroprotectant.

9. Acknowledgement

This study was supported by a grant to TK from the Promotion and Mutual Aid Corporation for Private Schools. The authors thank the Rotary Yoneyama Scholarship Association for the financial assistance to VVG.

10. References

Ali qureshi ali Syed G and Parvez SH; Oxidative Stress and Neurodegenerative Disorders 2004 Elsevier, role of selenium, iron, copper and zinc in parkinsonism edited By S. Hasan Parvez

Barberger-Gateau, P., C. Raffaitin, et al. (2007). "Dietary patterns and risk of dementia: the Three-City cohort study." *Neurology* 69(20): 1921-30.

Chen, N., P. Y. Chiu, et al. (2008). "Schisandrin B enhances cerebral mitochondrial antioxidant status and structural integrity, and protects against cerebral ischemia/reperfusion injury in rats." *Biol Pharm Bull* 31(7): 1387-91.

Chiu, P. Y. and K. M. Ko (2004). "Schisandrin B protects myocardial ischemia-reperfusion injury partly by inducing Hsp25 and Hsp70 expression in rats." *Mol Cell Biochem* 266(1-2): 139-44.

Chiu, P. Y., H. Y. Leung, et al. (2008). "Schisandrin B Enhances Renal Mitochondrial Antioxidant Status, Functional and Structural Integrity, and Protects against Gentamicin-Induced Nephrotoxicity in Rats." *Biol Pharm Bull* 31(4): 602-5.

Cooke, M. S., M. D. Evans, et al. (2003). "Oxidative DNA damage: mechanisms, mutation, and disease." *FASEB J* 17(10): 1195-214.

Das, A., M. Dikshit, et al. (2005). "Role of molecular isoforms of acetylcholinesterase in learning and memory functions." *Pharmacol Biochem Behav* 81(1): 89-99.

Deshmukh, R., V. Sharma, et al. (2009). "Amelioration of intracerebroventricular streptozotocin induced cognitive dysfunction and oxidative stress by vinpocetine -- a PDE1 inhibitor." *Eur J Pharmacol* 620(1-3): 49-56.

Evans, M. D., M. Dizdaroglu, et al. (2004). "Oxidative DNA damage and disease: induction, repair and significance." *Mutat Res* 567(1): 1-61.

Fernandez-Bachiller, M. I., C. Perez, et al. (2009). "Tacrine-melatonin hybrids as multifunctional agents for Alzheimer's disease, with cholinergic, antioxidant, and neuroprotective properties." *ChemMedChem* 4(5): 828-41.

Gerst, J. L., S. L. Siedlak, et al. (1999). "Role of oxidative stress in frontotemporal dementia." *Dement Geriatr Cogn Disord* 10 Suppl 1: 85-7.

Giridharan, V. V., R. A. Thandavarayan, et al. (2011). "Effect of Shengmai-san on cognitive performance and cerebral oxidative damage in BALB/c mice." *J Med Food* 14(6): 601-9.

Giridharan, V. V., R. A. Thandavarayan, et al. (2011). "Ocimum sanctum Linn. Leaf Extracts Inhibit Acetylcholinesterase and Improve Cognition in Rats with Experimentally Induced Dementia." *J Med Food* 14(9): 912-9.

Giridharan, V. V., R. A. Thandavarayan, et al. (2011). "Prevention of scopolamine-induced memory deficits by schisandrin B, an antioxidant lignan from Schisandra chinensis in mice." *Free Radic Res* 45(8): 950-8.

Giridharan, V. V., R. A. Thandavarayan, et al. (2011). " "Schisandrin B attenuates cisplatin-induced oxidative stress, genotoxicity and neurotoxicity through modulating NF-κB pathway in mice." Free Radic Res. 2011.

Gonzalez, L. E. and S. E. File (1997). "A five minute experience in the elevated plus-maze alters the state of the benzodiazepine receptor in the dorsal raphe nucleus." *J Neurosci* 17(4): 1505-11.

Guglielmotto, M., L. Giliberto, et al. "Oxidative stress mediates the pathogenic effect of different Alzheimer's disease risk factors." *Front Aging Neurosci* 2: 3.

Harris, E. D. (1992). "Regulation of antioxidant enzymes." *J Nutr* 122(3 Suppl): 625-6.

Ichikawa, H., L. Wang, et al. (2006). "Prevention of cerebral oxidative injury by post-ischemic intravenous administration of Shengmai San." *Am J Chin Med* 34(4): 591-600.

Kim, S. R., M. K. Lee, et al. (2004). "Dibenzocyclooctadiene lignans from Schisandra chinensis protect primary cultures of rat cortical cells from glutamate-induced toxicity." *J Neurosci Res* 76(3): 397-405.

Konishi, T. (2009). "Brain oxidative stress as basic target of antioxidant traditional oriental medicines." *Neurochem Res* 34(4): 711-6.

Kumar, A., N. Seghal, et al. (2006). "Differential effects of cyclooxygenase inhibitors on intracerebroventricular colchicine-induced dysfunction and oxidative stress in rats." *Eur J Pharmacol* 551(1-3): 58-66.

Leon, R. and J. Marco-Contelles (2011). "A step further towards multitarget drugs for Alzheimer and neuronal vascular diseases: targeting the cholinergic system, amyloid-beta aggregation and Ca(2+) dyshomeostasis." *Curr Med Chem* 18(4): 552-76.

Leong, P. K., P. Y. Chiu, et al. (2011). "Schisandrin B elicits a glutathione antioxidant response and protects against apoptosis via the redox-sensitive ERK/Nrf2 pathway in AML12 hepatocytes." *Free Radic Res* 45(4): 483-95.

Li, L., Q. Pan, et al. (2007). "Schisandrin B prevents doxorubicin-induced cardiotoxicity via enhancing glutathione redox cycling." *Clin Cancer Res* 13(22 Pt 1): 6753-60.

Li, Y., C. Xu, et al. (2005). "In vitro anti-Helicobacter pylori action of 30 Chinese herbal medicines used to treat ulcer diseases." *J Ethnopharmacol* 98(3): 329-33.

Lovell, M. A. and W. R. Markesbery (2007). "Oxidative DNA damage in mild cognitive impairment and late-stage Alzheimer's disease." *Nucleic Acids Res* 35(22): 7497-504.

Markesbery, W. R. and M. A. Lovell (2006). "DNA oxidation in Alzheimer's disease." *Antioxid Redox Signal* 8(11-12): 2039-45.

Massoulie, J., L. Pezzementi, et al. (1993). "Molecular and cellular biology of cholinesterases." *Prog Neurobiol* 41(1): 31-91.

McKhann, G., D. Drachman, et al. (1984). "Clinical diagnosis of Alzheimer's disease: report of the NINCDS-ADRDA Work Group under the auspices of Department of Health and Human Services Task Force on Alzheimer's Disease." *Neurology* 34(7): 939-44.

Pong, K. (2003). "Oxidative stress in neurodegenerative diseases: therapeutic implications for superoxide dismutase mimetics." *Expert Opin Biol Ther* 3(1): 127-39.

Song, J. X., X. Lin, et al. "Protective effects of dibenzocyclooctadiene lignans from Schisandra chinensis against beta-amyloid and homocysteine neurotoxicity in PC12 cells." *Phytother Res* 25(3): 435-43.

Srinivasan, V., S. R. Pandi-Perumal, et al. (2005). "Role of melatonin in neurodegenerative diseases." *Neurotox Res* 7(4): 293-318.

Stacchiotti, A., G. Li Volti, et al. (2011). "Different role of Schisandrin B on mercury-induced renal damage in vivo and in vitro." *Toxicology* 286(1-3): 48-57.

Strimpakos, A. S. and R. A. Sharma (2008). "Curcumin: preventive and therapeutic properties in laboratory studies and clinical trials." *Antioxid Redox Signal* 10(3): 511-45.

Sun, A. Y., Q. Wang, et al. (2010). "Resveratrol as a therapeutic agent for neurodegenerative diseases." *Mol Neurobiol* 41(2-3): 375-83.

Tabet, N. (2006). "Acetylcholinesterase inhibitors for Alzheimer's disease: anti-inflammatories in acetylcholine clothing!" *Age Ageing* 35(4): 336-8.

Tang, M. H., P. Y. Chiu, et al. (2003). "Hepatoprotective action of schisandrin B against carbon tetrachloride toxicity was mediated by both enhancement of mitochondrial glutathione status and induction of heat shock proteins in mice." *Biofactors* 19(1-2): 33-42.

Vorhees, C. V. and M. T. Williams (2006). "Morris water maze: procedures for assessing spatial and related forms of learning and memory." *Nat Protoc* 1(2): 848-58.

Walf, A. A. and C. A. Frye (2007). "The use of the elevated plus maze as an assay of anxiety-related behavior in rodents." *Nat Protoc* 2(2): 322-8.

Wang, B. and X. M. Wang (2009). "Schisandrin B protects rat cortical neurons against Abeta1-42-induced neurotoxicity." *Pharmazie* 64(7): 450-4.

Wang, B. L., J. P. Hu, et al. (2008). "Simultaneous quantification of four active schisandra lignans from a traditional Chinese medicine Schisandra chinensis(Wuweizi) in rat plasma using liquid chromatography/mass spectrometry." *J Chromatogr B Analyt Technol Biomed Life Sci* 865(1-2): 114-20.
Xuejiang, W., T. Magara, et al. (1999). "Prevention and repair of cerebral ischemia-reperfusion injury by Chinese herbal medicine, shengmai san, in rats." *Free Radic Res* 31(5): 449-55.
You, J. S., T. L. Pan, et al. (2006). "Schisandra chinensis protects against adriamycin-induced cardiotoxicity in rats." *Chang Gung Med J* 29(1): 63-70.
Zhu, M., K. F. Lin, et al. (1999). "Evaluation of the protective effects of Schisandra chinensis on Phase I drug metabolism using a CCl4 intoxication model." *J Ethnopharmacol* 67(1): 61-8.

Part 3

Brain Cancer

9

CREB Signaling in Neural Stem/Progenitor Cells: Implications for a Role in Brain Tumors

Theo Mantamadiotis[1,2], Nikos Papalexis[2] and Sebastian Dworkin[3]
[1]*Department of Pathology, The University of Melbourne,*
[2]*Laboratory of Physiology, Medical School, University of Patras,*
[3]*Central Clinical School, Monash University, Melbourne,*
[1,3]*Australia*
[2]*Greece*

1. Introduction

Since its discovery in the PC12 rat pheochromocytoma cell line (Montminy & Bilezikjian 1987) the cAMP Response Element Binding (CREB) protein has been implicated in a variety of neuronal responses such as excitation, long-term memory formation, neural cell proliferation and opiate tolerance. Its importance is underscored by the attention this factor has attracted in the neuroscience community, as evidenced by the thousands of citations in the academic bibliographic databases. CREB is a transcription factor which potentially regulates the transcription of hundreds or even thousands of genes in neurons. A variety of protein kinases possess the capability of driving CREB phosphorylation and activation, placing CREB at a hub of multiple intraneuronal signalling cascades. The array of neuronal functions attributed to CREB has expanded recently, with studies showing that CREB has role in neural stem/progenitor cell growth, differentiation and survival. This data, together with complementary studies in tissues outside the CNS showing that CREB activation has oncogenic effects has led to the hypothesis that CREB has an important role in brain tumour biology. Therefore, CREB is a factor which sits within a molecular network potentially integrating signalling events regulating neural stem cells and neurogenesis, neural cancer cells and other cells within brain tumors.

To gain an the understanding of the link between stem cells normally residing in the adult brain and the stem cells which can give rise to a brain tumour, it is important to introduce the concepts relating to the so-called 'cancer stem cell hypothesis'. Indeed, one of the most important advances in brain tumour biology has been the discovery that tumors can develop from cells with stem cell-like characteristics. The reason for the excitement is better understood when one considers the nature of treatments of typical cancers/tumors in a patient. The most relevant example to consider in the context of this chapter is the most common and deadly brain tumour, glioblastoma multiforme (high-grade glioma). Gliomas are difficult to treat and patients usually succumb within months to 1-2 years, even with multiple treatment approaches. Standard treatments rely on 'debulking' of the tumour(s), achieved by surgical excision and/or cytotoxic therapies, usually radiation and chemotherapy. Almost inevitably, this first treatment is followed by relatively rapid relapse and aggressive tumour recurrence. Considering the existence of glioma cancer stem cells, which give rise to the original tumour mass, it has become clear that these cells, which although few in number, probably lie at the periphery or even outside the main tumour mass and are also resistant to current cytotoxic therapies. Thus, surgery only removes the large tumour mass and cancer stem cells within, sparing other cancer stem cells outside the main tumour mass. These surviving cancer stem cells are able to give rise to the recurring/secondary tumors, which have also evolved to become more resistant to further treatments. Thus, much research has focussed on stem cell biology in the context of cancer and the processes which give rise to cancer stem cells or tumour initiating cells. Research on the mechanisms that play a role in neural birth and brain development are gaining traction in the understanding of brain tumour biology, since there must be common molecular genetic mechanisms operating in both normal/non-tumor neural stem cells and neural cancer stem cells. Indeed, once the parallel mechanisms are understood, then the differences will also become apparent. These differences will also provide the rational basis for therapeutic targeting of neural cancer stem cells.

Aside from contributing to furthering the understanding of the ongoing cellular plasticity of the brain, the knowledge that adult organs, including the brain, harbour stem and progenitor cells throughout the life of the organism has helped develop new concepts on what happens when these cells accumulate mutations in the context of diseases such as cancer. Indeed, the understanding of cancer stem cells has provided a new optimism in the development of novel strategies for cancer therapy (Schatton *et al.* 2009). The signalling networks operating in normal neural and brain tumour initiating cells involve complex molecular networks. At the hub of these networks are the transcription factors, which determine which genes are expressed, when they are expressed and how much of each corresponding mRNA is expressed. There are many transcription factors which have been identified as being important for neural stem cell function but research linking transcription factors regulating normal stem cells and cancer stem cells is still at an early stage. In fact, little is known about what distinguishes a cancer stem cell from a physiologically normal stem cell.

2. Neural stem cells and neurogenesis

The origins of the mammalian central nervous system lie within the neuroepithelium, a thin layer of developing nerve cells. Much of this early developmental period in vertebrates is dedicated to organising the structure of the brain. This organisation precedes a period of rapid cellular expansion, the peak of neurogenesis.

The discovery that neurogenesis persists in the adult vertebrate brain was contrary to the long-held dogma, oft quoted as Santiago Ramon y Cajal's statement referring to the central nervous system that "...nothing may be regenerated". Of course, the available methods over a century ago made it almost impossible to observe or measure the minute fraction of nerve cells undergoing cell division amongst the billions of postmitotic cells in an adult mammalian brain. Since Cajal's time there were sporadic but important reports on the existence of mitotic cells in mature adult mammalian brains (Allen 1912; Altman & Das 1965). The prevailing understanding of neurogenesis is that neural stem cells arise during embryogenesis, and a fraction of these persist into adulthood within discrete regions of adult brain ("neurogenic regions") (reviewed in (Abrous *et al.* 2005)). These cells are distinct from other, non-neural cell types in the brain (most notably microglia – the "immune cells" of the brain) which retain the ability to proliferate, but cannot generate cells of other neural lineages. Cells fulfilling the criteria of "stemness" (self-renewal, multipotentiality) have been identified in the brains of higher vertebrates, including humans (Eriksson *et al.* 1998). The best characterised neurogenic regions in higher vertebrates lie in the sub-ventricular zone of the lateral ventricles and the sub-granular zone of the hippocampus. The number of proliferating cells and newborn neurons in the dentate gyrus, olfactory bulb and sub-ventricular zone decreases with age (Altman & Das 1965; Kuhn *et al.* 1996), consistent with an age-dependent decline in neurogenic potential. As mentioned previously, there are many factors which regulate neurogenesis, including transcription factors. The CREB transcription factor has only recently been recognised to play an important role in this process. This factor is at the hub of multiple signalling cascades, which are active in neural stem cells and regulates the expression of a series of downstream target genes important for stem cell survival and growth (see Figure 1).

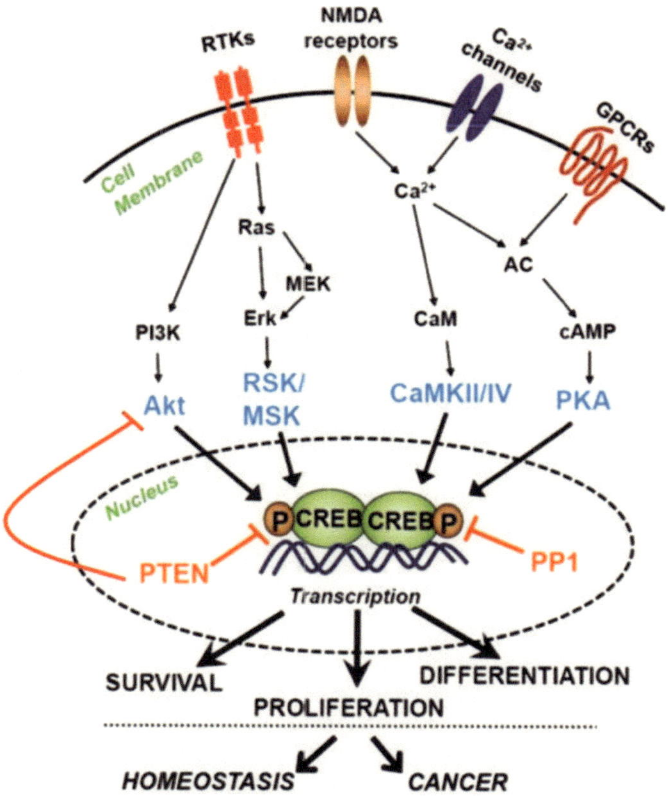

Fig. 1. Several pathways lead to CREB phosphorylation/activation to promote cell survival, proliferation and differentiation. In the context of neural stem cells and cancer, Receptor Tyrosine Kinases (RTKs) are important, since their ligands such as EGF and PDGF are growth factors necessary for cell survival and proliferation. However, the role of the other pathways shown, remain to be investigated in this context. Note that dephosphorylation of CREB via phosphatases occurs via the activity of PTEN and PP1. PTEN may be critical in the context of brain tumors and CREB signalling, as it is often mutated in gliomas.

In the context of neural stem cells and cancer, Receptor Tyrosine Kinases (RTKs) are important, since the ligands for these receptors such as EGF and PDGF are growth factors and necessary for cell survival and proliferation. However, the roles of the other pathways shown remain to be investigated in this context. Note that dephosphorylation of CREB via phosphatases occurs via the activity of PTEN and PP1. PTEN may be critical in the context of brain tumors and CREB signalling, as it is often mutated in glioma.

3. The CREB transcription factor family

Transcription factors are the terminal convergence points of many signalling pathways, these genes function as effector molecules to activate downstream target genes which in turn regulate NSPC proliferation, cell-cycle exit, induction of differentiation and survival (for a concise review see (Ahmed *et al.* 2009)). The precise cell stage at which a particular transcription factor is active determines its contribution to the cell's progression from immaturity to maturity.

CREB is a nuclear-localised basic leucine zipper superfamily transcription factor, acting as a conduit between upstream signalling kinases and downstream target-gene transcription. Three major isoforms of CREB are known (α, Δ and γ), all transcribed from the same gene, CREB1. Although the best characterised member of the CREB family is CREB itself (Montminy & Bilezikjian 1987), the family also includes CREM (Foulkes et al. 1991) and ATF1 (Hai *et al.* 1988), products of distinct genes. These transcription factors are able to homodimerize or heterodimerize with each other, bind to cyclic-AMP Response Element (CREs) sequences present in target gene promoters and are activated by serine-threonine kinases targeting the phosphorylation of their Kinase-Inducible Domain (KID). Thus, there is an inherent functional redundancy in the CREB transcription factor family which has been shown in mouse knockout studies, where CREB deletion results in an upregulation of CREM expression in an attempt to compensate for many of the cellular functions normally attributed to CREB (Blendy *et al.* 1996; Mantamadiotis *et al.* 2002). Phosphorylation of the KID then causes increased affinity to various transcriptional coactivators such as CREB-Binding Protein (CBP), p300 and the Transducers Of Regulated CREB activity (TORCs), which then leads to the assembly of the transcriptional machinery and transcription initiation. The CREB transcription factor family are potent transcriptional activators although there is some evidence that in certain contexts these factors are capable of repressing transcription (Rutberg *et al.* 1999).

3.1 CREB in NSPCs and neurogenesis

CREB's role in embryonic brain development and neurogenesis is conserved across at least two vertebrate species separated by over 300 million years of evolution, as studies in zebrafish embryos show that CREB has a role in developmental neurogenesis and in midbrain-hindbrain patterning (Dworkin *et al.* 2007). There is also evidence that CREB has a role in the regeneration of the simple nerve net in *Hydra* species and the more complex nervous system of the roundworm *Caenorhabditis elegans* (Chera *et al.* 2007; Ghosh-Roy *et al.* 2010).

In the developing mouse brain, the active phosphorylated form of CREB is seen in cells clustered in the neurogenic regions at E14.5, a time when the brain takes on recognisable neuro-anatomical features and neurogenesis peaks and becomes regionally localised. These regions include the ventricular zones of both the lateral and third ventricles and the olfactory bulb (Dworkin *et al.* 2009). Consistent with a role in neurogenesis, the activated, phosphorylated form of CREB, pCREB is enriched and restricted to the neurogenic zones of the adult mouse brain, whereas total (unphosphorylated and phosphorylated) CREB protein is present in almost all cells of the brain (Figure 2).

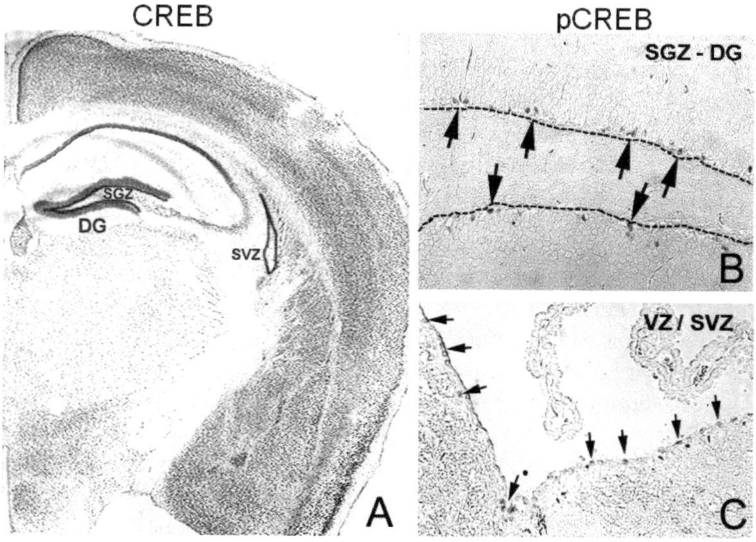

Fig. 2. Immunohistochemical analysis of CREB protein expression in mouse brain.
A) A coronal section of mouse brain showing the global expression of CREB protein (phosphorylated and unphosphorylated). The positive signals are evident as the dark nuclear staining in each cell/neuron. The neurogenic zones, SGZ (sub-granular zone) of the dentate gyrus (DG) located in the hippocampus and SVZ (sub-ventricular zone) are indicated by the dark lines and labels (x4 power). B) In contrast to total CREB protein expression, phospho-CREB expression is evident and restricted to the neurogenic sub-granular zone (SGZ) of the hippocampal dentate gyrus (DG), where some positive cells are indicated by the arrow heads. C) phospho-CREB expression is evident in the SVZ, where some positive cells are indicated by the arrow heads. (B & C x100).

Regulated transient CREB phosphorylation and de-phosphorylation is a well described mechanism by which neuronal activity is regulated in many regions of adult mouse brain (Lonze & Ginty 2002). Moreover, CREB is required for the survival of post-mitotic neurons in mouse brain (Ao *et al.* 2006; Dworkin *et al.* 2009; Giachino *et al.* 2005; Herold *et al.* 2010; Mantamadiotis *et al.* 2002; Riccio *et al.* 1999), while the role of CREB signalling in the proliferation and migration stages of immature neurons is less well defined. In a number of studies, the use of phospho-specific CREB antibodies demonstrate that constitutive CREB activation is restricted to cells in neurogenic regions (Bender *et al.* 2001; Dworkin *et al.* 2007; Dworkin *et al.* 2009; Fujioka *et al.* 2004; Gampe *et al.* 2011; Giachino *et al.* 2005; Herold *et al.* 2010; Nakagawa *et al.* 2002). In zebrafish, phosphorylated CREB is expressed throughout the highly proliferative embryonic brain but in the adult expression is restricted to cells in the proliferative zones (Dworkin *et al.* 2007), in patterns identical to those previously reported for proliferating cells (Grandel *et al.* 2006). Taken together, these data suggest a role for

CREB in proliferating cells in the post-natal adult vertebrate brain. Furthermore, pCREB is also expressed in zones of NSPC migration (Giachino *et al.* 2005), indicating it may also function in maintaining survival of migratory neuroblasts.

A number of CREB mouse mutants have been critical to the investigation of CREB function in vivo. Transgenic mice expressing a dominant-negative mutant CREB shows that CREB has a role in cell expansion and survival in the pituitary gland (Struthers *et al.* 1991) and seminiferous tubules of the testis (Scobey *et al.* 2001). CREB over-expression on the other hand results in increased cellular proliferation (Shankar & Sakamoto 2004; Zhu *et al.* 2004). Mice with germline deletion of all CREB isoforms show a decrease in the size of the corpus callosum and an increase in lateral ventricle area (Rudolph *et al.* 1998), consistent with a decrease in cellularity and displayed significant defects in brain development which were attributed to neurogenic defects (Dworkin *et al.* 2009).

Since loss of CREB leads to an upregulation of the related factor CREM as a compensatory mechanism for CREB loss, a more sophisticated approach was needed to assess the role of CREB signalling loss. Therefore, mice were generated with a germline deletion of CREM and lacking CREB specifically in neural cells. These brain-specific compound CREB-CREM mutant mice displayed severe neuronal death (Mantamadiotis *et al.* 2002), stressing the importance of CREB signalling in neuronal survival. Further studies on on NSPCs derived from CREB-null mice displayed severe defects in survival, cellular expansion and neurosphere forming potential (Dworkin *et al.* 2009). An important question on whether CREB is also important for neural expansion comes from studies in mice where a transcriptionally constitutive active fusion of the CREB DNA-binding domain with the transactivation domain of Herpes Simplex Virus, VP-16-CREB has demonstrated that CREB-dependent genes contribute to neurogenesis (Zhu *et al.* 2004). Similarly, a constitutively active CREB mutant leads to an overproduction of neural cells in zebrafish embryos while a dominant-negative CREB mutant which is able to silence kinase-induced CREB activation, has the opposite effect and inhibits neurogenesis (Dworkin *et al.* 2007).

The upstream or downstream factors associated with the CREB-dependent mechanisms promoting proliferation are not well understood. However, activation of the PI3K/Akt pathway by FGF-2 in cultured adult hippocampal NSPCs resulted in increased CREB phosphorylation and increased progenitor proliferation and decreased differentiation, as did over-expression of wild-type CREB (Peltier *et al.* 2007). Furthermore, increasing cGMP, Akt and GSK3β activity, upstream signals, which phosphorylate CREB, in adult SVZ-derived neurospheres increased NSPC proliferation, whereas down-regulating these signals resulted in decreased proliferation (Peltier *et al.* 2007). Recent work also shows that CREB-dependent NSPC proliferation and neurogenesis is mediated via EGF-induced activation of both PKA (Iguchi *et al.* 2011) and ERK (Gampe *et al.* 2011). All the above mentioned studies were performed in animal model organisms or NSPCs derived from these. So far there are no reports on the role of the CREB pathway in human NSPCs but recent work shows that CREB is activated and functional in neurogenic cells in the adult primate (Japanese macacque) brain (Boneva & Yamashima 2011).

3.2 CREB's oncogenic properties

There are numerous reports in cell, animal and human tissue studies showing a positive correlation between the level of CREB expression and activation and malignancy. A role for

CREB-mediated transcription in cancer was first reported through the identification of a chromosomal translocation t(12;22)(q13;q12) in clear cell sarcomas of soft tissue to give a fusion protein EWS-ATF1 (Zucman *et al.* 1993). This chimaeric protein, consisting of the N-terminal region of EWS (Ewing's Sarcoma) fused with the C-terminal DNA-binding domain of the CREB-related protein ATF1, generates a constitutively active transcription activator capable of binding to the promoters of CREB/ATF1 target genes, which in turn promote tumour development and growth. More recently, a EWS-CREB1 fusion was discovered in a clear cell sarcoma variant (Antonescu *et al.* 2006) and angiomatoid fibrous histiocytomas (Rossi *et al.* 2007).

CREB has been implicated in contributing to the progression of several other tumour types (Conkright & Montminy 2005; Rosenberg *et al.* 2002). Analysis of prostate tumors from patients demonstrated that pCREB expression was restricted to poorly-differentiated prostate cancers and bone metastatic tissue but not to non-tumour benign prostate glands (Wu *et al.* 2007). Increased mRNA levels of CREB are also a feature of breast cancer tissue compared to non-tumour mammary tissue and the level of CREB expression correlated with disease progression and survival (Chhabra *et al.* 2007). In non-small-cell lung cancer the expression levels of CREB and pCREB were elevated in tumour compared to adjacent normal tissues and increased CREB expression correlated with poor patient survival (Seo *et al.* 2008). Human ovarian tumors also exhibit increased CREB expression and ovarian tumour cell lines in which CREB expression is silenced display significantly reduced proliferation (Linnerth *et al.* 2008). Some of the best studies implicating CREB in cancer development come from evidence showing that CREB has a role in the development of bone marrow malignancies. The oncogenic virus human T-cell leukemia virus type 1 (HTLV-1) is strongly associated with T-cell leukemia (ATL) [29, 30]. T-cell oncogenic transformation mediated by the HTLV-1 Tax oncoprotein requires intact CREB signalling (Smith & Greene 1991). Moreover, increased CREB and pCREB expression is seen in bone marrow from patients with ALL (acute lymphoid leukemia) and AML (acute myeloid leukemia) compared to that from healthy patients (Crans-Vargas *et al.* 2002). In addition, CREB expression and in some cases increased CREB gene copy number correlates with disease stage in leukemia patients where CREB overexpression is associated with accelerated relapse and event-free survival (Crans-Vargas *et al.* 2002; Pigazzi *et al.* 2007; Shankar *et al.* 2005). Finally, CREB also appears to regulate malignant melanoma biology by promoting tumour cell survival and metastasis (Jean & Bar-Eli 2000; Melnikova *et al.* 2010).

How CREB regulates tumour growth is still a question that remains unanswered. An obvious approach to unravel the underlying CREB-mediated oncogenic mechanisms is to determine the array of "cancer-associated" genes which CREB directly regulates at the level of transcription. Several genes known to be directly regulated by CREB are implicated in tumourigenesis and uncontrolled proliferation. CREB directly regulates several cell-cycle control genes known to be aberrantly expressed in hyper-proliferative disorders, including *cyclin D1* (Pradeep *et al.* 2004), *cyclin A1* and *A2* (Desdouets *et al.* 1995)(Shankar and Sakamoto, 2004), *bcl-2* (Wilson *et al.* 1996), *HEC1* (a cell-cycle regulatory protein which localizes to the kinetochore in mitosis and is implicated in cancer progression (H. Y. Cheng *et al.* 2007) and *cyclin D2*. Increased *cyclin D2* transcription following CREB transactivation has been implicated in regulating the proliferation of lymphocytes, putatively through phosphorylation of CREB by PI3K and PKA (Assanah *et al.* 2006). In cultured mouse embryonic fibroblasts (MEFs), phosphorylation of CREB by LiCl increases cyclin D2 expression, whereas inhibition of the CREB-cyclin D2 pathway by the tumour-suppressor phosphatase PTEN decreases the abundance of cyclin D2 mRNA and protein (Huang *et al.* 2007), indicating that CREB-mediated regulation of cyclin D2 may be a conserved partnership regulating proliferation. VEGF was also increased in tandem with increased CREB signalling in metastatic prostate cancer derived from human bone (He *et al.* 2007), strongly supporting a direct role for CREB in mediating cellular proliferation and possibly metastasis. In human brain tumour derived cell lines there is evidence that CREB can be activated by prostaglandin E_2 via the PKA pathway to stimulate cell proliferation (Bidwell *et al.* 2010). Thus, data from cell lines, animal models and importantly patient tumour samples, indicate that CREB not only serves as a diagnostic marker but also has a role in promoting and supporting the development tumors in a variety of cell and tissue types. In the next section we discuss the evidence that suggests CREB may also be an important factor in brain tumour development and growth.

4. Converging evidence for the involvement of CREB in brain cancer

Various studies using brain tumour cell lines suggest that signalling pathways involving CREB activation are important for tumour cell growth and differentiation (Bidwell *et al.* 2010; Golan *et al.* 2011; Kim *et al.* 2010; Morioka *et al.* 2010). To date there has been no evidence linking CREB to brain cancer development or progression in vivo, although a number of recent findings linking CREB activity to PTEN and growth factors, together with the knowledge of CREB's role in NSPC biology, lend support to the view that CREB is an important factor in brain tumour signalling pathways. Of note, recent data shows that CREB is a protein target of PTEN phosphatase activity and that PTEN loss induces CREB-dependent gene expression and cell growth (Boneva & Yamashima 2011). PTEN is a tumour suppressor gene frequently mutated in many cancers including the most aggressive forms of brain cancer, glioblastoma multiforme and related astrocytomas. Indeed PTEN expression appears to directly affect glioblastoma growth as well as glioma-initiating cell proliferation and self-renewal (R. B. Cheng *et al.* 2011). Thus, PTEN loss-of-function mutations would lead to loss of CREB deactivation, allowing the over activation of CREB-dependent cell survival and growth signals in brain cancer stem cells or brain tumour initiating cells (BTICs). Other important signalling pathways in patient brain tumour cells are the epidermal growth factor receptor (EGFR) and the platelet-derived growth factor receptor (PDGFR) pathways (Brennan *et al.* 2009). EGFR activation is important for glioma stem/progenitor cell growth and resistance to anti-cancer treatments (Murat *et al.* 2008). EGF is able to induce CREB in NSPCs in vivo (Gampe *et al.* 2011); most likely acting through EGFR induced CREB activation via the Ras-MAPK dependent kinase, RSK-2 (Xing *et al.* 1996). Furthermore, there is evidence that CREB is activated in human glioma cells lines and that inhibition of CREB leads to reduced survival of glioma cells (Malla *et al.* 2010). This study also shows that PDGFR-dependent PI3K/Akt signals which converge upon CREB are important for tumour invasiveness, a process which BTICs use to migrate and generate metastatic tumors.

Data from the Human Protein Atlas (www.proteinatlas.org) shows that CREB is highly expressed in all glioma patient samples tested (24 cases) and consistent with the mouse data, human brain also shows robust CREB expression in neurogenic zones (Figure 3). Data from our own laboratory shows that human glioma tumour tissue (40 cases) exhibits robust pCREB expression compared to only weak staining in non-tumour tissue controls (unpublished data). This implies that the CREB pathway is overactive in human glioma cells, thereby driving the survival and growth of these cells. More interest is the potential role that CREB may be playing in the glioma stem cells, which are the cellular source of the tumour and which may also be responsible for the relapse of tumour growth following therapy. Data from primary mouse NSPCs shows that CREB is required for the expression of various growth and survival factors including BDNF, NGF, PACAP and Bcl-2 (Dworkin *et al.* 2009). It is likely that the expression of growth and survival factors will be dependent upon CREB-dependent transcription.

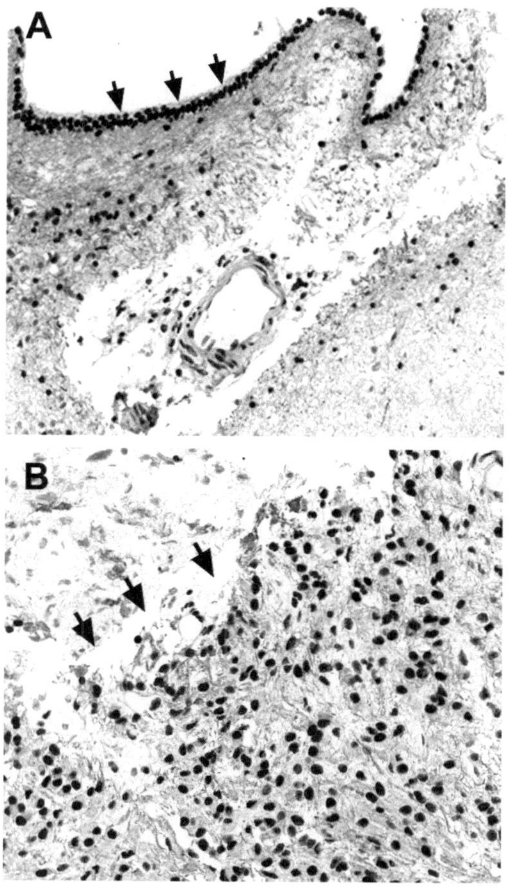

Fig. 3. CREB expression is human brain.
A) CREB expression is enriched in the human brain SVZ, as seen by the intense nuclear staining of cells lining the ventricular space (indicated by arrow heads). B) Intense CREB expression is clearly evident in human high grade glioma. The non-tumour cells show weak staining (behind the arrow heads). According to the Human Protein Atlas data, 100% (24 cases) of brain tumour samples tested exhibited strong CREB expression (Uhlen et al., *Nat Biotechnol.* 2010 28(12):1248-50 and http://www.proteinatlas.org). The images were from the Human Protein Atlas database.

5. Conclusion

In conclusion, there is significant emerging experimental data implicating the CREB signalling pathway in the development and maintenance of brain tumors. Investigation of the CREB signalling pathway and transcriptome in glioma cell lines, BTICs and new animal models will shed light on the importance of this pathway in glioma biology. This knowledge will provide an opportunity to investigate novel drug targeting approaches in glioma treatment, targeting CREB itself or an upstream or downstream component of the CREB-pathway. Opinions on whether widely expressed factors which are critical to cell function are good targets vary widely and have evolved over the last decades. CREB may well prove to be a good anti-tumour target in the brain, as tumors seem to express high levels of the activated phosphorylated form. This is in contrast with the physiologically normal adult brain which only transiently exhibits pCREB expression only in discreet nuclei responsible for a specific neuronal response (eg. the suprachiasmatic nucleus in response to visual light stimulation). This observation together with the ever advancing drug delivery technologies may allow targeting of CREB in brain tumors with minimal toxicity to neurons outside the tumour.

6. Acknowledgements

This work has been supported by an FP7 Marie Curie IRG (IRG231032/Neurogencreb), Karatheodori Grant (2010-4735 Uni Patras) and the Department of Pathology, The University of Melbourne.

7. References

Abrous, D.N., M. Koehl & M. Le Moal. (2005). Adult neurogenesis: From precursors to network and physiology. *Physiol Rev* 85, no. 2: 523-69.

Ahmed, S., H.T. Gan, C.S. Lam, A. Poonepalli, S. Ramasamy, Y. Tay, M. Tham & Y.H. Yu. (2009). Transcription factors and neural stem cell self-renewal, growth and differentiation. *Cell Adh Migr* 3, no. 4: 412-24.

Allen, E. (1912). The cessation of mitosis in the central nervous system of the albino rat. *J Comp Neurol.* 22, no. 6: 547–68.

Altman, J. & G.D. Das. (1965). Autoradiographic and histological evidence of postnatal hippocampal neurogenesis in rats. *J Comp Neurol* 124, no. 3: 319-35.

Antonescu, C.R., K. Nafa, N.H. Segal, P. Dal Cin & M. Ladanyi. (2006). Ews-Creb1: A recurrent variant fusion in clear cell sarcoma--association with gastrointestinal location and absence of melanocytic differentiation. *Clin Cancer Res* 12, no. 18: 5356-62.

Ao, H., S.W. Ko & M. Zhuo. (2006). CREB activity maintains the survival of cingulate cortical pyramidal neurons in the adult mouse brain. *Mol Pain* 2: 15.

Assanah, M., R. Lochhead, A. Ogden, J. Bruce, J. Goldman & P. Canoll. (2006). Glial progenitors in adult white matter are driven to form malignant gliomas by platelet-derived growth factor-expressing retroviruses. *J Neurosci* 26, no. 25: 6781-90.

Bender, R.A., J.C. Lauterborn, C.M. Gall, W. Cariaga & T.Z. Baram. (2001). Enhanced CREB phosphorylation in immature dentate gyrus granule cells precedes neurotrophin expression and indicates a specific role of creb in granule cell differentiation. *Eur J Neurosci* 13, no. 4: 679-86.

Bidwell, P., K. Joh, H.A. Leaver & M.T. Rizzo. (2010). Prostaglandin e2 activates camp response element-binding protein in glioma cells via a signaling pathway involving pka-dependent inhibition of erk. *Prostaglandins Other Lipid Mediat* 91, no. 1-2: 18-29.

Blendy, J.A., K.H. Kaestner, W. Schmid, P. Gass & G. Schutz. (1996). Targeting of the creb gene leads to up-regulation of a novel creb mrna isoform. *Embo J* 15, no. 5: 1098-106.

Boneva, N.B. & T. Yamashima. (2011). New insights into "gpr40-CREB interaction inadult neurogenesis" specific for primates. *Hippocampus*.

Brennan, C., H. Momota, D. Hambardzumyan, T. Ozawa, A. Tandon, A. Pedraza & E. Holland. (2009). Glioblastoma subclasses can be defined by activity among signal transduction pathways and associated genomic alterations. *PLoS One* 4, no. 11: e7752.

Cheng, H.Y., J.W. Papp, O. Varlamova, H. Dziema, B. Russell, J.P. Curfman, T. Nakazawa, K. Shimizu, H. Okamura, S. Impey & K. Obrietan. (2007). Microrna modulation of circadian-clock period and entrainment. *Neuron* 54, no. 5: 813-29.

Cheng, R.B., R.J. Ma, Z.K. Wang, S.J. Yang, X.Z. Lin, H. Rong & Y. Ma. (2011). Pten status is related to cell proliferation and self-renewal independent of cd133 phenotype in the glioma-initiating cells. *Mol Cell Biochem* 349, no. 1-2: 149-57.

Chera, S., K. Kaloulis & B. Galliot. (2007). The camp response element binding protein (CREB) as an integrative hub selector in metazoans: Clues from the hydra model system. *Biosystems* 87, no. 2-3: 191-203.

Chhabra, A., H. Fernando, G. Watkins, R.E. Mansel & W.G. Jiang. (2007). Expression of transcription factor Creb1 in human breast cancer and its correlation with prognosis. *Oncol Rep* 18, no. 4: 953-8.

Conkright, M.D. & M. Montminy. (2005). CREB: The unindicted cancer co-conspirator. *Trends Cell Biol* 15, no. 9: 457-9.

Crans-Vargas, H.N., E.M. Landaw, S. Bhatia, G. Sandusky, T.B. Moore & K.M. Sakamoto. (2002). Expression of cyclic adenosine monophosphate response-element binding protein in acute leukemia. *Blood* 99, no. 7: 2617-9.

Desdouets, C., G. Matesic, C.A. Molina, N.S. Foulkes, P. Sassone-Corsi, C. Brechot & J. Sobczak-Thepot. (1995). Cell cycle regulation of cyclin a gene expression by the cyclic amp-responsive transcription factors CREB and CREM. *Mol Cell Biol* 15, no. 6: 3301-9.

Dworkin, S., J.K. Heath, T.A. Dejong-Curtain, B.M. Hogan, G.J. Lieschke, J. Malaterre, R.G. Ramsay & T. Mantamadiotis. (2007). CREB activity modulates neural cell proliferation, midbrain-hindbrain organization and patterning in zebrafish. *Dev Biol* 307, no. 1: 127-41.

Dworkin, S., J. Malaterre, F. Hollande, P.K. Darcy, R.G. Ramsay & T. Mantamadiotis. (2009). Camp response element binding protein is required for mouse neural progenitor cell survival and expansion. *Stem Cells* 27, no. 6: 1347-57.

Eriksson, P.S., E. Perfilieva, T. Bjork-Eriksson, A.M. Alborn, C. Nordborg, D.A. Peterson & F.H. Gage. (1998). Neurogenesis in the adult human hippocampus. *Nat Med* 4, no. 11: 1313-7.

Foulkes, N.S., E. Borrelli & P. Sassone-Corsi. (1991). Crem gene: Use of alternative DNA-binding domains generates multiple antagonists of camp-induced transcription. *Cell* 64, no. 4: 739-49.

Fujioka, T., A. Fujioka & R.S. Duman. (2004). Activation of camp signaling facilitates the morphological maturation of newborn neurons in adult hippocampus. *J Neurosci* 24, no. 2: 319-28.

Gampe, K., M.S. Brill, S. Momma, M. Gotz & H. Zimmermann. (2011). Egf induces CREB and ERK activation at the wall of the mouse lateral ventricles. *Brain Res* 1376: 31-41.

Ghosh-Roy, A., Z. Wu, A. Goncharov, Y. Jin & A.D. Chisholm. (2010). Calcium and cyclic amp promote axonal regeneration in caenorhabditis elegans and require dlk-1 kinase. *J Neurosci* 30, no. 9: 3175-83.

Giachino, C., S. De Marchis, C. Giampietro, R. Parlato, I. Perroteau, G. Schutz, A. Fasolo & P. Peretto. (2005). Camp response element-binding protein regulates differentiation and survival of newborn neurons in the olfactory bulb. *J Neurosci* 25, no. 44: 10105-18.

Golan, M., G. Schreiber & S. Avissar. (2011). Antidepressants elevate GDNF expression and release from c6 glioma cells in a beta-arrestin1-dependent, CREB interactive pathway. *Int J Neuropsychopharmacol*: 1-12.

Grandel, H., J. Kaslin, J. Ganz, I. Wenzel & M. Brand. (2006). Neural stem cells and neurogenesis in the adult zebrafish brain: Origin, proliferation dynamics, migration and cell fate. *Dev Biol* 295, no. 1: 263-77.

Hai, T.W., F. Liu, E.A. Allegretto, M. Karin & M.R. Green. (1988). A family of immunologically related transcription factors that includes multiple forms of atf and ap-1. *Genes Dev* 2, no. 10: 1216-26.

He, G., D. Wu, A. Sun, Y. Xue, Z. Jin, H. Qiu, M. Miao, X. Tang, Z. Fu & Z. Chen. (2007). Cytcd79a expression in acute leukemia with t(8;21): Biphenotypic or myeloid leukemia? *Cancer Genet Cytogenet* 174, no. 1: 76-7.

Herold, S., R. Jagasia, K. Merz, K. Wassmer & D.C. Lie. (2010). CREB signalling regulates early survival, neuronal gene expression and morphological development in adult subventricular zone neurogenesis. *Mol Cell Neurosci.*

Huang, W., H.Y. Chang, T. Fei, H. Wu & Y.G. Chen. (2007). Gsk3 beta mediates suppression of cyclin d2 expression by tumor suppressor pten. *Oncogene* 26, no. 17: 2471-82.

Iguchi, H., T. Mitsui, M. Ishida, S. Kanba & J. Arita. (2011). Camp response element-binding protein (CREB) is required for epidermal growth factor (EGF)-induced cell proliferation and serum response element activation in neural stem cells isolated from the forebrain subventricular zone of adult mice. *Endocr J.*

Jean, D. & M. Bar-Eli. (2000). Regulation of tumor growth and metastasis of human melanoma by the CREB transcription factor family. *Mol Cell Biochem* 212, no. 1-2: 19-28.

Kim, Y.H., H.S. Joo & D.S. Kim. (2010). Nitric oxide induction of ire1-alpha-dependent CREB phosphorylation in human glioma cells. *Nitric Oxide* 23, no. 2: 112-20.

Kuhn, H.G., H. Dickinson-Anson & F.H. Gage. (1996). Neurogenesis in the dentate gyrus of the adult rat: Age-related decrease of neuronal progenitor proliferation. *J Neurosci* 16, no. 6: 2027-33.

Linnerth, N.M., J.B. Greenaway, J.J. Petrik & R.A. Moorehead. (2008). Camp response element-binding protein is expressed at high levels in human ovarian adenocarcinoma and regulates ovarian tumor cell proliferation. *Int J Gynecol Cancer* 18, no. 6: 1248-57.

Lonze, B.E. & D.D. Ginty. (2002). Function and regulation of CREB family transcription factors in the nervous system. *Neuron* 35, no. 4: 605-23.

Malla, R., S. Gopinath, K. Alapati, C.S. Gondi, M. Gujrati, D.H. Dinh, S. Mohanam & J.S. Rao. (2010). Downregulation of upar and cathepsin b induces apoptosis via regulation of bcl-2 and bax and inhibition of the PI3K/Akt pathway in gliomas. *PLoS One* 5, no. 10: e13731.

Mantamadiotis, T., T. Lemberger, S.C. Bleckmann, H. Kern, O. Kretz, A. Martin Villalba, F. Tronche, C. Kellendonk, D. Gau, J. Kapfhammer, C. Otto, W. Schmid & G. Schutz. (2002). Disruption of CREB function in brain leads to neurodegeneration. *Nat Genet* 31, no. 1: 47-54.

Melnikova, V.O., A.S. Dobroff, M. Zigler, G.J. Villares, R.R. Braeuer, H. Wang, L. Huang & M. Bar-Eli. (2010). CREB inhibits ap-2alpha expression to regulate the malignant phenotype of melanoma. *PLoS One* 5, no. 8: e12452.

Montminy, M.R. & L.M. Bilezikjian. (1987). Binding of a nuclear protein to the cyclic-amp response element of the somatostatin gene. *Nature* 328, no. 6126: 175-8.

Morioka, N., T. Sugimoto, M. Tokuhara, T. Dohi & Y. Nakata. (2010). Noradrenaline induces clock gene per1 mrna expression in c6 glioma cells through beta(2)-adrenergic receptor coupled with protein kinase a - camp response element binding protein (pka-creb) and src-tyrosine kinase - glycogen synthase kinase-3beta (src-gsk-3beta). *J Pharmacol Sci* 113, no. 3: 234-45.

Murat, A., E. Migliavacca, T. Gorlia, W.L. Lambiv, T. Shay, M.F. Hamou, N. De Tribolet, L. Regli, W. Wick, M.C. Kouwenhoven, J.A. Hainfellner, F.L. Heppner, P.Y. Dietrich, Y. Zimmer, J.G. Cairncross, R.C. Janzer, E. Domany, M. Delorenzi, R. Stupp & M.E. Hegi. (2008). Stem cell-related "self-renewal" signature and high epidermal growth factor receptor expression associated with resistance to concomitant chemoradiotherapy in glioblastoma. *J Clin Oncol* 26, no. 18: 3015-24.

Nakagawa, S., J.E. Kim, R. Lee, J. Chen, T. Fujioka, J. Malberg, S. Tsuji & R.S. Duman. (2002). Localization of phosphorylated camp response element-binding protein in immature neurons of adult hippocampus. *J Neurosci* 22, no. 22: 9868-76.

Peltier, J., A. O'neill & D.V. Schaffer. (2007). PI3K/Akt and CREB regulate adult neural hippocampal progenitor proliferation and differentiation. *Dev Neurobiol* 67, no. 10: 1348-61.

Pigazzi, M., E. Ricotti, G. Germano, D. Faggian, M. Arico & G. Basso. (2007). Camp response element binding protein (CREB) overexpression has been described as critical for leukemia progression. *Haematologica* 92, no. 10: 1435-7.

Pradeep, A., C. Sharma, P. Sathyanarayana, C. Albanese, J.V. Fleming, T.C. Wang, M.M. Wolfe, K.M. Baker, R.G. Pestell & B. Rana. (2004). Gastrin-mediated activation of cyclin d1 transcription involves beta-catenin and CREB pathways in gastric cancer cells. *Oncogene* 23, no. 20: 3689-99.

Riccio, A., S. Ahn, C.M. Davenport, J.A. Blendy & D.D. Ginty. (1999). Mediation by a CREB family transcription factor of ngf-dependent survival of sympathetic neurons. *Science* 286, no. 5448: 2358-61.

Rosenberg, D., L. Groussin, E. Jullian, K. Perlemoine, X. Bertagna & J. Bertherat. (2002). Role of the PKA-regulated transcription factor CREB in development and tumorigenesis of endocrine tissues. *Ann N Y Acad Sci* 968: 65-74.

Rossi, S., K. Szuhai, M. Ijszenga, H.J. Tanke, L. Zanatta, R. Sciot, C.D. Fletcher, A.P. Dei Tos & P.C. Hogendoorn. (2007). Ewsr1-creb1 and EWSR1-ATF1 fusion genes in angiomatoid fibrous histiocytoma. *Clin Cancer Res* 13, no. 24: 7322-8.

Rudolph, D., A. Tafuri, P. Gass, G.J. Hammerling, B. Arnold & G. Schutz. (1998). Impaired fetal T cell development and perinatal lethality in mice lacking the camp response element binding protein. *Proc Natl Acad Sci U S A* 95, no. 8: 4481-6.

Rutberg, S.E., T.L. Adams, M. Olive, N. Alexander, C. Vinson & S.H. Yuspa. (1999). Cre DNA binding proteins bind to the AP-1 target sequence and suppress AP-1 transcriptional activity in mouse keratinocytes. *Oncogene* 18, no. 8: 1569-79.

Schatton, T., N.Y. Frank & M.H. Frank. (2009). Identification and targeting of cancer stem cells. *Bioessays* 31, no. 10: 1038-49.

Scobey, M., S. Bertera, J. Somers, S. Watkins, A. Zeleznik & W. Walker. (2001). Delivery of a cyclic adenosine 3',5'-monophosphate response element-binding protein (CREB) mutant to seminiferous tubules results in impaired spermatogenesis. *Endocrinology* 142, no. 2: 948-54.

Seo, H.S., D.D. Liu, B.N. Bekele, M.K. Kim, K. Pisters, S.M. Lippman, Wistuba, Ii & J.S. Koo. (2008). Cyclic amp response element-binding protein overexpression: A feature associated with negative prognosis in never smokers with non-small cell lung cancer. *Cancer Res* 68, no. 15: 6065-73.

Shankar, D.B., J.C. Cheng, K. Kinjo, N. Federman, T.B. Moore, A. Gill, N.P. Rao, E.M. Landaw & K.M. Sakamoto. (2005). The role of CREB as a proto-oncogene in hematopoiesis and in acute myeloid leukemia. *Cancer Cell* 7, no. 4: 351-62.

Shankar, D.B. & K.M. Sakamoto. (2004). The role of cyclic-amp binding protein (CREB) in leukemia cell proliferation and acute leukemias. *Leuk Lymphoma* 45, no. 2: 265-70.

Smith, M.R. & W.C. Greene. (1991). Type I human T cell leukemia virus Tax protein transforms rat fibroblasts through the Cyclic adenosine monophosphate response element binding protein/Activating transcription factor pathway. *J Clin Invest* 88, no. 3: 1038-42.

Struthers, R.S., W.W. Vale, C. Arias, P.E. Sawchenko & M.R. Montminy. (1991). Somatotroph hypoplasia and dwarfism in transgenic mice expressing a non-phosphorylatable CREB mutant. *Nature* 350, no. 6319: 622-4.

Uhlen, M, P.Oksvold, L.Fagerberg, E.Lundberg, K.Jonasson, M.Forsberg, M.Zwahlen, C.Kampf, K.Wester, S.Hober, H.Wernerus, L.Björling & F.Ponten. (2010) Towards a knowledge-based Human Protein Atlas. *Nat Biotechnol.* 28:1248-50.

Wilson, B.E., E. Mochon & L.M. Boxer. (1996). Induction of Bcl-2 expression by phosphorylated creb proteins during B-cell activation and rescue from apoptosis. *Mol Cell Biol* 16, no. 10: 5546-56.

Wu, D., H.E. Zhau, W.C. Huang, S. Iqbal, F.K. Habib, O. Sartor, L. Cvitanovic, F.F. Marshall, Z. Xu & L.W. Chung. (2007). Camp-responsive element-binding protein regulates vascular endothelial growth factor expression: Implication in human prostate cancer bone metastasis. *Oncogene* 26, no. 35: 5070-7.

Xing, J., D.D. Ginty & M.E. Greenberg. (1996). Coupling of the Ras-MAPK pathway to gene activation by Rsk2, a growth factor-regulated CREB kinase. *Science* 273, no. 5277: 959-63.

Zhu, D.Y., L. Lau, S.H. Liu, J.S. Wei & Y.M. Lu. (2004). Activation of camp-response-element-binding protein (CREB) after focal cerebral ischemia stimulates neurogenesis in the adult dentate gyrus. *Proc Natl Acad Sci U S A* 101, no. 25: 9453-7.

Zucman, J., O. Delattre, C. Desmaze, A.L. Epstein, G. Stenman, F. Speleman, C.D. Fletchers, A. Aurias & G. Thomas. (1993). EWS and ATF-1 gene fusion induced by t(12;22) translocation in malignant melanoma of soft parts. *Nat Genet* 4, no. 4: 341-5.

Part 4

Brain Imaging

10

MRI Techniques and New Animal Models for Imaging the Brain

Elodie Chaillou[1], Yves Tillet[1] and Frédéric Andersson[2]
[1]*Reproductive and Behavioural Physiology INRA, CNRS UMR 6175, University François Rabelais of Tours, EFCE, IFR135, Nouzilly,*
[2]*IFR135 Functional Imaging, Tours, France*

1. Introduction

This chapter describes how large animal models can be used to improve our knowledge of neuroscience and brain disorders. Various animal models have been used in Magnetic Resonance Imaging (MRI). Rodents and non-human primates are the most commonly used, but they present a number of drawbacks; for example, the rodent brain is smaller than that of humans and thus a higher spatial resolution is required. In addition, there are significant differences between human and rodent brain morphology: for example the rodent brain is smooth, whereas that of the human is gyrencephalic. By contrast, the brain of large mammals such as the pig, sheep or goat is gyrencephalic and has greater similarities with the human brain (Lind et al. 2007). The Göttingen minipig is increasingly used in experimental neuroscience, to investigate brain disorders and is a suitable alternative model to non-human primates for economic, ethical and genetical homogeneity reasons.

Functional imaging studies usually use the haemodynamic response to neuronal activity which induces the Blood Oxygenation Level Dependent (BOLD) effect. In general, BOLD functional MRI (fMRI) paradigms use block design protocols for stimulus presentation to study cognitive processes. However, due to a number of constraints (immobilization, conscious animals, etc.) these experimental paradigms are often unsuitable for animal models. The development of MRI apparatus for animals offers new MR imaging techniques to study brain functionality, including neuronal tracing by manganese-enhanced MRI, pharmacological MRI or MR Spectroscopy (MRS). Toxicity and acquisition time can make some of these techniques unsuitable for humans, and animal models could be used to overcome these problems and improve the signal-to-noise ratio.

The first part of this chapter describes MRI techniques that can be used as alternatives to typical block-design paradigms with large animal models, illustrated by a number of research examples. The second part explores the state of knowledge about the functioning of the central nervous system and its involvement in major functions and behaviour of farm animals such as the pig and sheep. We discuss the relevance of these animal models for human research into brain disorders.

2. MRI techniques

MRI is a non-invasive and *in vivo* technique, both essential features for biomedical research. It enables repeated measures to be carried out and also the longitudinal study of phenomena such as development, ageing, and the influence of environmental factors and physiopathology. MRI can also provide information about structural anatomy, functional activity, cerebral blood flow and water diffusion.

MRI uses a high magnetic field (B0) that aligns the magnetic spin of hydrogen atoms in the tissue in a low energy configuration. The spins are then excited out of equilibrium by a radiofrequency pulse. During the relaxation phase (return to equilibrium), time constants T1 (longitudinal magnetization) and T2 (transverse magnetization) can be measured. These values are used to construct MR images, as relaxation times differ across tissues.

One important advantage of MRI is its high spatial resolution associated with a higher grey/white matter contrast than in X-ray imaging. Due to these properties, cerebral structures can easily be identified. Depending on the animal model, the expected grey/white contrast, and the sequence of acquisition, it is possible to obtain an in-plane resolution of less than one millimetre and as low as tens of micrometres. Moreover, with its ability to perform rapid imaging (e.g. Echo Planar Imaging, EPI), MRI can also be used to obtain dynamic and thus functional imaging.

The most commonly used MRI techniques and their underlying principles are described below, illustrated by a number of studies.

2.1 Pratical issues, anaesthesia, and immobilization of animals

The brains of small ruminants and other mammals with a bodyweight of less than 150 kg (sheep, pigs, dogs, etc.) can be studied using conventional clinical scanners. Depending on the morphological specificities of the mammals involved (size, shape, presence of horns, etc.), surface or knee coils can be used.

The brain functions of healthy subjects can be studied using fMRI under non-invasive conditions and without injection of exogenous markers (e.g. radio-isotope). Recent advances have led to the possibility of imaging brain activity during cognitive processing, revealing the neural bases of various cognitive processes such as language (Vigneau et al. 2006), memory (Wager & Smith 2003), emotion (Sabatinelli et al. 2011), social cognition (Van Overwalle 2009) and neural network dysfunctions associated with various brain disorders (Ragland et al. 2007, Vocks et al. 2010). The method is based on localizing variations in blood flow or metabolism rates under basal or stimulated conditions. The method requires short-duration acquisition with repeated stimulations; the subject has to be immobile, which may require anaesthesia.

The question of anaesthesia has been raised for clinical applications with children (Orhan et al. 2011) and also for experimental applications, with large and small animals. The impact of various anaesthetics under different brain functioning conditions has been compared (for reviews: Boly et al. 2004, Gyulai 2004, Heinke & Schwarzbauer 2002), showing the importance of the type of anaesthetic (volatile e.g. halothane, isoflurane, or systemic e.g. propofol, ketamine), and the dose (low doses with analgesic effect without loss of consciousness, or higher doses with loss of ability to respond to commands). The impact of anaesthesia varies according to these factors and can be specific to a particular brain area (Fig. 1).

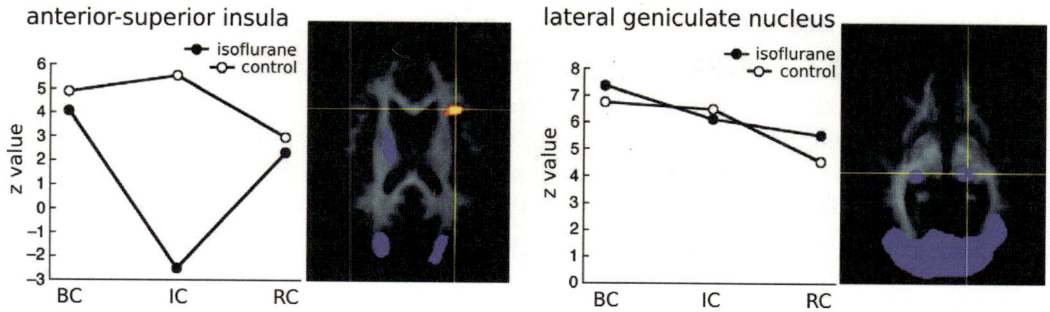

Fig. 1. Quantitative analysis of isoflurane-related changes in task-induced brain activation. Representative voxels were selected in two different regions (left: anterior-superior insula; right: lateral geniculate nucleus). The plots show the group-specific z-values for each group (isoflurane, control) and condition (BC=baseline condition; IC=isoflurane condition; RC=recovery condition). Comparing the corresponding time courses of the isoflurane and control groups reveals a significant isoflurane-related decrease (z>3.1 corresponding to P<0.001) in the anterior-superior insula, but not in the lateral geniculate nucleus. (Adapted from Heinke & Schwarzbauer 2002).

It is clear that the neural processes involved in cognitive functions cannot be studied under deep general anaesthesia; human brain activations induced by noxious, auditory or visual stimulations decrease in a dose-dependent manner after analgesia by ketamine (Rogers et al. 2004), and after sedation by propofol (Plourde et al. 2006, Purdon et al. 2009). In these studies, the authors described a decrease in BOLD in certain regions, but not in the primary cortical areas. Experimental studies with immobilized or anaesthetized animals have used new MRI paradigms with longer acquisition times or pharmacological agents, unsuitable for use with humans. For example, in a rat exposed to hypercapnia, brain activations were higher in conscious animals than those anaesthesia with isoflurane (Sicard et al. 2003). Conversely, the networks of vision, motor or auditory sensitivity described in the resting state persisted regardless of the depth or type of general anaesthesia (Hutchison et al. 2010), and no difference between anaesthetics was found after visual stimulation in dogs (Willis et al. 2001). Several MRI paradigms in anaesthetized animals have been developed to map brain activation induced by serotonin infusion in the baboon (Wey et al. 2010) and cat (Henderson et al. 2002) or brain connectivity in the rat (Pawela et al. 2009, Zhao et al. 2008).

Alternative functional MRI methods for paradigms requiring conscious animals, which comply with ethical standards of experimentation with large animal models, can be used to explore the organization and functioning of the brain (see section 3).

2.2 Structural studies

MRI allows brain images to be obtained with a very high spatial resolution (<0.5mm) and high grey/white matter contrast. Cortical and subortical structures can be easily segmented and their volumes can be determined precisely. Thus, both qualitative and quantitative studies can be conducted. T1-weighted images are mainly used for anatomical studies, but MRI can generate images based on numerous sequences and modalities, obtaining different contrast images (T2, T2*). Among T2-based sequences, Fluid Attenuated Inversion Recovery (FLAIR) enables an easier identification of white matter lesions by suppressing the signal from cerebro-spinal fluid (CSF).

2.2.1 Idenfication of structures – Qualitative studies

Schmidt and colleagues demonstrated that MRI is a useful tool for identifying and studying in detail anatomical cerebral structures in small ruminants (Fig. 2) (Schmidt et al. 2011). Using a conventional 1 Tesla MR scanner, they compared the brains of small ruminants with those of dogs and observed several distinct features (deep depression of the insula, pronounced gyri, larger diencephalon, and dominant positions of the visual and olfactory systems). Using a 4.7 Tesla MR scanner, Saikali and colleagues (Saikali et al. 2010) built a high-resolution (0.1x0.15x0.1mm) 3D atlas of the pig brain, including more than 100 cerebral and cerebellar regions. Although this atlas was constructed *post mortem* from one hemisphere, it can help to identify different structures.

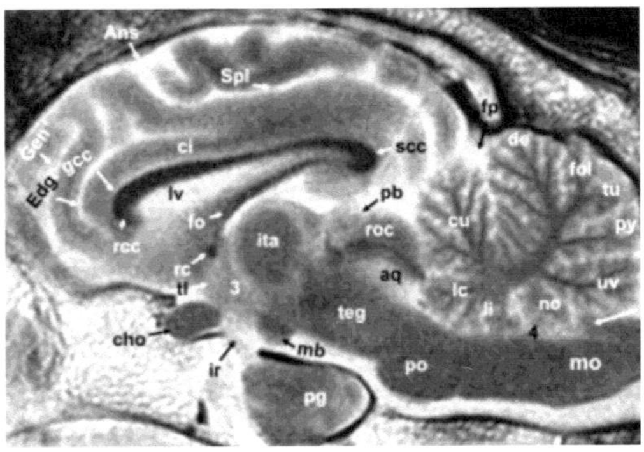

Fig. 2. T2-weighted mid-sagittal MRI of a sheep brain. (Adapted form Schmidt et al. 2011). Ans, ansate sulcus; aq, mesencephalic aqueduct; cho, optic chiasm; ci, cingulated gyrus; cu, culmen; de, declive; Edg, endogenual sulcus; fo, fornix; fol, folium; fp, primary fissure; fs, secondary fissure; gcc, genu of the corpus callosum; Gen, genual sulcus; ir, infundibular recess; ita, interthalamic adhesion; li, lingula; lc, central lobule; lv, lateral ventricle; mb, mamillary body; mo, medulla oblongata; no, nodulus; ob, obex; pb, pineal body; po, pons; py, pyramis; rc, rostral commissure; rcc, rostrum of the corpus callosum; roc, rostral colliculus; scc, splenium of the corpus callosum; Spl, splenial sulcus; teg, tegmentum of the mesencephalon; tu, tuber vermis; uv, uvula; 3, third ventricle; 4, fourth ventricle.

2.2.2 Morphometry – Quantitative studies

As mentioned above, MRI can be used for morphometric measures due to its high spatial resolution and grey/white matter contrast. Furthemore, as MRI is a non-invasive *in vivo* technique, it can be a valuable tool in longitudinal studies, revealing variations in the volume of cerebral structures. For example, it has been shown that an oestrogenic anabolic agent (zeranol) enhances the growth of the pituitary gland of rams (Carroll et al. 2007).

One limitation of morphometric studies is the anatomical variability between individuals. Most morphometric analysis methods in humans include a spatial normalisation step to overcome this problem. This involves a spatial transformation that places each individual brain in a standard, common space. This step requires a template of a standard target brain, which is constructed from several brains via linear affine coregistrations (see Collins et al. 1994 for method). Several templates (and atlases) have been constructed and are available, but most of them concern non-human primates (Black et al. 2004, McLaren et al. 2009) and rodents (Schweinhardt et al. 2003). As mentioned above, a high-resolution atlas of the pig brain has been constructed (Saikali et al. 2010). The same researchers also built a 3D probabilistic pig brain atlas of the deep brain structures using *ex vivo* adult Large White pig brains. The DaNex study group has also computed a template of the average brain of the Göttingen minipig and a probabilistic atlas including 34 regions (Watanabe et al. 2001).

With the possibility of spatial normalisation, focal variations in brain anatomy can be studied by Voxel Based Morphometry (VBM). VBM is a statistical analysis method that consists in voxel-wise comparisons of the local concentration of grey (or white) matter. VBM includes various steps such as spatial normalisation and segmentation (white matter, grey matter and cerebro-spinal fluid). Voxel-wise statistical tests are then performed on these tissue maps to identify group-wise differences or longitudinal changes based on the General Linear Model (GLM) (Ashburner & Friston 2000). For example, a longitudinal paradigm has revealed that training induces grey/white matter volume changes in macaques (Quallo et al. 2009). VBM can also highlight phenotypic variations. It has been demonstrated that MRI, and particularly VBM, can be successfully used to test the heritability of cerebral anatomy in baboons (Rogers et al. 2007).

2.3 Magnetic Resonance Spectroscopy

Magnetic Resonance Spectroscopy (MRS) is widely used in both clinical and preclinical research for the *in vivo* study of cerebral metabolism and the quantification of numerous metabolites (Fig. 3). This quantification is computed from the MR spectrum (intensity of the resonance interaction against the frequency of the chemical compound). The frequency of each compound is linked to its chemical shift which is affected by the chemical environment of the hydrogen atoms. The area under the peak provides a measure of the relative abundance of the corresponding compound. Among the detectable peaks, creatine is used as a relative control value because its concentration remains relatively constant. For example, choline and lactate are considered as markers for brain tumours, while N-Acetylaspartate is used as a marker of neuronal integrity. The spectrum is usually acquired in one voxel (single voxel spectroscopy) and the size of this volume of interest (VOI) is around 1 cm^3. As acquisition time is not necessarily a constraint in animals, a smaller VOI size could be expected with a similar signal-to-noise ratio.

A limitation of MRS is that it uses metabolite ratios for quantification. This may produce ambiguous results whenever several metabolite levels vary simultaneously. An absolute quantification method has been developed (Barantin et al. 1997) called ERETIC (Electric REference To access In vivo Concentrations). It uses a synthetic reference signal which is synthesized as an amplitude modulated radio-frequency pulse, and is injected during the acquisition of the spectrum.

Due to their brain size, small animal brains require higher spatial resolution than for human brains to obtain similar acquisitions. In the macaque, MR spectroscopy has been performed successfully with a spatial resolution of 0.05 cm^3 (Gonen et al. 2008). These authors used multivoxel spectroscopy to compute 2D or 3D maps of spectra and to distinguish brain regions according to their metabolite content.

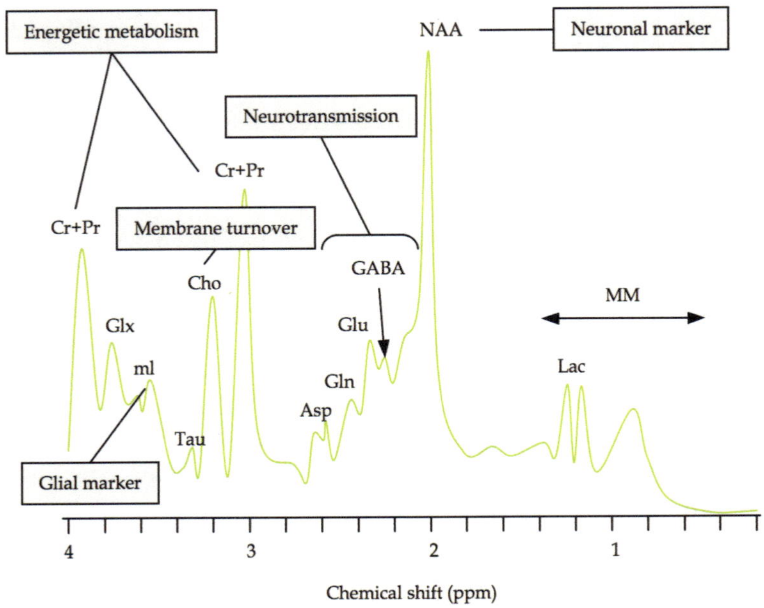

Fig. 3. Example of MR spectrum. Cr: Creatine, PCr: Phosphocreatine, Glx: Glutamate + Glutamine, ml: Myo-inositol, Tau: Taurine, Cho: Choline, Asp: Aspartate, Glu: Glutamate, Gln: Glutamine, NAA: N-Acetylaspartate, Lac: Lactate, MM: Macromolecules.

2.4 Contrast agents

The role of contrast agents is to improve the contrast-to-noise ratio and the spatial sensitivity of the MR signal. They are used in structural and functional studies. Several types of contrast agents have been proposed, some of them directly injected into blood vessels and others used to label cells that are subsequently injected. The use of several contrast agents is limited, especially in humans, due to their putative toxicity.

2.4.1 Gadolinium

Gadolinium (Gd) is a lanthanide metal with paramagnetic properties. However, as a free ion, Gd is highly toxic for mammals, so chelated Gd compounds are used as contrast agents. These agents enhance MRI by shortening the T1 relaxation time. In clinical examinations, Gd is widely used in MR angiography to enhance vessels. It is also commonly used for the exploration of brain tumours and blood-brain-barrier (BBB) integrity. Gd is a marker for BBB breakdown because it is restricted to the intravascular space when the BBB is not disrupted. Wuerfel and colleagues found that Gd-enhanced MRI could be successfully used to explore BBB changes *in-vivo* during the development of neuroinflammation (Wuerfel et al. 2010). A number of studies have also demonstrated the possibility of labelling and tracking cardio-vascular stem cells (Adler et al. 2009).

2.4.2 Manganese-Enchanced Magnetic Resonance Imaging (MEMRI)

Manganese ions (Mn^{2+}) are paramagnetic and enhance MRI contrast mainly by shortening the T1 relaxation time in tissue. Divalent Mn^{2+} is a calcium analogue and enters neurons through voltage-gated Ca^{2+} channels. Due to these two properties, Mn^{2+} is a unique contrast agent for tracing axonal pathways and neuronal connections in the central nervous system (for review see Silva & Bock 2008). Injections of low concentrations of Mn^{2+} into a specific cerebral structure produce significant contrast enhancement along the known relative pathways (Watanabe et al. 2004). Jelsing and colleagues demonstrated in the Göttingen minipig that *in vivo* tracking with MEMRI is very sensitive and corresponds closely to histological labelling (Jelsing et al. 2006).

However, use of MEMRI remains limited because of the neurotoxicity of the Mn^{2+} ion at high concentrations (Shukakidze et al. 2003). Only one agent, Mn-dipyridoxyl-diphosphate, is used in human clinical imaging of the liver.

2.4.3 Inorganic nanoparticles

The main inorganic contrast agents in use are SuperParamagnetic Iron Oxide (SPIO) and Ultrasmall SuperParamagnetic Iron Oxide (USPIO) particles. They vary in size from 20-140nm for SPIO to 60-150nm for USPIO. When placed in a magnetic field, iron oxide particles induce local inhomogeneities, shortening T2 relaxation time. Iron oxide particles produce hypointensity on T2 and T2* weighted images and hyperintensity on T1-weighted images. The signal changes induced by iron oxide particles on T1 and T2 relaxation times are linked to the particle size and the compartment of the particles (extra/intracellular). The toxicity of nanoparticles seems to be limited, but their effect on stem cells is still discussed (Farrell et al. 2008, Muldoon et al. 2005, Schlorf et al. 2010).

Several works have also demonstrated that Monocrystalline Iron Oxide Nanocompounds (MION) can be used in functional studies in animals (Leite et al. 2002). Their main advantage is the specificity of fMRI signal change induced by MION which is only influenced by cerebral blood volume, whereas the BOLD signal is also influenced by cerebral blood flow (CBF) and the metabolic rate of oxygen.

An alternative way of using iron oxyde particles is cellular MRI. This technique allows to transplant and to follow labelled cells. Numerous studies have shown that *in vitro* neural stem and progenitor cells can be loaded with iron oxyde particles (for review Couillard-Despres & Aigner 2011). It has been suggested that this method has a very low detection threshold (Kustermann et al. 2008). One limitation of this method is that the detected contrast on MR images refers only to the particles and not to the labelled cells themselves. This could lead to non-specific observations due to the lack of information on type or viability of cells.

2.5 Diffusion Imaging and Diffusion Tensor Imaging

Diffusion MRI produces *in vivo* images of water diffusion (Le Bihan et al. 1986). Since water diffusion is affected by the microarchitecture of cerebral tissue, in particular the white matter, it can be used to study the organization of neural pathways. Measurement of diffusion provides a non-invasive imaging method to estimate cellular integrity and pathology, and to investigate disease-related changes in neuropathological processes that cannot be observed directly. Several measures can be computed, such as the average diffusivity, apparent diffusion coefficient (ADC), and the fraction of anisotropy (FA) that corresponds to the degree of anisotropy of the diffusion process. These variables are influenced by factors such as fibre diameter or degree of myelination. Whole brain FA changes may be linked to numerous neuropathological mechanisms including neuronal loss, astrogliosis, myelin pallor and diffuse astrocytosis.

Diffusion tensor imaging (DTI) is an advanced method that produces images of the direction and the magnitude of water diffusion. DTI can be used to study white-matter fibre architecture and the influence of experience, disease or other factors on the white-matter fibre networks. Based on DTI data and the FA value of each voxel for several directions, different algorithms can be used to compute the location of white matter fibres and to perform tractography of the neural pathways. DTI can be considered as a functional imaging technique since it provides information about white matter tracts which carry functional information between brain regions.

2.6 Functional Magnetic Resonance Imaging (fMRI)

fMRI enables the measurement of BOLD changes associated with neuronal electrical activity. The BOLD effect is due to a local variation of desoxyhemoglobin concentration (acting as an endogenous contrast agent) which induces a T2* modification and a variation of the MR signal. fMRI uses EPI sequences that produces low spatial resolution images but with a relatively high sampling rate (typically 1-3 seconds). A time course of the MR signal (T2*) for each voxel can be computed. Neuronal activity induces a BOLD effect that affects the time course which is known as the haemodynamic response function. The relationship between neuronal activity and the BOLD effect is a combination of several physiological changes (cerebral blood flow, cerebral blood volumes, cerebral metabolic oxygen consumption, etc.) and is a subject of current research (Ekstrom 2010, Logothetis 2002).

When the effect of stimuli is assumed to be high, it can be examined by comparing the BOLD signal with and without stimulus presentation (Ferris et al. 2001, Makiranta et al. 2002). The size of the effect can then be estimated by computing the percentage of signal change ([average response over the stimulation period - average response over the control period]/[average response over the control period]) (Fig. 4). As the effect of the stimuli may be too weak to be observed with this method, block design paradigms have been developed.

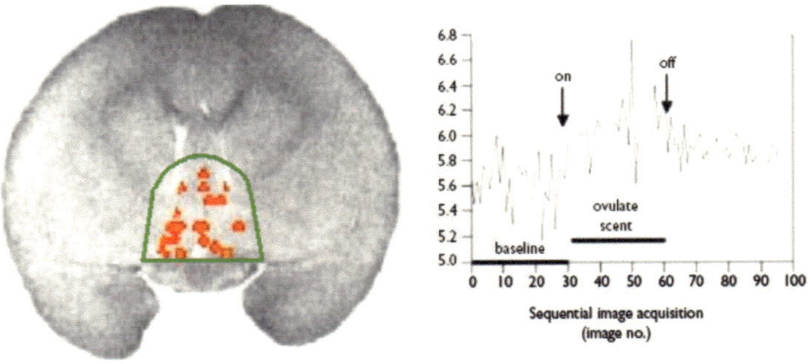

Fig. 4. Enhancement of BOLD signal in the preoptic area of male marmosets exposed to the scent of peri-ovulatory females. Red spots correspond to regions with a significant increase in the percentage of signal change during stimulus presentation. The average changes in signal in the region of interest (in green) are shown in the time course data. (Adapted from Ferris et al. 2001).

2.6.1 Typical activation studies: Block designs

Typical activation studies use block designs and analysis based on the general linear model (GLM). This method is used to make inferences about the effects of the stimuli by decomposing data into effects and errors, and computes statistical maps related to the effects of the stimuli (see Monti 2011 for principles). This kind of study is widely used in human and non-human primates, but due to a required subjects's involvement, typical fMRI activation paradigms have only been used in a few studies in large animals such as pigs or sheep (Fang et al. 2005b, Fang et al. 2005c, Fang et al. 2006, Opdam et al. 2002). Due to the constraints mentioned above (see 2.1), this kind of experimental paradigm will not be discussed further in this chapter.

2.6.2 Other experimental paradigms

The constraints relating to typical block activation paradigms can be avoided by analyzing the data with model-free methods. These do not require any presentation of stimuli and are thus also called data-driven analyses. One method widely used to identify brain networks is correlation analysis which is the most straightforward way to examine the functional connections of brain regions. It consists in computing correlations between the time course of the MR signal in one particular region (known as the seed region) against the time courses of all other regions, providing a connectivity map relative to the seed region. Numerous studies have used this method to explore the resting-state network in humans (van den Heuvel & Hulshoff Pol 2010 for review), non-human primates (Vincent et al. 2007) and rats (Zhang et al. 2010). One of the limitations of this method is that the functional connectivity map refers to a specific region and does not provide a whole-brain analysis.

Another data-driven approach is independent component analysis (ICA) whose goal is to recover independent sources given only observations. ICA transforms the observed signals into components and maximizes independency of these resulting components (see McKeown et al. 1998 for principles). In other words, ICA identifies functionally connected brain networks which covary independently of other regions. ICA has been used to explore resting state and functional connectivity in arousal states in humans, non-human primates (Moeller et al. 2009) and rodents (Hutchison et al. 2010).

2.6.3 Arterial Spin Labelling (ASL)

The ASL method measures CBF by providing cerebral perfusion maps without requiring a contrast agent. This approach uses magnetically labelled endogenous blood water as a freely diffusible tracer. The first studies were conducted in 1992 (Williams et al. 1992) and since then various improvements have been proposed. The principle of ASL is to sequencially acquire brain volumes and to obtain time series composed of tag images in which arterial blood is magnetically labelled (by apllying a 180 degre radiofrequency inversion pulse) and control images in which the inflowing blood is not labelled. First, the arterial blood water is tagged in a region that is proximal to the imaging region, and after a period of time the image of the region is acquired. The procedure is then repeated without the tagging step. This pattern of alternate acquisition is repeated several times. The difference between the control and tagged images provides a volume containing values proportional to the perfusion.

Because ASL measures CBF and uses rapid imaging sequences, activation studies similar to BOLD fMRI can be performed. The advantage of ASL-fMRI is that the ASL signal is thought to be only associated with CBF in capillaries, while the BOLD effect results from numerous haemodynamic changes in nearby veins. However, ASL-fMRI has a lower signal-to-noise ratio, lower spatial and temporal resolutions, and can be less sensitive to stimuli.

In this section, we have described the different MRI techniques and their applications (Table 1). As there are a number of drawbacks to the use of rodents and non-human primates, commonly used in MRI investigations (see introduction), we propose the use of large animals (sheep, pigs) as alternative models. In the following section, we will outline the main advantages of using these models for a better understanding of cerebral functioning and related brain disorders.

		Resolution
MORPHOLOGY	**Structural Imaging**	<0.2mm^3
	Grey/white matter volumes	
	Long-term modifications	
	Diffusion Imaging	<10mm^3
	Architecture of white matter tracts	
	Long-term modifications	
FUNCTION	**Blood Oxygenation Level-Dependent Effect**	<5mm^3
	Neural bases of cognitive processes	
	Neural networks	
	Arterial Spin Labelling	<20mm^3
	Perfusion maps	
	Neural bases of cognitive processes	
METABOLISM	**MR Spectroscopy**	<1cm^3
	Metabolite distribution	
	Neuronal death	
	Neurogenesis	

Table 1. Summary of the main MRI applications.

3. Animal models

The central nervous system of farm animals has been studied to understand the regulation of major functions such as reproduction and food intake, with the aim of improving yields. Researchers soon found that these models could also be used to improve understanding of the brain (Lind et al. 2007, Sauleau et al. 2009) and the neurobiological regulation of various functions (Lehman et al. 2002, Malpaux et al. 2002, Skinner et al. 1997). The next section presents data obtained in large animals (pigs and sheep) providing fundamental knowledge about brain functioning and the central control of various functions and behaviours.

3.1 Brain injury

Large animals are commonly used as experimental models for human-infant research into brain disorders (pig, Lind et al. 2007), sudden infant death syndrome (pig, Tong et al. 1995), head injury (Lehman et al. 2002), brain injury induced by hypoxia (pig, Foster et al. 2001;

sheep, Laurini et al. 1999) or by preterm birth (sheep, Patural et al. 2010 , Pladys et al. 2008, Riddle et al. 2006), and neurobehavioural topics (pig, Friess et al. 2007). They can also be used for xenografts in Parkinson's disease (Molenaar et al. 1997). Some of these studies have focused on neuronal activation induced by hypercapnia in the dorsal vagal complex of piglets (Ruggiero et al. 1999, Sica et al. 1999) and on cyto-architectural modifications induced by hypoxia/ischaemia (HI), such as neuronal necrosis in the piglet hippocampus (Foster et al. 2001), while others have investigated cell degeneration in the cerebral cortex of fetal lambs (Riddle et al. 2006).

With regard to the development of MRI techniques, some authors have combined these approaches with histological methods. For example, Fang and collaborators studied the development of the pig brain (Fang et al. 2005a) and compared nociceptive and motor stimulations at different ages (Fang et al. 2005b). They demonstrated the usefulness of fMRI in non-anaesthetized piglets to identify differences in brain activation induced by pain stimulation and passive movement (Fang et al. 2005b). Immunohistochemistry enabled the authors to propose a hypothesis of functional brain maturation to explain the effect of age on brain activation measured by fMRI (Fang et al. 2005a). It has also been demonstrated that the volumetric analysis of brain lesions by MRI reveals the impact of traumatic brain injury in a similar way to histological approaches (Grate et al. 2003; Fig. 5). The use of MRI has been validated to detect HI injury in preterm fetal sheep, although detection was limited to injury in deep structures (Fraser et al. 2007). These studies demonstrate first how MRI and histology are complementary methods for understanding brain functioning, and secondly, that MRI produces similar results to histology while offering a more ethical approach.

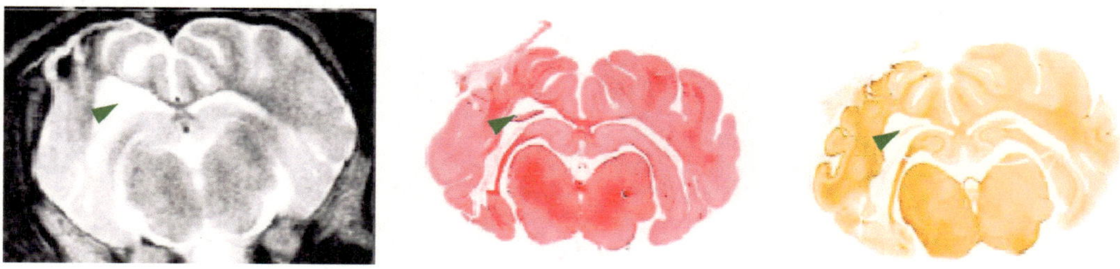

Fig. 5. Serial T2-weighted MR images, histological section stained with hematoxylin and eosin, and adjacent section stained with an antibody against glial fibrillary acidic protein obtained at one-month post-injury in a one-month old piglet subjected to scaled focal brain injury. Note that the traumatic brain lesion (green arrow) is found whatever the method (adapted from Grate et al. 2003).

In the case of HI-induced brain injury in newborn piglets, magnetic resonance spectroscopy (MRS) has been used to monitor the cerebral metabolite ratio *in vivo* (Björkman et al. , Li et al. 2010, Vial et al. 2004). Björkman and colleagues measured the severity of the brain injury with EEG, ADC, MRS and neuropathological analysis. They observed correlations between these measures (Björkman et al. 2010).

MRI methods have also been used with large animal models in studies on epilepsy (sheep: Opdam et al. 2002), to develop new chemotherapeutic strategies such as local injection in the fourth ventricle (pig, Sandberg et al. 2008), and to test the toxicity of chemotherapeutic treatment on normal brain tissue close to the injection site (Makiranta et al. 2002). In sheep, MRI has validated *in vivo* ultra-sound transcranial brain surgery (Pernot et al. 2007).

3.2 Cerebrospinal fluid functionality

The ewe has commonly been used in neuroendocrinology studies as an animal model for neuroanatomical research (Lehman et al. 2002) into the neuroendocrine mechanisms of reproduction (Malpaux et al. 2002), or to study the effect of drugs on the central nervous system (Parry 1976). In this large animal model, CSF content can be analysed in real-time by continuous sampling over several days in conscious and unstressed animals at different stages of development (Dziegielewska et al. 1980, Tricoire et al. 2003).

Studies conducted in sheep have demonstrated that the gonadotropin releasing hormone (GnRH) pulses measured in the CSF are coincident with those measured in the hypophyseal portal blood and with the luteinizing hormone pulses measured in jugular blood (Skinner et al. 1997). Similar observations have been made for the melatonin (MLT) concentration measured in the jugular vein and CSF which vary with day-night rhythm (Skinner & Malpaux, 1999). It has been demonstrated in sheep that the CSF content varies according to the cerebroventricular compartment (Fig. 6, GnRH, Caraty & Skinner 2008; MLT, Malpaux et al. 2002, Tricoire et al. 2003), light-dark cycles (Skinner & Malpaux 1999, Thiery & Malpaux 2003, Thiery et al. 2003, Thiery et al. 2006, Thiery et al. 2009) and ageing (Chen et al. 2010a, Chen et al. 2010b). These findings suggest that the CSF is an active medium which could play a role in regulating various functions (Malpaux et al. 2002, Skipor & Thiery 2008).

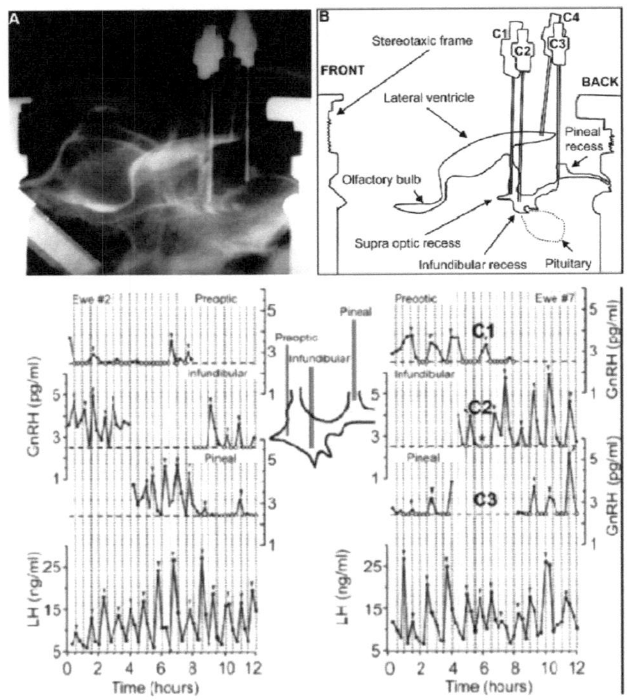

Fig. 6. A: Lateral X-ray image, and B: diagram showing the placement of the four cannulae implanted in the supraoptic (C1), infundibular (C2) and pineal (C3) recesses and in the lateral ventricle. C: Examples of GnRH concentration profiles in the CSF harvested simultaneously from the different cannulae (C1, C2, C3) with the corresponding LH secretion in the peripheral blood. (Adapted from Caraty & Skinner 2008).

One of the hypotheses regarding the variations in CSF content linked to season or ageing concerns variations in the BBB permeability, as demonstrated in sheep (Chen et al. 2010b, Lagaraine et al. 2011). BBB involvement and dysfunction in brain disorders has been extensively documented (de Vries et al. 1997, Forster 2008, Hawkins & Davis 2005, Strbian et al. 2008). Using MRI methods it is possible to study the BBB and its permeability in physiological or pathological paradigms (Hjort et al. 2008, Israeli et al. 2011, Wuerfel et al. 2010), and also to develop new therapeutic strategies (Liu et al. 2010).

Another hypothesis about the CSF-brain-endocrine interaction concerns tanycytes, which are ependymal cells of the third ventricle (Rodriguez et al. 2005). Their putative involvement in photoperiodic regulations has been described in the hamster (Ebling 2010). Apart from their physiological role, they are also implicated in brain disorders, as some chordoid gliomas could have a tanycytic origin (Sato et al. 2003). These tanycytoma are differentiated from other intracranial neoplasms by their specific location in the hypothalamus (Lieberman et al. 2003).

We therefore suggest that large animals such as pigs and sheep are relevant animal models, as the CSF content is easily measurable (e.g. in sheep), the permeability of the BBB can be investigated physiologically through day-night cycles (e.g. in sheep) and pharmacologically using ultrasound (e.g. in pigs, Xie et al. 2008).

3.3 Neurogenesis, cell proliferation

Evidence of adult neurogenesis was first presented in 1965 (Altman & Das 1965). It is now thought to play a role in different functions (Aimone et al. 2010) such as memory (Deng et al. 2010), in sensory systems such as olfaction (Whitman & Greer 2009), and in mental health disorders (Eisch et al. 2008), epilepsy (Rakhade & Jensen 2009) and Alzheimer's disease (Lazarov & Marr 2010).

In sheep, cell proliferation, evaluated by bromodeoxyuridine (BrdU) incorporation, has been observed in the dentate gyrus of the hippocampus of ewes exposed to a novel male (Hawken et al. 2009). Using BrdU incorporation and cellular biomarkers such as doublecortin or glial fibrillary acid protein (for review Sierra et al. 2011), it has been demonstrated that cell proliferation is down-regulated in the subventricular zone, the dentate gyrus and the main olfactory bulb at parturition and during interactions with the young (Brus et al. 2010, Fig. 7A). These authors suggest that cell proliferation could play a role in maternal behaviour via the olfactory and memory neuronal systems. New neurogenesis sites that could be involved in photoperiodic neuroendocrine systems have also been described in the hypothalamus (Migaud et al. 2010, Fig. 7B).

MRS is a promising method for visualizing and studying endogenous neural progenitor cells (Ramm et al. 2009, Sierra et al. 2011). *In vivo* imaging needs to be developed in humans (Couillard-Despres & Aigner 2011) to study adult neurogenesis (Couillard-Despres et al. 2011).

Based on current knowledge and available tools, we suggest that large animal models such as sheep can be used to validate the development of MRI techniques and to understand the role of neurogenesis through longitudinal *in vivo* studies.

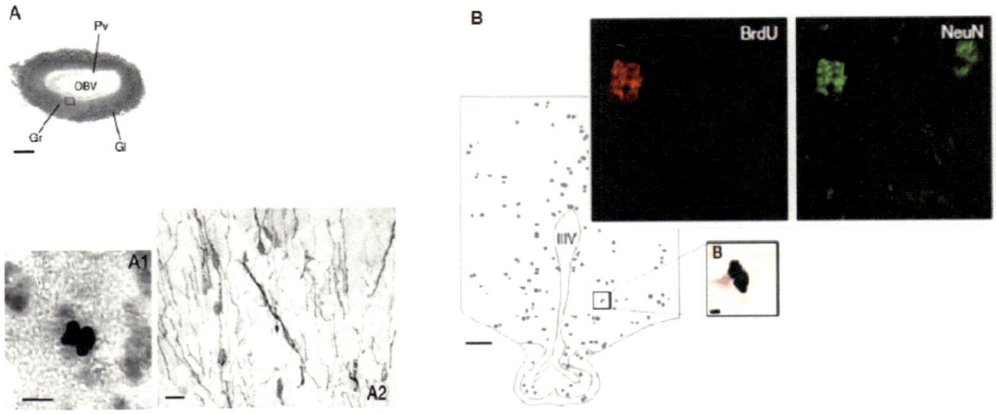

Fig. 7. BrdU integrated cells observed in the olfactory bulb (A, A1) and the hypothalamus (B) of adult sheep. In the main olfactory bulb, positive mature neuroblasts (A2) were observed in the same area as the BrdU incorporated cells (A1) at parturition (adapted from Brus et al., 2010). Constitutive cell proliferation observed in the adult sheep hypothalamus (B, BrdU in red), the new cells differentiated into mature neurons (NeuN in green) (adapted from Migaud et al. 2010).

3.4 Neurobiological regulation

3.4.1 Feeding behavior

The role of the central nervous system in regulating appetite and food intake has been extensively studied (for review Berthoud 2006, Kalra et al. 1999, Schwartz 2006). Regulatory systems in central areas include the hypothalamic system (ventromedian nucleus, arcuate nucleus, etc.), the caudal brainstem (area postrema, nucleus of the tractus solitary, etc.) and cortical structures (prefrontal cortex, amygdala, hippocampus, etc.). At the hypothalamic level, numerous neuropeptides have been identified as major orexigens (neuropeptide Y, galanin, etc.) or anorexigens (cholecystokinin, somatostatin, etc.), most of them regulated by hormones such as insulin or leptin. In sheep, similar factors have been observed to regulate food intake (Baile & McLaughlin 1987, Chaillou et al. 2000, Della-Fera & Baile 1984) or to be regulated by nutrition (Chaillou & Tillet 2005, Zieba et al. 2008). The same factors have been described in the pig (Baldwin et al. 1990a, Baldwin et al. 1990b, Baldwin & Sukhchai 1996, Czaja et al. 2002, Czaja et al. 2007, Parrott et al. 1986), and similarities have also been found in humans for preferences for sweet food (Houpt et al. 1979) and for energy metabolism (Spurlock & Gabler 2008). All these observations support the idea that the pig can be used as a model for human studies (Johansen et al. 2001).

Knowledge about the central regulation of feeding behaviour has been documented using techniques including central injections of neuropeptides or hormones, comparison of neuropeptide expression levels in different nutritional states, and more recently by MRI (Van Vugt 2010). MRI has been used in human studies of the cognitive component of eating disorders such as anorexia nervosa (volumetric MRI, Muhlau et al. 2007; fMRI, Vocks et al. 2010) or nutritional disorders such as obesity (fMRI, Killgore & Yurgelun-Todd 2010).

We suggest that large animal models could be used to study the putative consequences on human brain functioning of nutritional disorders such as obesity. For example, functional connectivity, measured by MRI, is impaired in obese human subjects, and is correlated with metabolic indicators such as insulin (Kullmann et al. 2011). It would be interesting to compare the impact of different neuropeptides or diets on brain activation that could be measured at different times of life, but this type of protocol would be difficult to standardize in humans. The effects of gastric bypass surgery on hypothalamic functional connectivity and on various indicators (inflammatory and metabolic) have been studied in obese human subjects (van de Sande-Lee et al. 2011). Similar protocols could be designed in the pig, making it easier to select animals and to set up a sham-surgery control group, and could be used to study the long-term effects of surgery. Other studies could investigate interactions between nutrition and other functions such as reproduction, or to evaluate the putative sensorial effects induced by cognitive perturbations during prenatal, perinatal or childhood periods. For example, a recent brain imaging investigation using PET scan compared the cerebral blood flow of lean and obese minipigs (Val-Laillet et al. 2011).

3.4.2 Reproduction

Reproduction is controlled by the central nervous system, more particularly by the hypothalamus where the neuronal population containing GnRH is located. This neuropeptide is the key factor in the regulation of the hypothalamic- pituitary-gonadal axis. It is released in a pulsatile fashion into the hypophyseal portal blood. Numerous studies have been performed using a sheep model, as the oestrus cycle of ewes has the same temporal pattern as the menstrual cycle of women, and because it is possible to sample blood from the hypophyseal portal system of the ewe (Caraty et al. 1982). Many peripheral hormones from the gonads act on distinct neuronal populations in the brain to regulate the neuronal activity of GnRH neurons. The neuronal network controlling reproduction in sheep has been extensively described, and the neuroendocrine factors regulating this network are known (steroids, neuropeptides, monoamines, etc; for reviews Herbison 1995, Herbison 2006, Tillet 1995). All these data have contributed to our knowledge of the central control of reproduction in mammals, and more particularly in humans.

However, one outstanding difficulty concerns the precise description of the temporal activations and interactions between the different neuronal partners in regulating the menstrual cycle and puberty. MRI could help to overcome this difficulty and a number of studies have already been performed in humans to investigate the interaction between the neuronal population and the feedback effect of gonadal hormones. At puberty, when the gonads start to produce hormones and particularly steroids, MRI methods have been used to determine how steroids (oestrogen and testosterone) act on brain development and plasticity (Jernigan et al. 2011). Another field of study has focused on brain functioning in women during the menstrual cycle. Throughout the cycle, the ovaries produce successively increasing levels of oestradiol and progesterone (Goodman & Inskeep 2006), concomitant with changes in functional cerebral asymmetries (Weis & Hausmann 2010) which are potentially due to variations in functional connectivity (Weis et al. 2010). These hormonal variations during the menstrual cycle or caused by hormonal contraceptives affect the volumes of grey matter (Pletzer et al. 2010) and modify the activation induced by negative emotion in the amygdala and hippocampus as demonstrated by fMRI (Andreano & Cahill 2010), and hormonal variations also affect food perception in interaction with feeding disorders (Van Vugt 2010). These data have been obtained under clinical conditions and it is clearly impossible to extend these human studies for obvious ethical reasons. The female sheep is an excellent model to understand the central effect of steroids on brain functioning and can also be very useful for developing treatment strategies for central or pituitary infertility in humans, and for investigating central effects of new therapeutics and contraceptives.

3.4.3 Social behaviour

For all species, the neuronal networks involved in social behaviour combine autonomic regulatory and sensorial integrative structures. In the case of sexual and maternal behaviour, partner recognition results from the interaction between the olfactory system (the main sense involved in social recognition), and the neuroendocrine circuit involved in oestrus for sexual behaviour and parturition for maternal behaviour (Gelez & Fabre-Nys 2006, Levy & Keller 2009, Poindron et al. 2007). Similarly, olfaction is important in establishing maternal recognition by the lamb, and the development of the mother-young bond is reinforced by oro-gastro-intestinal stimulation (Nowak et al. 2007). However, while olfaction is the first proximal sense used (i) by the mother and infant to establish a bond, and (ii) by the male and female to identify social partners, visual (Kendrick et al. 2001) and auditory (Sebe et al. 2007, Sebe et al. 2008) factors are also involved in the expression of social preferences.

In order to understand social behavioural, disorders, and the establishment of social bonds, we need to study the sensory systems and how they interact with the neuroendocrine system. Neuroanatomical approaches require a large sample and complex protocols. MRI techniques can be used to show how the brain discriminates social sensory indices or is activated by social neuroendocrine factors. For example, the BOLD signal of conscious non-human primate males exposed to the scent of peri-ovulatory females is greater than when exposed to the scent of ovariectomized females in various hypothalamic (Ferris et al. 2001; see above Fig. 4) and cortical areas (Ferris et al. 2004). With regard to the formation of a maternal bond in the rat, it has been shown that suckling activates similar brain areas to those activated by a central injection of oxytocin (Febo et al. 2005), a neuropeptide involved in social attachment (Young et al. 2008) and maternal behaviour (Levy et al. 2004).

Ungulates are similar to humans in the preference shown by the mother for her own offspring, a process known as maternal selectivity (Poindron et al. 2007). This suggests that ewes could provide an interesting model to investigate disorders of maternal behaviour. For example, the impact of the offspring's odour on variations in cerebral blood flow could be compared between selective, maternal, and non-maternal ewes. Functional connectivity MRI could also be used to describe the dynamic functional interactions between the cortical structures involved in sensory integration and deep structures such as the hypothalamus or amygdala, since the neuroanatomical connections between these neuronal systems are known in sheep (Levy et al. 1999, Meurisse et al. 2009).

3.4.4 Emotional reaction

Animals' emotional reactions can be described through behavioural and physiological responses. These are regulated in mammals by numerous neuronal networks: the corticotrope axis (Herman et al. 2003), the brainstem, and the periaqueductal grey matter that regulates motor responses (Keay & Bandler 2001, LeDoux 2000). These deep structures interact with the prefrontal cortex and the amygdala (Herman et al. 2003, Keay & Bandler 2001, LeDoux 2000) and are all involved with neurochemical factors as cortico-releasing factor (CRF), and serotoninergic and dopaminergic systems (Charney 2004, Rotzinger et al.

2010). In large animals, such as sheep or pigs, similar neurobiological factors have been found to be involved in emotional responses, especially in stressful situations. Invasive neurobiological approaches based on functional neuroanatomy (sheep: da Costa et al. 2004, Rivalland et al. 2007, Vellucci & Parrott 1994), intracerebroventricular pharmacology (pigs: Johnson et al. 1994, Salak-Johnson et al. 2004), and neurochemical brain content (e.g. in pigs, Kanitz et al. 2003, Loijens et al. 2002, Piekarzewska et al. 1999, Piekarzewska et al. 2000, Zanella et al. 1996) have demonstrated the involvement of neuropeptides such as CRF and enkephalins in different brain areas including the hypothalamus, brainstem and cortices. While neuroanatomical methods have been used to describe the immunoreactive content of brain areas (in sheep: Tillet 1995; in pigs: Kineman et al. 1989, Leshin et al. 1996, Niblock et al. 2005, Rowniak et al. 2008; in large mammals: Tillet & Kitahama, 1998), and neuronal-tracing methods have been used to describe the interconnections between some of these brain areas (sheep: Qi et al. 2008, Rivalland et al. 2006, Tillet et al. 1993, Tillet et al. 2000; pigs: Chaillou et al. 2009), no dynamic functional information is available about the functional interactions among these different factors. The use of MRI techniques could be an interesting way of gaining a better understanding of the neuronal circuits of animal emotion and other functions.

MRI has been used in humans to develop knowledge of the neuroscience of emotions (Junghofer et al. 2006), describing the neuronal circuit in order to demonstrate the impact of pathological emotional behaviour (e.g. posttraumatic stress disorder) on hippocampal volume (Wang et al. 2010) or the effects of antidepressants in major depression (Bellani et al. 2011). These studies have all focused on cortical structures. The posterior hypothalamic area has been shown to play a major role in seasonal affective disorder (SAD) (Vandewalle et al. 2011). The sheep has been proposed as a model for SAD as it is a photoperiodic mammal. More information is now available in sheep about emotional states (Guesdon et al. 2011) and how they can be modified (Doyle et al. 2011, Erhard et al. 2004, Greiveldinger et al. 2007, Vandenheede & Bouissou 1998). For example, it has been suggested that the serotoninergic pathway is involved in the affective state of sheep (Doyle et al. 2011). MRI techniques could be used to investigate the impact of various neurobiological factors on emotional state, as shown in pharmacological models of depression (Michael-Titus et al. 2008). More interestingly, we propose the use of large animal models to study the long-term effects of strong acute emotion in prenatal or perinatal life on brain development and behaviour. For example in the pig, prenatal stresses have been shown to affect ontogeny of the corticotrope axis (Kanitz et al. 2003) and behaviour (Jarvis et al. 2006).

We suggest that large animal models can be used to validate and/or study the impact of non-pharmacological clinical treatments that are now used in mood and anxiety disorders (Ressler & Mayberg 2007), using standardized protocols that are inappropriate to conduct in humans.

4. Conclusion

This chapter described various MRI methods and their use in exploring brain anatomy and functioning in large animal models. We discussed the way these models can be used to study brain injury such as hypoxia/ischaemia, and the different compartments of the central nervous system (e.g. CSF) or neurobiological control (e.g. food intake).

The brain circumvolutions, the brain size and development as well as the neurobiological regulations are the most evident arguments to justify the interest for large animal models for human brain studies. These models also present many advantages for studying the dynamic functional interactions between brain structures using functional connectivity MRI, to understand the interaction between different brain functions with fMRI and for conducting standardized longitudinal studies that are not feasible in human studies. They can also be used to test new surgical procedures and the impact of treatment on healthy brain tissue and behaviour.

5. References

Adler, E. D.; Bystrup, A.; Briley-Saebo, K. C.; Mani, V.; Young, W.; Giovanonne, S.; Altman, P.; Kattman, S. J.; Frank, J. A.; Weinmann, H. J.; Keller, G. M. & Fayad, Z. A. (2009). In vivo detection of embryonic stem cell-derived cardiovascular progenitor cells using Cy3-labeled Gadofluorine M in murine myocardium. *JACC Cardiovasc Imaging*, Vol. 2, No 9, pp. 1114-1122, ISSN 1876-7591

Aimone, J. B.; Deng, W. & Gage, F. H. (2010). Adult neurogenesis: integrating theories and separating functions. *Trends Cogn Sci*, Vol. 14, No 7, pp. 325-337, ISSN 1879-307X

Altman, J. & Das, G. D. (1965). Autoradiographic and histological evidence of postnatal hippocampal neurogenesis in rats. *J Comp Neurol*, Vol. 124, No 3, pp. 319-335, ISSN 0021-9967

Andreano, J. M. & Cahill, L. (2010). Menstrual cycle modulation of medial temporal activity evoked by negative emotion. *Neuroimage*, Vol. 53, No 4, pp. 1286-1293, ISSN 1095-9572

Ashburner, J. & Friston, K. J. (2000). Voxel-based morphometry-the methods. *Neuroimage*, Vol. 11, No 6 Pt 1, pp. 805-821, ISSN 1053-8119

Baile, C. A. & McLaughlin, C. L. (1987). Mechanisms controlling feed intake in ruminants: a review. *J Anim Sci*, Vol. 64, No 3, pp. 915-922, ISSN 0021-8812

Baldwin, B. A.; de la Riva, C. & Ebenezer, I. S. (1990a). Effects of intracerebroventricular injection of dynorphin, leumorphin and alpha neo-endorphin on operant feeding in pigs. *Physiol Behav*, Vol. 48, No 6, pp. 821-824, ISSN 0031-9384

Baldwin, B. A.; Ebenezer, I. S. & De La Riva, C. (1990b). Effects of intracerebroventricular injection of muscimol or GABA on operant feeding in pigs. *Physiol Behav*, Vol. 48, No 3, pp. 417-421, ISSN 0031-9384

Baldwin, B. A. & Sukhchai, S. (1996). Intracerebroventricular injection of CCK reduces operant sugar intake in pigs. *Physiol Behav*, Vol. 60, No 1, pp. 231-233, ISSN 0031-9384

Barantin, L.; Le Pape, A. & Akoka, S. (1997). A new method for absolute quantitation of MRS metabolites. *Magn Reson Med*, Vol. 38, No 2, pp. 179-182, ISSN 0740-3194

Bellani, M.; Dusi, N.; Yeh, P. H.; Soares, J. C. & Brambilla, P. (2011). The effects of antidepressants on human brain as detected by imaging studies. Focus on major depression. *Prog Neuropsychopharmacol Biol Psychiatry*, Vol. 35, No 7, pp. 1544-1552, ISSN 1878-4216

Berthoud, H. R. (2006). Homeostatic and non-homeostatic pathways involved in the control of food intake and energy balance. *Obesity*, Vol. 14 Suppl 5, No pp. 197S-200S, ISSN 1930-7381

Björkman, S. T.; Miller, S. M.; Rose, S. E.; Burke, C. & Colditz, P. B. Seizures are associated with brain injury severity in a neonatal model of hypoxia-ischemia. *Neuroscience*, Vol. 166, No 1, pp. 157-167, ISSN 1873-7544

Black, K. J.; Koller, J. M.; Snyder, A. Z. & Perlmutter, J. S. (2004). Atlas template images for nonhuman primate neuroimaging: baboon and macaque. *Methods Enzymol*, Vol. 385, No pp. 91-102, ISSN 0076-6879

Boly, M.; Faymonville, M. E.; Peigneux, P.; Lambermont, B.; Damas, P.; Del Fiore, G.; Degueldre, C.; Franck, G.; Luxen, A.; Lamy, M.; Moonen, G.; Maquet, P. & Laureys, S. (2004). Auditory processing in severely brain injured patients: differences between the minimally conscious state and the persistent vegetative state. *Arch Neurol*, Vol. 61, No 2, pp. 233-238, ISSN 0003-9942

Brus, M.; Meurisse, M.; Franceschini, I.; Keller, M. & Levy, F. (2010). Evidence for cell proliferation in the sheep brain and its down-regulation by parturition and interactions with the young. *Horm Behav*, Vol. 58, No 5, pp. 737-746, ISSN 1095-6867

Caraty, A.; Orgeur, P. & Thiery, J. C. (1982). Demonstration of the pulsatile secretion of LH-RH into hypophysial portal blood of ewes using an original technic for multiple samples. *C R Seances Acad Sci III*, Vol. 295, No 2, pp. 103-106, ISSN 0249-6313

Caraty, A. & Skinner, D. C. (2008). Gonadotropin-releasing hormone in third ventricular cerebrospinal fluid: endogenous distribution and exogenous uptake. *Endocrinology*, Vol. 149, No 10, pp. 5227-5234, ISSN 0013-7227

Carroll, J. A.; Walker, M. A.; Hartsfield, S. M.; McArthur, N. H. & Welsh, T. H., Jr. (2007). Visual documentation of ovine pituitary gland development with magnetic resonance imaging following zeranol treatment. *Lab Anim*, Vol. 41, No 1, pp. 120-127, ISSN 0023-6772

Chaillou, E.; Baumont, R.; Tramu, G. & Tillet, Y. (2000). Effect of feeding on Fos protein expression in sheep hypothalamus with special reference to the supraoptic and paraventricular nuclei: an immunohistochemical study. *Eur J Neurosci*, Vol. 12, No 12, pp. 4515-4524, ISSN 0953-816X

Chaillou, E. & Tillet, Y. (2005). Nutrition and hypothalamic neuropeptides in sheep: histochemical studies. *Histol Histopathol*, Vol. 20, No 4, pp. 1209-1225, ISSN 0213-3911

Chaillou, E.; Tillet, Y. & Malbert, C. H. (2009). Organisation of the catecholaminergic system in the vagal motor nuclei of pigs: a retrograde fluorogold tract tracing study combined with immunohistochemistry of catecholaminergic synthesizing enzymes. *J Chem Neuroanat*, Vol. 38, No 4, pp. 257-265, ISSN 1873-6300

Charney, D. S. (2004). Psychobiological mechanisms of resilience and vulnerability: implications for successful adaptation to extreme stress. *Am J Psychiatry*, Vol. 161, No 2, pp. 195-216, ISSN 0002-953X

Chen, C. P.; Chen, R. L. & Preston, J. E. (2010a). The influence of cerebrospinal fluid turnover on age-related changes in cerebrospinal fluid protein concentrations. *Neurosci Lett*, Vol. 476, No 3, pp. 138-141, ISSN 1872-7972

Chen, R. L.; Chen, C. P. & Preston, J. E. (2010b). Elevation of CSF albumin in old sheep: relations to CSF turnover and albumin extraction at blood-CSF barrier. *J Neurochem*, Vol. 113, No 5, pp. 1230-1239, ISSN 1471-4159

Collins, D. L.; Neelin, P.; Peters, T. M. & Evans, A. C. (1994). Automatic 3D intersubject registration of MR volumetric data in standardized Talairach space. *J Comput Assist Tomogr*, Vol. 18, No 2, pp. 192-205, ISSN 0363-8715

Couillard-Despres, S. & Aigner, L. (2011). In vivo imaging of adult neurogenesis. *Eur J Neurosci*, Vol. 33, No 6, pp. 1037-1044, ISSN 1460-9568

Couillard-Despres, S.; Vreys, R.; Aigner, L. & Van der Linden, A. (2011). In vivo monitoring of adult neurogenesis in health and disease. *Front Neurosci*, Vol. 5, No 67 pp. 1-10, ISSN 1662-453X

Czaja, K.; Lakomy, M.; Sienkiewicz, W.; Kaleczyc, J.; Pidsudko, Z.; Barb, C. R.; Rampacek, G. B. & Kraeling, R. R. (2002). Distribution of neurons containing leptin receptors in the hypothalamus of the pig. *Biochem Biophys Res Commun*, Vol. 298, No 3, pp. 333-337, ISSN 0006-291X

Czaja, K.; Barb, C. R. & Kraeling, R. R. (2007). Hypothalamic neurons innervating fat tissue in the pig express leptin receptor immunoreactivity. *Neurosci Lett*, Vol. 425, No 1, pp. 6-11, ISSN 0304-3940

da Costa, A. P.; Leigh, A. E.; Man, M. S. & Kendrick, K. M. (2004). Face pictures reduce behavioural, autonomic, endocrine and neural indices of stress and fear in sheep. *Proc Biol Sci*, Vol. 271, No 1552, pp. 2077-2084, ISSN 0962-8452

de Vries, H. E.; Kuiper, J.; de Boer, A. G.; Van Berkel, T. J. & Breimer, D. D. (1997). The blood-brain barrier in neuroinflammatory diseases. *Pharmacol Rev*, Vol. 49, No 2, pp. 143-155, ISSN 0031-6997

Della-Fera, M. A. & Baile, C. A. (1984). Control of feed intake in sheep. *J Anim Sci*, Vol. 59, No 5, pp. 1362-1368, ISSN 0021-8812

Deng, W.; Aimone, J. B. & Gage, F. H. (2010). New neurons and new memories: how does adult hippocampal neurogenesis affect learning and memory? *Nat Rev Neurosci*, Vol. 11, No 5, pp. 339-350, ISSN 1471-0048

Doyle, R. E.; Hinch, G. N.; Fisher, A. D.; Boissy, A.; Henshall, J. M. & Lee, C. (2011). Administration of serotonin inhibitor p-Chlorophenylalanine induces pessimistic-like judgement bias in sheep. *Psychoneuroendocrinology*, Vol. 36, No 2, pp. 279-288, ISSN 1873-3360

Dziegielewska, K. M.; Evans, C. A.; Fossan, G.; Lorscheider, F. L.; Malinowska, D. H.; Mollgard, K.; Reynolds, M. L.; Saunders, N. R. & Wilkinson, S. (1980). Proteins in cerebrospinal fluid and plasma of fetal sheep during development. *J Physiol*, Vol. 300, pp. 441-455, ISSN 0022-3751

Ebling, F. J. (2010). Photoperiodic regulation of puberty in seasonal species. *Mol Cell Endocrinol*, Vol. 324, No 1-2, pp. 95-101, ISSN 1872-8057

Eisch, A. J.; Cameron, H. A.; Encinas, J. M.; Meltzer, L. A.; Ming, G. L. & Overstreet-Wadiche, L. S. (2008). Adult neurogenesis, mental health, and mental illness: hope or hype? *J Neurosci*, Vol. 28, No 46, pp. 11785-11791, ISSN 1529-2401

Ekstrom, A. (2010). How and when the fMRI BOLD signal relates to underlying neural activity: the danger in dissociation. *Brain Res Rev*, Vol. 62, No 2, pp. 233-244, ISSN 1872-6321

Erhard, H. W.; Boissy, A.; Rae, M. T. & Rhind, S. M. (2004). Effects of prenatal undernutrition on emotional reactivity and cognitive flexibility in adult sheep. *Behav Brain Res*, Vol. 151, No 1-2, pp. 25-35, ISSN 0166-4328

Fang, M.; Li, J.; Gong, X.; Antonio, G.; Lee, F.; Kwong, W. H.; Wai, S. M. & Yew, D. T. (2005a). Myelination of the pig's brain: a correlated MRI and histological study. *Neurosignals*, Vol. 14, No 3, pp. 102-108, ISSN 1424-862X

Fang, M.; Lorke, D. E.; Li, J.; Gong, X.; Yew, J. C. & Yew, D. T. (2005b). Postnatal changes in functional activities of the pig's brain: a combined functional magnetic resonance imaging and immunohistochemical study. *Neurosignals*, Vol. 14, No 5, pp. 222-233, ISSN 1424-862X

Fang, M.; Zhang, L.; Li, J.; Wang, C.; Chung, C. H.; Wai, S. M. & Yew, D. T. (2005c). The postnatal development of the cerebellum - a fMRI and silver study. *Cell Mol Neurobiol*, Vol. 25, No 6, pp. 1043-1050, ISSN 0272-4340

Fang, M.; Li, J.; Rudd, J. A.; Wai, S. M.; Yew, J. C. & Yew, D. T. (2006). fMRI mapping of cortical centers following visual stimulation in postnatal pigs of different ages. *Life Sci*, Vol. 78, No 11, pp. 1197-1201, ISSN 0024-3205

Farrell, E.; Wielopolski, P.; Pavljasevic, P.; van Tiel, S.; Jahr, H.; Verhaar, J.; Weinans, H.; Krestin, G.; O'Brien, F. J.; van Osch, G. & Bernsen, M. (2008). Effects of iron oxide incorporation for long term cell tracking on MSC differentiation in vitro and in vivo. *Biochem Biophys Res Commun*, Vol. 369, No 4, pp. 1076-1081, ISSN 1090-2104

Febo, M.; Numan, M. & Ferris, C. F. (2005). Functional magnetic resonance imaging shows oxytocin activates brain regions associated with mother-pup bonding during suckling. *J Neurosci*, Vol. 25, No 50, pp. 11637-11644, ISSN 1529-2401

Ferris, C. F.; Snowdon, C. T.; King, J. A.; Duong, T. Q.; Ziegler, T. E.; Ugurbil, K.; Ludwig, R.; Schultz-Darken, N. J.; Wu, Z.; Olson, D. P.; Sullivan Jr, J. M.; Tannenbaum, P. L. & Vaughan, J. T. (2001). Functional imaging of brain activity in conscious monkeys responding to sexually arousing cues. *Neuroreport*, Vol. 12, No 10, pp. 2231-2236, ISSN 0959-4965

Ferris, C. F.; Snowdon, C. T.; King, J. A.; Sullivan, J. M., Jr.; Ziegler, T. E.; Olson, D. P.; Schultz-Darken, N. J.; Tannenbaum, P. L.; Ludwig, R.; Wu, Z.; Einspanier, A.; Vaughan, J. T. & Duong, T. Q. (2004). Activation of neural pathways associated with sexual arousal in non-human primates. *J Magn Reson Imaging*, Vol. 19, No 2, pp. 168-175, ISSN 1053-1807

Forster, C. (2008). Tight junctions and the modulation of barrier function in disease. *Histochem Cell Biol*, Vol. 130, No 1, pp. 55-70, ISSN 0948-6143

Foster, K. A.; Colditz, P. B.; Lingwood, B. E.; Burke, C.; Dunster, K. R. & Roberts, M. S. (2001). An improved survival model of hypoxia/ischaemia in the piglet suitable for neuroprotection studies. *Brain Res*, Vol. 919, No 1, pp. 122-131, ISSN 0006-8993

Fraser, M.; Bennet, L.; Helliwell, R.; Wells, S.; Williams, C.; Gluckman, P.; Gunn, A. J. & Inder, T. (2007). Regional specificity of magnetic resonance imaging and histopathology following cerebral ischemia in preterm fetal sheep. *Reprod Sci*, Vol. 14, No 2, pp. 182-191, ISSN 1933-7205

Friess, S. H.; Ichord, R. N.; Owens, K.; Ralston, J.; Rizol, R.; Overall, K. L.; Smith, C.; Helfaer, M. A. & Margulies, S. S. (2007). Neurobehavioral functional deficits following closed head injury in the neonatal pig. *Exp Neurol*, Vol. 204, No 1, pp. 234-243, ISSN 0014-4886

Gelez, H. & Fabre-Nys, C. (2006). Role of the olfactory systems and importance of learning in the ewes' response to rams or their odors. *Reprod Nutr Dev*, Vol. 46, No 4, pp. 401-415, ISSN 0926-5287

Gonen, O.; Liu, S.; Goelman, G.; Ratai, E. M.; Pilkenton, S.; Lentz, M. R. & Gonzalez, R. G. (2008). Proton MR spectroscopic imaging of rhesus macaque brain in vivo at 7T. *Magn Reson Med*, Vol. 59, No 4, pp. 692-699, ISSN 0740-3194

Goodman, R. L. & Inskeep, E. I. (2006). Neuroendocrine control of the ovarian cycle of the sheep. In: Neill JD ed. Knobil and Neill's Physiology of Reproduction r. edition, New-York, pp. 2389-2449, 978-0-12-515400-0 978-0-12-515400-0

Grate, L. L.; Golden, J. A.; Hoopes, P. J.; Hunter, J. V. & Duhaime, A. C. (2003). Traumatic brain injury in piglets of different ages: techniques for lesion analysis using histology and magnetic resonance imaging. *J Neurosci Methods*, Vol. 123, No 2, pp. 201-206, ISSN 0165-0270

Greiveldinger, L.; Veissier, I. & Boissy, A. (2007). Emotional experience in sheep: predictability of a sudden event lowers subsequent emotional responses. *Physiol Behav*, Vol. 92, No 4, pp. 675-683, ISSN 0031-9384

Guesdon, V.; Ligout, S.; Delagrange, P.; Spedding, M.; Levy, F.; Laine, A. L.; Malpaux, B. & Chaillou, E. (2011). Multiple exposures to familiar conspecific withdrawal is a novel robust stress paradigm in ewes. *Physiol Behav*, Vol. 105, No 2, pp. 203-208, ISSN 1873-507X

Gyulai, F. E. (2004). Anesthetics and cerebral metabolism. *Curr Opin Anaesthesiol*, Vol. 17, No 5, pp. 397-402, ISSN 0952-7907

Hawken, P. A.; Jorre, T. J.; Rodger, J.; Esmaili, T.; Blache, D. & Martin, G. B. (2009). Rapid induction of cell proliferation in the adult female ungulate brain (Ovis aries) associated with activation of the reproductive axis by exposure to unfamiliar males. *Biol Reprod*, Vol. 80, No 6, pp. 1146-1151, ISSN 0006-3363

Hawkins, B. T. & Davis, T. P. (2005). The blood-brain barrier/neurovascular unit in health and disease. *Pharmacol Rev*, Vol. 57, No 2, pp. 173-185, ISSN 0031-6997

Heinke, W. & Schwarzbauer, C. (2002). In vivo imaging of anaesthetic action in humans: approaches with positron emission tomography (PET) and functional magnetic resonance imaging (fMRI). *Br J Anaesth*, Vol. 89, No 1, pp. 112-122, ISSN 0007-0912

Henderson, L. A.; Yu, P. L.; Frysinger, R. C.; Galons, J. P.; Bandler, R. & Harper, R. M. (2002). Neural responses to intravenous serotonin revealed by functional magnetic resonance imaging. *J Appl Physiol*, Vol. 92, No 1, pp. 331-342, ISSN 8750-7587

Herbison, A. E. (1995). Neurochemical identity of neurones expressing oestrogen and androgen receptors in sheep hypothalamus. *J Reprod Fertil Suppl*, Vol. 49, No pp. 271-283, ISSN 0449-3087

Herbison, A. E. (2006). Physiology of the Gonadotropin-releasing hormone neuronal network. Ìn: Neill JD ed. Knobil and Neill's Physiology of Reproduction r. edition, New-York, pp. 1415-1482, 978-0-12-515400-0 978-0-12-515400-0

Herman, J. P.; Figueiredo, H.; Mueller, N. K.; Ulrich-Lai, Y.; Ostrander, M. M.; Choi, D. C. & Cullinan, W. E. (2003). Central mechanisms of stress integration: hierarchical circuitry controlling hypothalamo-pituitary-adrenocortical responsiveness. *Front Neuroendocrinol*, Vol. 24, No 3, pp. 151-180, ISSN 0091-3022

Hjort, N.; Wu, O.; Ashkanian, M.; Solling, C.; Mouridsen, K.; Christensen, S.; Gyldensted, C.; Andersen, G. & Ostergaard, L. (2008). MRI detection of early blood-brain barrier disruption: parenchymal enhancement predicts focal hemorrhagic transformation after thrombolysis. *Stroke*, Vol. 39, No 3, pp. 1025-1028, ISSN 1524-4628

Houpt, K. A.; Houpt, T. R. & Pond, W. G. (1979). The pig as a model for the study of obesity and of control of food intake: a review. *Yale J Biol Med*, Vol. 52, No 3, pp. 307-329, ISSN 0044-0086

Hutchison, R. M.; Mirsattari, S. M.; Jones, C. K.; Gati, J. S. & Leung, L. S. (2010). Functional networks in the anesthetized rat brain revealed by independent component analysis of resting-state FMRI. *J Neurophysiol*, Vol. 103, No 6, pp. 3398-3406, ISSN 1522-1598

Israeli, D.; Tanne, D.; Daniels, D.; Last, D.; Shneor, R.; Guez, D.; Landau, E.; Roth, Y.; Ocherashvilli, A.; Bakon, M.; Hoffman, C.; Weinberg, A.; Volk, T. & Mardor, Y.

(2011). The application of MRI for depiction of subtle blood brain barrier disruption in stroke. *Int J Biol Sci*, Vol. 7, No 1, pp. 1-8, ISSN 1449-2288

Jarvis, S.; Moinard, C.; Robson, S. K.; Baxter, E.; Ormandy, E.; Douglas, A. J.; Seckl, J. R.; Russell, J. A. & Lawrence, A. B. (2006). Programming the offspring of the pig by prenatal social stress: neuroendocrine activity and behaviour. *Horm Behav*, Vol. 49, No 1, pp. 68-80, ISSN 0018-506X

Jelsing, J.; Hay-Schmidt, A.; Dyrby, T.; Hemmingsen, R.; Uylings, H. B. & Pakkenberg, B. (2006). The prefrontal cortex in the Gottingen minipig brain defined by neural projection criteria and cytoarchitecture. *Brain Res Bull*, Vol. 70, No 4-6, pp. 322-336, ISSN 0361-9230

Jernigan, T. L.; Baare, W. F.; Stiles, J. & Madsen, K. S. (2011). Postnatal brain development: structural imaging of dynamic neurodevelopmental processes. *Prog Brain Res*, Vol. 189, pp. 77-92, ISSN 1875-7855

Johansen, T.; Hansen, H. S.; Richelsen, B. & Malmlof, R. (2001). The obese Gottingen minipig as a model of the metabolic syndrome: dietary effects on obesity, insulin sensitivity, and growth hormone profile. *Comp Med*, Vol. 51, No 2, pp. 150-155, ISSN 1532-0820

Johnson, R. W.; von Borell, E. H.; Anderson, L. L.; Kojic, L. D. & Cunnick, J. E. (1994). Intracerebroventricular injection of corticotropin-releasing hormone in the pig: acute effects on behavior, adrenocorticotropin secretion, and immune suppression. *Endocrinology*, Vol. 135, No 2, pp. 642-648, ISSN 0013-7227

Junghofer, M.; Peyk, P.; Flaisch, T. & Schupp, H. T. (2006). Neuroimaging methods in affective neuroscience: selected methodological issues. *Prog Brain Res*, Vol. 156, pp. 123-143, ISSN 0079-6123

Kalra, S. P.; Dube, M. G.; Pu, S.; Xu, B.; Horvath, T. L. & Kalra, P. S. (1999). Interacting appetite-regulating pathways in the hypothalamic regulation of body weight. *Endocr Rev*, Vol. 20, No 1, pp. 68-100, ISSN 0163-769X

Kanitz, E.; Otten, W.; Tuchscherer, M. & Manteuffel, G. (2003). Effects of prenatal stress on corticosteroid receptors and monoamine concentrations in limbic areas of suckling piglets (Sus scrofa) at different ages. *J Vet Med A Physiol Pathol Clin Med*, Vol. 50, No 3, pp. 132-139, ISSN 0931-184X

Keay, K. A. & Bandler, R. (2001). Parallel circuits mediating distinct emotional coping reactions to different types of stress. *Neurosci Biobehav Rev*, Vol. 25, No 7-8, pp. 669-678, ISSN 0149-7634

Kendrick, K. M.; da Costa, A. P.; Leigh, A. E.; Hinton, M. R. & Peirce, J. W. (2001). Sheep don't forget a face. *Nature*, Vol. 414, No 6860, pp. 165-166, ISSN 0028-0836

Killgore, W. D. & Yurgelun-Todd, D. A. (2010). Sex differences in cerebral responses to images of high versus low-calorie food. *Neuroreport*, Vol. 21, No 5, pp. 354-358, ISSN 1473-558X

Kineman, R. D.; Kraeling, R. R.; Crim, J. W.; Leshin, L. S.; Barb, C. R. & Rampacek, G. B. (1989). Localization of proopiomelanocortin (POMC) immunoreactive neurons in the forebrain of the pig. *Biol Reprod*, Vol. 40, No 5, pp. 1119-1126, ISSN 0006-3363

Kullmann, S.; Heni, M.; Veit, R.; Ketterer, C.; Schick, F.; Haring, H. U.; Fritsche, A. & Preissl, H. (2011). The obese brain: Association of body mass index and insulin sensitivity with resting state network functional connectivity. *Hum Brain Mapp*, doi: 10.1002/hbm.21268. ISSN 1097-0193

Kustermann, E.; Himmelreich, U.; Kandal, K.; Geelen, T.; Ketkar, A.; Wiedermann, D.; Strecker, C.; Esser, J.; Arnhold, S. & Hoehn, M. (2008). Efficient stem cell labeling for MRI studies. *Contrast Media Mol Imaging*, Vol. 3, No 1, pp. 27-37, ISSN 1555-4317

Lagaraine, C.; Skipor, J.; Szczepkowska, A.; Dufourny, L. & Thiery, J. C. (2011). Tight junction proteins vary in the choroid plexus of ewes according to photoperiod. *Brain Res*, Vol. 1393, No pp. 44-51, ISSN 1872-6240

Laurini, R. N.; Arbeille, B.; Gemberg, C.; Akoka, S.; Locatelli, A.; Lansac, J. & Arbeille, P. (1999). Brain damage and hypoxia in an ovine fetal chronic cocaine model. *Eur J Obstet Gynecol Reprod Biol*, Vol. 86, No 1, pp. 15-22, ISSN 0301-2115

Lazarov, O. & Marr, R. A. (2010). Neurogenesis and Alzheimer's disease: at the crossroads. *Exp Neurol*, Vol. 223, No 2, pp. 267-281, ISSN 1090-2430

Le Bihan, D.; Breton, E.; Lallemand, D.; Grenier, P.; Cabanis, E. & Laval-Jeantet, M. (1986). MR imaging of intravoxel incoherent motions: application to diffusion and perfusion in neurologic disorders. *Radiology*, Vol. 161, No 2, pp. 401-407, ISSN 0033-8419

LeDoux, J. E. (2000). Emotion circuits in the brain. *Annu Rev Neurosci*, Vol. 23, No pp. 155-184, ISSN 0147-006X

Lehman, M. N.; Coolen, L. M.; Goodman, R. L.; Viguie, C.; Billings, H. J. & Karsch, F. J. (2002). Seasonal plasticity in the brain: the use of large animal models for neuroanatomical research. *Reprod Suppl*, Vol. 59, No pp. 149-165, ISSN 1477-0415

Leite, F. P.; Tsao, D.; Vanduffel, W.; Fize, D.; Sasaki, Y.; Wald, L. L.; Dale, A. M.; Kwong, K. K.; Orban, G. A.; Rosen, B. R.; Tootell, R. B. & Mandeville, J. B. (2002). Repeated fMRI using iron oxide contrast agent in awake, behaving macaques at 3 Tesla. *Neuroimage*, Vol. 16, No 2, pp. 283-294, ISSN 1053-8119

Leshin, L. S.; Kraeling, R. R.; Kineman, R. D.; Barb, C. R. & Rampacek, G. B. (1996). Immunocytochemical distribution of catecholamine-synthesizing neurons in the hypothalamus and pituitary gland of pigs: tyrosine hydroxylase and dopamine-beta-hydroxylase. *J Comp Neurol*, Vol. 364, No 1, pp. 151-168, ISSN 0021-9967

Levy, F.; Meurisse, M.; Ferreira, G.; Thibault, J. & Tillet, Y. (1999). Afferents to the rostral olfactory bulb in sheep with special emphasis on the cholinergic, noradrenergic and serotonergic connections. *J Chem Neuroanat*, Vol. 16, No 4, pp. 245-263, ISSN 0891-0618

Levy, F.; Keller, M. & Poindron, P. (2004). Olfactory regulation of maternal behavior in mammals. *Horm Behav*, Vol. 46, No 3, pp. 284-302, ISSN 0018-506X

Levy, F. & Keller, M. (2009). Olfactory mediation of maternal behavior in selected mammalian species. *Behav Brain Res*, Vol. 200, No 2, pp. 336-345, ISSN 1872-7549

Li, Y. K.; Liu, G. R.; Zhou, X. G. & Cai, A. Q. (2010). Experimental hypoxic-ischemic encephalopathy: comparison of apparent diffusion coefficients and proton magnetic resonance spectroscopy. *Magn Reson Imaging*, Vol. 28, No 4, pp. 487-494, ISSN 1873-5894

Lieberman, K. A.; Wasenko, J. J.; Schelper, R.; Swarnkar, A.; Chang, J. K. & Rodziewicz, G. S. (2003). Tanycytomas: a newly characterized hypothalamic-suprasellar and ventricular tumor. *AJNR Am J Neuroradiol*, Vol. 24, No 10, pp. 1999-2004, ISSN 0195-6108

Lind, N. M.; Moustgaard, A.; Jelsing, J.; Vajta, G.; Cumming, P. & Hansen, A. K. (2007). The use of pigs in neuroscience: modeling brain disorders. *Neurosci Biobehav Rev*, Vol. 31, No 5, pp. 728-751, ISSN 0149-7634

Liu, H. L.; Hua, M. Y.; Yang, H. W.; Huang, C. Y.; Chu, P. C.; Wu, J. S.; Tseng, I. C.; Wang, J. J.; Yen, T. C.; Chen, P. Y. & Wei, K. C. (2010). Magnetic resonance monitoring of focused ultrasound/magnetic nanoparticle targeting delivery of therapeutic agents to the brain. *Proc Natl Acad Sci U S A*, Vol. 107, No 34, pp. 15205-15210, ISSN 1091-6490

Logothetis, N. K. (2002). The neural basis of the blood-oxygen-level-dependent functional magnetic resonance imaging signal. *Philos Trans R Soc Lond B Biol Sci*, Vol. 357, No 1424, pp. 1003-1037, ISSN 0962-8436

Loijens, L. W.; Janssens, C. J.; Schouten, W. G. & Wiegant, V. M. (2002). Opioid activity in behavioral and heart rate responses of tethered pigs to acute stress. *Physiol Behav*, Vol. 75, No 5, pp. 621-626, ISSN 0031-9384

Makiranta, M. J.; Lehtinen, S.; Jauhiainen, J. P.; Oikarinen, J. T.; Pyhtinen, J. & Tervonen, O. (2002). MR perfusion, diffusion and BOLD imaging of methotrexate-exposed swine brain. *J Magn Reson Imaging*, Vol. 15, No 5, pp. 511-519, ISSN 1053-1807

Malpaux, B.; Tricoire, H.; Mailliet, F.; Daveau, A.; Migaud, M.; Skinner, D. C.; Pelletier, J. & Chemineau, P. (2002). Melatonin and seasonal reproduction: understanding the neuroendocrine mechanisms using the sheep as a model. *Reprod Suppl*, Vol. 59, No pp. 167-179, ISSN 1477-0415

McKeown, M. J.; Makeig, S.; Brown, G. G.; Jung, T. P.; Kindermann, S. S.; Bell, A. J. & Sejnowski, T. J. (1998). Analysis of fMRI data by blind separation into independent spatial components. *Hum Brain Mapp*, Vol. 6, No 3, pp. 160-188, ISSN 1065-9471

McLaren, D. G.; Kosmatka, K. J.; Oakes, T. R.; Kroenke, C. D.; Kohama, S. G.; Matochik, J. A.; Ingram, D. K. & Johnson, S. C. (2009). A population-average MRI-based atlas collection of the rhesus macaque. *Neuroimage*, Vol. 45, No 1, pp. 52-59, ISSN 1095-9572

Meurisse, M.; Chaillou, E. & Levy, F. (2009). Afferent and efferent connections of the cortical and medial nuclei of the amygdala in sheep. *J Chem Neuroanat*, Vol. 37, No 2, pp. 87-97, ISSN 1873-6300

Michael-Titus, A. T.; Albert, M.; Michael, G. J.; Michaelis, T.; Watanabe, T.; Frahm, J.; Pudovkina, O.; van der Hart, M. G.; Hesselink, M. B.; Fuchs, E. & Czeh, B. (2008). SONU20176289, a compound combining partial dopamine D(2) receptor agonism with specific serotonin reuptake inhibitor activity, affects neuroplasticity in an animal model for depression. *Eur J Pharmacol*, Vol. 598, No 1-3, pp. 43-50, ISSN 0014-2999

Migaud, M.; Batailler, M.; Segura, S.; Duittoz, A.; Franceschini, I. & Pillon, D. (2010). Emerging new sites for adult neurogenesis in the mammalian brain: a comparative study between the hypothalamus and the classical neurogenic zones. *Eur J Neurosci*, Vol. 32, No 12, pp. 2042-2052, ISSN 1460-9568

Moeller, S.; Nallasamy, N.; Tsao, D. Y. & Freiwald, W. A. (2009). Functional connectivity of the macaque brain across stimulus and arousal states. *J Neurosci*, Vol. 29, No 18, pp. 5897-5909, ISSN 1529-2401

Molenaar, G. J.; Hogenesch, R. I.; Sprengers, M. E. & Staal, M. J. (1997). Ontogenesis of embryonic porcine ventral mesencephalon in the perspective of its potential use as a xenograft in Parkinson's disease. *J Comp Neurol*, Vol. 382, No 1, pp. 19-28, ISSN 0021-9967

Monti, M. M. (2011). Statistical Analysis of fMRI Time-Series: A Critical Review of the GLM Approach. *Front Hum Neurosci*, Vol. 5, No 28 pp. 1-13, ISSN 1662-5161

Muhlau, M.; Gaser, C.; Ilg, R.; Conrad, B.; Leibl, C.; Cebulla, M. H.; Backmund, H.; Gerlinghoff, M.; Lommer, P.; Schnebel, A.; Wohlschlager, A. M.; Zimmer, C. & Nunnemann, S. (2007). Gray matter decrease of the anterior cingulate cortex in anorexia nervosa. *Am J Psychiatry*, Vol. 164, No 12, pp. 1850-1857, ISSN 0002-953X

Muldoon, L. L.; Sandor, M.; Pinkston, K. E. & Neuwelt, E. A. (2005). Imaging, distribution, and toxicity of superparamagnetic iron oxide magnetic resonance nanoparticles in the rat brain and intracerebral tumor. *Neurosurgery*, Vol. 57, No 4, pp. 785-796; discussion 785-796, ISSN 1524-4040

Niblock, M. M.; Luce, C. J.; Belliveau, R. A.; Paterson, D. S.; Kelly, M. L.; Sleeper, L. A.; Filiano, J. J. & Kinney, H. C. (2005). Comparative anatomical assessment of the piglet as a model for the developing human medullary serotonergic system. *Brain Res Brain Res Rev*, Vol. 50, No 1, pp. 169-183, ISSN

Nowak, R.; Keller, M.; Val-Laillet, D. & Levy, F. (2007). Perinatal visceral events and brain mechanisms involved in the development of mother-young bonding in sheep. *Horm Behav*, Vol. 52, No 1, pp. 92-98, ISSN 0018-506X

Opdam, H. I.; Federico, P.; Jackson, G. D.; Buchanan, J.; Abbott, D. F.; Fabinyi, G. C.; Syngeniotis, A.; Vosmansky, M.; Archer, J. S.; Wellard, R. M. & Bellomo, R. (2002). A sheep model for the study of focal epilepsy with concurrent intracranial EEG and functional MRI. *Epilepsia*, Vol. 43, No 8, pp. 779-787, ISSN 0013-9580

Orhan, M. E.; Bilgin, F.; Kilickaya, O.; Atim, A. & Kurt, E. (2011). Nitrous oxide anesthesia in children for MRI: a comparison with isofl urane and halothane. *Turk J Med Sci*, Vol. 41, No 3, pp. 387-396, ISSN 1303-6165

Parrott, R. F.; Heavens, R. P. & Baldwin, B. A. (1986). Stimulation of feeding in the satiated pig by intracerebroventricular injection of neuropeptide Y. *Physiol Behav*, Vol. 36, No 3, pp. 523-525, ISSN 0031-9384

Parry, G. J. (1976). An animal model for the study of drugs in the central nervous system. *Proc Aust Assoc Neurol*, Vol. 13, pp. 83-88, ISSN 0084-7224

Patural, H.; St-Hilaire, M.; Pichot, V.; Beuchee, A.; Samson, N.; Duvareille, C. & Praud, J. P. (2010). Postnatal autonomic activity in the preterm lamb. *Res Vet Sci*, Vol. 89, No 2, pp. 242-249, ISSN 1532-2661

Pawela, C. P.; Biswal, B. B.; Hudetz, A. G.; Schulte, M. L.; Li, R.; Jones, S. R.; Cho, Y. R.; Matloub, H. S. & Hyde, J. S. (2009). A protocol for use of medetomidine anesthesia in rats for extended studies using task-induced BOLD contrast and resting-state functional connectivity. *Neuroimage*, Vol. 46, No 4, pp. 1137-1147, ISSN 1095-9572

Pernot, M.; Aubry, J. F.; Tanter, M.; Boch, A. L.; Marquet, F.; Kujas, M.; Seilhean, D. & Fink, M. (2007). In vivo transcranial brain surgery with an ultrasonic time reversal mirror. *J Neurosurg*, Vol. 106, No 6, pp. 1061-1066, ISSN 0022-3085

Piekarzewska, A.; Sadowski, B. & Rosochacki, S. J. (1999). Alterations of brain monoamine levels in pigs exposed to acute immobilization stress. *Zentralbl Veterinarmed A*, Vol. 46, No 4, pp. 197-207, ISSN 0514-7158

Piekarzewska, A. B.; Rosochacki, S. J. & Sender, G. (2000). The effect of acute restraint stress on regional brain neurotransmitter levels in stress-susceptible pietrain pigs. *J Vet Med A Physiol Pathol Clin Med*, Vol. 47, No 5, pp. 257-269, ISSN 0931-184X

Pladys, P.; Arsenault, J.; Reix, P.; Rouillard Lafond, J.; Moreau-Bussiere, F. & Praud, J. P. (2008). Influence of prematurity on postnatal maturation of heart rate and arterial pressure responses to hypoxia in lambs. *Neonatology*, Vol. 93, No 3, pp. 197-205, ISSN 1661-7819

Pletzer, B.; Kronbichler, M.; Aichhorn, M.; Bergmann, J.; Ladurner, G. & Kerschbaum, H. H. (2010). Menstrual cycle and hormonal contraceptive use modulate human brain structure. *Brain Res*, Vol. 1348, pp. 55-62, ISSN 1872-6240

Plourde, G.; Belin, P.; Chartrand, D.; Fiset, P.; Backman, S. B.; Xie, G. & Zatorre, R. J. (2006). Cortical processing of complex auditory stimuli during alterations of consciousness with the general anesthetic propofol. *Anesthesiology*, Vol. 104, No 3, pp. 448-457, ISSN 0003-3022

Poindron, P.; Levy, F. & Keller, M. (2007). Maternal responsiveness and maternal selectivity in domestic sheep and goats: the two facets of maternal attachment. *Dev Psychobiol*, Vol. 49, No 1, pp. 54-70, ISSN 0012-1630

Purdon, P. L.; Pierce, E. T.; Bonmassar, G.; Walsh, J.; Harrell, P. G.; Kwo, J.; Deschler, D.; Barlow, M.; Merhar, R. C.; Lamus, C.; Mullaly, C. M.; Sullivan, M.; Maginnis, S.; Skoniecki, D.; Higgins, H. A. & Brown, E. N. (2009). Simultaneous electroencephalography and functional magnetic resonance imaging of general anesthesia. *Ann N Y Acad Sci*, Vol. 1157, pp. 61-70, ISSN 1749-6632

Qi, Y.; Iqbal, J.; Oldfield, B. J. & Clarke, I. J. (2008). Neural connectivity in the mediobasal hypothalamus of the sheep brain. *Neuroendocrinology*, Vol. 87, No 2, pp. 91-112, ISSN 1423-0194

Quallo, M. M.; Price, C. J.; Ueno, K.; Asamizuya, T.; Cheng, K.; Lemon, R. N. & Iriki, A. (2009). Gray and white matter changes associated with tool-use learning in macaque monkeys. *Proc Natl Acad Sci U S A*, Vol. 106, No 43, pp. 18379-18384, ISSN 1091-6490

Ragland, J. D.; Yoon, J.; Minzenberg, M. J. & Carter, C. S. (2007). Neuroimaging of cognitive disability in schizophrenia: search for a pathophysiological mechanism. *Int Rev Psychiatry*, Vol. 19, No 4, pp. 417-427, ISSN 0954-0261

Rakhade, S. N. & Jensen, F. E. (2009). Epileptogenesis in the immature brain: emerging mechanisms. *Nat Rev Neurol*, Vol. 5, No 7, pp. 380-391, ISSN 1759-4766

Ramm, P.; Couillard-Despres, S.; Plotz, S.; Rivera, F. J.; Krampert, M.; Lehner, B.; Kremer, W.; Bogdahn, U.; Kalbitzer, H. R. & Aigner, L. (2009). A nuclear magnetic resonance biomarker for neural progenitor cells: is it all neurogenesis? *Stem Cells*, Vol. 27, No 2, pp. 420-423, ISSN 1549-4918

Ressler, K. J. & Mayberg, H. S. (2007). Targeting abnormal neural circuits in mood and anxiety disorders: from the laboratory to the clinic. *Nat Neurosci*, Vol. 10, No 9, pp. 1116-1124, ISSN 1097-6256

Riddle, A.; Luo, N. L.; Manese, M.; Beardsley, D. J.; Green, L.; Rorvik, D. A.; Kelly, K. A.; Barlow, C. H.; Kelly, J. J.; Hohimer, A. R. & Back, S. A. (2006). Spatial heterogeneity in oligodendrocyte lineage maturation and not cerebral blood flow predicts fetal ovine periventricular white matter injury. *J Neurosci*, Vol. 26, No 11, pp. 3045-3055, ISSN 1529-2401

Rivalland, E. T.; Tilbrook, A. J.; Turner, A. I.; Iqbal, J.; Pompolo, S. & Clarke, I. J. (2006). Projections to the preoptic area from the paraventricular nucleus, arcuate nucleus and the bed nucleus of the stria terminalis are unlikely to be involved in stress-induced suppression of GnRH secretion in sheep. *Neuroendocrinology*, Vol. 84, No 1, pp. 1-13, ISSN 0028-3835

Rivalland, E. T.; Clarke, I. J.; Turner, A. I.; Pompolo, S. & Tilbrook, A. J. (2007). Isolation and restraint stress results in differential activation of corticotrophin-releasing hormone and arginine vasopressin neurons in sheep. *Neuroscience*, Vol. 145, No 3, pp. 1048-1058, ISSN 0306-4522

Rodriguez, E. M.; Blazquez, J. L.; Pastor, F. E.; Pelaez, B.; Pena, P.; Peruzzo, B. & Amat, P. (2005). Hypothalamic tanycytes: a key component of brain-endocrine interaction. *Int Rev Cytol*, Vol. 247, No pp. 89-164, ISSN 0074-7696

Rogers, J.; Kochunov, P.; Lancaster, J.; Shelledy, W.; Glahn, D.; Blangero, J. & Fox, P. (2007). Heritability of brain volume, surface area and shape: an MRI study in an extended pedigree of baboons. *Hum Brain Mapp*, Vol. 28, No 6, pp. 576-583, ISSN 1065-9471

Rogers, R.; Wise, R. G.; Painter, D. J.; Longe, S. E. & Tracey, I. (2004). An investigation to dissociate the analgesic and anesthetic properties of ketamine using functional magnetic resonance imaging. *Anesthesiology*, Vol. 100, No 2, pp. 292-301, ISSN 0003-3022

Rotzinger, S.; Lovejoy, D. A. & Tan, L. A. (2010). Behavioral effects of neuropeptides in rodent models of depression and anxiety. *Peptides*, Vol. 31, No 4, pp. 736-756, ISSN 1873-5169

Rowniak, M.; Robak, A.; Bogus-Nowakowska, K.; Kolenkiewicz, M.; Bossowska, A.; Wojtkiewicz, J.; Skobowiat, C. & Majewski, M. (2008). Somatostatin-like immunoreactivity in the amygdala of the pig. *Folia Histochem Cytobiol*, Vol. 46, No 2, pp. 229-238, ISSN 1897-5631

Ruggiero, D. A.; Gootman, P. M.; Ingenito, S.; Wong, C.; Gootman, N. & Sica, A. L. (1999). The area postrema of newborn swine is activated by hypercapnia: relevance to sudden infant death syndrome? *J Auton Nerv Syst*, Vol. 76, No 2-3, pp. 167-175, ISSN 0165-1838

Sabatinelli, D.; Fortune, E. E.; Li, Q.; Siddiqui, A.; Krafft, C.; Oliver, W. T.; Beck, S. & Jeffries, J. (2011). Emotional perception: meta-analyses of face and natural scene processing. *Neuroimage*, Vol. 54, No 3, pp. 2524-2533, ISSN 1095-9572

Saikali, S.; Meurice, P.; Sauleau, P.; Eliat, P. A.; Bellaud, P.; Randuineau, G.; Verin, M. & Malbert, C. H. (2010). A three-dimensional digital segmented and deformable brain atlas of the domestic pig. *J Neurosci Methods*, Vol. 192, No 1, pp. 102-109, ISSN 1872-678X

Salak-Johnson, J. L.; Anderson, D. L. & McGlone, J. J. (2004). Differential dose effects of central CRF and effects of CRF astressin on pig behavior. *Physiol Behav*, Vol. 83, No 1, pp. 143-150, ISSN 0031-9384

Sandberg, D. I.; Crandall, K. M.; Petito, C. K.; Padgett, K. R.; Landrum, J.; Babino, D.; He, D.; Solano, J.; Gonzalez-Brito, M. & Kuluz, J. W. (2008). Chemotherapy administration directly into the fourth ventricle in a new piglet model. Laboratory Investigation. *J Neurosurg Pediatr*, Vol. 1, No 5, pp. 373-380, ISSN 1933-0707

Sato, K.; Kubota, T.; Ishida, M.; Yoshida, K.; Takeuchi, H. & Handa, Y. (2003). Immunohistochemical and ultrastructural study of chordoid glioma of the third ventricle: its tanycytic differentiation. *Acta Neuropathol*, Vol. 106, No 2, pp. 176-180, ISSN 0001-6322

Sauleau, P.; Lapouble, E.; Val-Laillet, D. & Malbert, C. H. (2009). The pig model in brain imaging and neurosurgery. *Animal*, Vol. 3, No 8, pp. 1138-1151, ISSN 1751-7111

Schlorf, T.; Meincke, M.; Kossel, E.; Gluer, C. C.; Jansen, O. & Mentlein, R. (2010). Biological properties of iron oxide nanoparticles for cellular and molecular magnetic resonance imaging. *Int J Mol Sci*, Vol. 12, No 1, pp. 12-23, ISSN 1422-0067

Schmidt, M. J.; Langen, N.; Klumpp, S.; Nasirimanesh, F.; Shirvanchi, P.; Ondreka, N. & Kramer, M. (2011). A study of the comparative anatomy of the brain of domestic ruminants using magnetic resonance imaging. *Vet J*, doi: 10.1016/j.tvjl.2010.12.026 ISSN 1532-2971

Schwartz, G. J. (2006). Integrative capacity of the caudal brainstem in the control of food intake. *Philos Trans R Soc Lond B Biol Sci*, Vol. 361, No 1471, pp. 1275-1280, ISSN 0962-8436

Schweinhardt, P.; Fransson, P.; Olson, L.; Spenger, C. & Andersson, J. L. (2003). A template for spatial normalisation of MR images of the rat brain. *J Neurosci Methods*, Vol. 129, No 2, pp. 105-113, ISSN 0165-0270

Sebe, F.; Nowak, R.; Poindron, P. & Aubin, T. (2007). Establishment of vocal communication and discrimination between ewes and their lamb in the first two days after parturition. *Dev Psychobiol*, Vol. 49, No 4, pp. 375-386, ISSN 0012-1630

Sebe, F.; Aubin, T.; Boue, A. & Poindron, P. (2008). Mother-young vocal communication and acoustic recognition promote preferential nursing in sheep. *J Exp Biol*, Vol. 211, pp. 3554-3562, ISSN 0022-0949

Shukakidze, A.; Lazriev, I. & Mitagvariya, N. (2003). Behavioral impairments in acute and chronic manganese poisoning in white rats. *Neurosci Behav Physiol*, Vol. 33, No 3, pp. 263-267, ISSN 0097-0549

Sica, A. L.; Gootman, P. M. & Ruggiero, D. A. (1999). CO(2)-induced expression of c-fos in the nucleus of the solitary tract and the area postrema of developing swine. *Brain Res*, Vol. 837, No 1-2, pp. 106-116, ISSN 0006-8993

Sicard, K.; Shen, Q.; Brevard, M. E.; Sullivan, R.; Ferris, C. F.; King, J. A. & Duong, T. Q. (2003). Regional cerebral blood flow and BOLD responses in conscious and anesthetized rats under basal and hypercapnic conditions: implications for functional MRI studies. *J Cereb Blood Flow Metab*, Vol. 23, No 4, pp. 472-481, ISSN 0271-678X

Sierra, A.; Encinas, J. M. & Maletic-Savatic, M. (2011). Adult human neurogenesis: from microscopy to magnetic resonance imaging. *Front Neurosci*, Vol. 5, No 47 pp. 1-18, ISSN 1662-453X

Silva, A. C. & Bock, N. A. (2008). Manganese-enhanced MRI: an exceptional tool in translational neuroimaging. *Schizophr Bull*, Vol. 34, No 4, pp. 595-604, ISSN 0586-7614

Skinner, D. C.; Caraty, A.; Malpaux, B. & Evans, N. P. (1997). Simultaneous measurement of gonadotropin-releasing hormone in the third ventricular cerebrospinal fluid and hypophyseal portal blood of the ewe. *Endocrinology*, Vol. 138, No 11, pp. 4699-4704, ISSN 0013-7227

Skinner, D. C. & Malpaux, B. (1999). High melatonin concentrations in third ventricular cerebrospinal fluid are not due to Galen vein blood recirculating through the choroid plexus. *Endocrinology*, Vol. 140, No 10, pp. 4399-4405, ISSN 0013-7227

Skipor, J. & Thiery, J. C. (2008). The choroid plexus-cerebrospinal fluid system: undervaluated pathway of neuroendocrine signaling into the brain. *Acta Neurobiol Exp*, Vol. 68, No 3, pp. 414-428, ISSN 0065-1400

Spurlock, M. E. & Gabler, N. K. (2008). The development of porcine models of obesity and the metabolic syndrome. *J Nutr*, Vol. 138, No 2, pp. 397-402, ISSN 1541-6100

Strbian, D.; Durukan, A.; Pitkonen, M.; Marinkovic, I.; Tatlisumak, E.; Pedrono, E.; Abo-Ramadan, U. & Tatlisumak, T. (2008). The blood-brain barrier is continuously open for several weeks following transient focal cerebral ischemia. *Neuroscience*, Vol. 153, No 1, pp. 175-181, ISSN 0306-4522

Thiery, J. C. & Malpaux, B. (2003). Seasonal regulation of reproductive activity in sheep: modulation of access of sex steroids to the brain. *Ann N Y Acad Sci*, Vol. 1007, pp. 169-175, ISSN 0077-8923

Thiery, J. C.; Robel, P.; Canepa, S.; Delaleu, B.; Gayrard, V.; Picard-Hagen, N. & Malpaux, B. (2003). Passage of progesterone into the brain changes with photoperiod in the ewe. *Eur J Neurosci*, Vol. 18, No 4, pp. 895-901, ISSN 0953-816X

Thiery, J. C.; Lomet, D.; Schumacher, M.; Liere, P.; Tricoire, H.; Locatelli, A.; Delagrange, P. & Malpaux, B. (2006). Concentrations of estradiol in ewe cerebrospinal fluid are modulated by photoperiod through pineal-dependent mechanisms. *J Pineal Res*, Vol. 41, No 4, pp. 306-312, ISSN 0742-3098

Thiery, J. C.; Lomet, D.; Bougoin, S. & Malpaux, B. (2009). Turnover rate of cerebrospinal fluid in female sheep: changes related to different light-dark cycles. *Cerebrospinal Fluid Res*, Vol. 6, pp. 9, ISSN 1743-8454

Tillet, Y.; Batailler, M. & Thibault, J. (1993). Neuronal projections to the medial preoptic area of the sheep, with special reference to monoaminergic afferents: immunohistochemical and retrograde tract tracing studies. *J Comp Neurol*, Vol. 330, No 2, pp. 195-220, ISSN 0021-9967

Tillet, Y. (1995). Distribution of neurotransmitters in the sheep brain. *J Reprod Fertil Suppl*, Vol. 49, pp. 199-220, ISSN 0449-3087

Tillet Y. & Kitahama K. (1998). Distribution of central catecholaminergic neurons : a comparison between ungulates, humans and other species. *Histol Histopathol*, Vol.13, pp. 1163-1177, ISSN 0213-3911

Tillet, Y.; Batailler, M.; Thiery, J. C. & Thibault, J. (2000). Neuronal projections to the lateral retrochiasmatic area of sheep with special reference to catecholaminergic afferents: immunohistochemical and retrograde tract-tracing studies. *J Chem Neuroanat*, Vol. 19, No 1, pp. 47-67, ISSN 0891-0618

Tong, S.; Ingenito, S.; Anderson, J. E.; Gootman, N.; Sica, A. L. & Gootman, P. M. (1995). Development of a swine animal model for the study of sudden infant death syndrome. *Lab Anim Sci*, Vol. 45, No 4, pp. 398-403, ISSN 0023-6764

Tricoire, H.; Moller, M.; Chemineau, P. & Malpaux, B. (2003). Origin of cerebrospinal fluid melatonin and possible function in the integration of photoperiod. *Reprod Suppl*, Vol. 61, pp. 311-321, ISSN 1477-0415

Val-Laillet, D.; Layec, S.; Guerin, S.; Meurice, P. & Malbert, C. H. (2011). Changes in brain activity after a diet-induced obesity. *Obesity (Silver Spring)*, Vol. 19, No 4, pp. 749-756, ISSN 1930-7381

van de Sande-Lee, S.; Pereira, F. R.; Cintra, D. E.; Fernandes, P. T.; Cardoso, A. R.; Garlipp, C. R.; Chaim, E. A.; Pareja, J. C.; Geloneze, B.; Li, L. M.; Cendes, F. & Velloso, L. A. (2011). Partial reversibility of hypothalamic dysfunction and changes in brain activity after body mass reduction in obese subjects. *Diabetes*, Vol. 60, No 6, pp. 1699-1704, ISSN 1939-327X

van den Heuvel, M. P. & Hulshoff Pol , H. E. (2010). Exploring the brain network: a review on resting-state fMRI functional connectivity. *Eur Neuropsychopharmacol*, Vol. 20, No 8, pp. 519-534, ISSN 1873-7862

Van Overwalle, F. (2009). Social cognition and the brain: a meta-analysis. *Hum Brain Mapp*, Vol. 30, No 3, pp. 829-858, ISSN 1097-0193

Van Vugt, D. A. (2010). Brain imaging studies of appetite in the context of obesity and the menstrual cycle. *Hum Reprod Update*, Vol. 16, No 3, pp. 276-292, ISSN 1460-2369

Vandenheede, M. & Bouissou, M. F. (1998). Effects of an enriched environment of subsequent fear reactions of lambs and ewes. *Dev Psychobiol*, Vol. 33, No 1, pp. 33-45, ISSN 0012-1630

Vandewalle, G.; Hebert, M.; Beaulieu, C.; Richard, L.; Daneault, V.; Garon, M. L.; Leblanc, J.; Grandjean, D.; Maquet, P.; Schwartz, S.; Dumont, M.; Doyon, J. & Carrier, J. (2011). Abnormal Hypothalamic Response to Light in Seasonal Affective Disorder. *Biol Psychiatry*, doi: 10.1016/j.biopsych.2011.06.022 ISSN 1873-2402

Vellucci, S. V. & Parrott, R. F. (1994). Expression of c-fos in the ovine brain following different types of stress, or central administration of corticotrophin-releasing hormone. *Exp Physiol*, Vol. 79, No 2, pp. 241, ISSN 0958-0670

Vial, F.; Serriere, S.; Barantin, L.; Montharu, J.; Nadal-Desbarats, L.; Pourcelot, L. & Seguin, F. (2004). A newborn piglet study of moderate hypoxic-ischemic brain injury by 1H-MRS and MRI. *Magn Reson Imaging*, Vol. 22, No 4, pp. 457-465, ISSN 0730-725X

Vigneau, M.; Beaucousin, V.; Herve, P. Y.; Duffau, H.; Crivello, F.; Houde, O.; Mazoyer, B. & Tzourio-Mazoyer, N. (2006). Meta-analyzing left hemisphere language areas: phonology, semantics, and sentence processing. *Neuroimage*, Vol. 30, No 4, pp. 1414-1432, ISSN 1053-8119

Vincent, J. L.; Patel, G. H.; Fox, M. D.; Snyder, A. Z.; Baker, J. T.; Van Essen, D. C.; Zempel, J. M.; Snyder, L. H.; Corbetta, M. & Raichle, M. E. (2007). Intrinsic functional architecture in the anaesthetized monkey brain. *Nature*, Vol. 447, No 7140, pp. 83-86, ISSN 1476-4687

Vocks, S.; Busch, M.; Gronemeyer, D.; Schulte, D.; Herpertz, S. & Suchan, B. (2010). Neural correlates of viewing photographs of one's own body and another woman's body in anorexia and bulimia nervosa: an fMRI study. *J Psychiatry Neurosci*, Vol. 35, No 3, pp. 163-176, ISSN 1488-2434

Wager, T. D. & Smith, E. E. (2003). Neuroimaging studies of working memory: a meta-analysis. *Cogn Affect Behav Neurosci*, Vol. 3, No 4, pp. 255-274, ISSN 1530-7026

Wang, Z.; Neylan, T. C.; Mueller, S. G.; Lenoci, M.; Truran, D.; Marmar, C. R.; Weiner, M. W. & Schuff, N. (2010). Magnetic resonance imaging of hippocampal subfields in posttraumatic stress disorder. *Arch Gen Psychiatry*, Vol. 67, No 3, pp. 296-303, ISSN 1538-3636

Watanabe, H.; Andersen, F.; Simonsen, C. Z.; Evans, S. M.; Gjedde, A. & Cumming, P. (2001). MR-based statistical atlas of the Gottingen minipig brain. *Neuroimage*, Vol. 14, No 5, pp. 1089-1096, ISSN 1053-8119

Watanabe, T.; Frahm, J. & Michaelis, T. (2004). Functional mapping of neural pathways in rodent brain in vivo using manganese-enhanced three-dimensional magnetic resonance imaging. *NMR Biomed*, Vol. 17, No 8, pp. 554-568, ISSN 0952-3480

Weis, S. & Hausmann, M. (2010). Sex hormones: modulators of interhemispheric inhibition in the human brain. *Neuroscientist*, Vol. 16, No 2, pp. 132-138, ISSN 1089-4098

Weis, S.; Hausmann, M.; Stoffers, B. & Sturm, W. (2010). Dynamic changes in functional cerebral connectivity of spatial cognition during the menstrual cycle. *Hum Brain Mapp*, Vol. 32, No 10, pp. 1544-1556, ISSN 1097-0193

Wey, H. Y.; Li, J.; Szabo, C. A.; Fox, P. T.; Leland, M. M.; Jones, L. & Duong, T. Q. (2010). BOLD fMRI of visual and somatosensory-motor stimulations in baboons. *Neuroimage*, Vol. 52, No 4, pp. 1420-1427, ISSN 1095-9572

Whitman, M. C. & Greer, C. A. (2009). Adult neurogenesis and the olfactory system. *Prog Neurobiol*, Vol. 89, No 2, pp. 162-175, ISSN 1873-5118

Williams, D. S.; Detre, J. A.; Leigh, J. S. & Koretsky, A. P. (1992). Magnetic resonance imaging of perfusion using spin inversion of arterial water. *Proc Natl Acad Sci U S A*, Vol. 89, No 1, pp. 212-216, ISSN 0027-8424

Willis, C. K.; Quinn, R. P.; McDonell, W. M.; Gati, J.; Partlow, G. & Vilis, T. (2001). Functional MRI activity in the thalamus and occipital cortex of anesthetized dogs induced by monocular and binocular stimulation. *Can J Vet Res*, Vol. 65, No 3, pp. 188-195, ISSN 0830-9000

Wuerfel, E.; Infante-Duarte, C.; Glumm, R. & Wuerfel, J. T. (2010). Gadofluorine M-enhanced MRI shows involvement of circumventricular organs in neuroinflammation. *J Neuroinflammation*, Vol. 7, pp. 70, ISSN 1742-2094

Xie, F.; Boska, M. D.; Lof, J.; Uberti, M. G.; Tsutsui, J. M. & Porter, T. R. (2008). Effects of transcranial ultrasound and intravenous microbubbles on blood brain barrier permeability in a large animal model. *Ultrasound Med Biol*, Vol. 34, No 12, pp. 2028-2034, ISSN 1879-291X

Young, K. A.; Liu, Y. & Wang, Z. (2008). The neurobiology of social attachment: A comparative approach to behavioral, neuroanatomical, and neurochemical studies. *Comp Biochem Physiol C Toxicol Pharmacol*, Vol. 148, No 4, pp. 401-410, ISSN 1532-0456

Zanella, A. J.; Broom, D. M.; Hunter, J. C. & Mendl, M. T. (1996). Brain opioid receptors in relation to stereotypies, inactivity, and housing in sows. *Physiol Behav*, Vol. 59, No 4-5, pp. 769-775, ISSN 0031-9384

Zhang, N.; Rane, P.; Huang, W.; Liang, Z.; Kennedy, D.; Frazier, J. A. & King, J. (2010). Mapping resting-state brain networks in conscious animals. *J Neurosci Methods*, Vol. 189, No 2, pp. 186-196, ISSN 1872-678X

Zhao, F.; Zhao, T.; Zhou, L.; Wu, Q. & Hu, X. (2008). BOLD study of stimulation-induced neural activity and resting-state connectivity in medetomidine-sedated rat. *Neuroimage*, Vol. 39, No 1, pp. 248-260, ISSN 1053-8119

Zieba, D. A.; Szczesna, M.; Klocek-Gorka, B. & Williams, G. L. (2008). Leptin as a nutritional signal regulating appetite and reproductive processes in seasonally-breeding ruminants. *J Physiol Pharmacol*, Vol. 59 Suppl 9, pp. 7-18, ISSN 1899-1505